5.1.96
$148

BERRY LIBRARY
CINCINNATI STATE
3520 CENTRAL PARKWAY
CINCINNATI, OHIO 45223-2690

D1294522

HERRY LIBRARY
CINCINNATI STATE
3520 CENTRAL PARKWAY
CINCINNATI, OHIO 45223-2690

Solar Energy Conversion

THE SOLAR CELL

(SECOND EDITION)

621.31244
N525
1995

Solar
Energy
Conversion
THE SOLAR CELL
(SECOND EDITION)

Richard C. Neville

College of Engineering & Technology
Northern Arizona University
Flagstaff, AZ, U.S.A.

1995

ELSEVIER

Amsterdam – Lausanne – New York – Oxford – Shannon – Tokyo

ELSEVIER SCIENCE B.V.
Sara Burgerhartstraat 25
P.O. Box 211, 1000 AE Amsterdam, The Netherlands

ISBN: 0 444 89818 2

© 1995 Elsevier Science B.V. All rights reserved.

No part of this publication may be reproduced, stored in a retrieval system or transmitted in any form or by any means, electronic, mechanical, photocopying, recording or otherwise, without the prior written permission of the publisher, Elsevier Science B.V., Copyright & Permissions Department, P.O. Box 521, 1000 AM Amsterdam, The Netherlands.

Special regulations for readers in the U.S.A. – This publication has been registered with the Copyright Clearance Center Inc. (CCC), Salem, Massachusetts. Information can be obtained from the CCC about conditions under which photocopies of parts of this publication may be made in the U.S.A. All other copyright questions, including photocopying outside of the U.S.A., should be referred to the copyright owner, Elsevier Science B.V., unless otherwise specified.

No responsibility is assumed by the publisher for any injury and/or damage to persons or property as a matter of products liability, negligence or otherwise, or from any use or operation of any methods, products, instructions or ideas contained in the material herein.

This book is printed on acid-free paper.

Printed in The Netherlands.

PREFACE

That the human race faced an energy crisis became painfully obvious during the 1970s. Since that time the blatant obviousness of the problem has waned, but the underlying technical and political problems have not disappeared. Humanity continues to increase in number (the world population is, at present, in excess of five billion) and, despite major efforts towards improving the efficiency of energy consumption, the overall per capita use of energy continues to increase.

Projections concerning the human population and its energy requirements during the next century estimate populations in excess of seven billion and energy consumption per person in excess of 40,000 kilowatt hours per year (approximately twice the current rate). This increasing energy use must be viewed in the light of the finite availability of conventional energy sources. When done so the energy crisis can be seen to be all too real for any long term comfort.

A frequently mentioned solution to the problem of increasing requirements for energy and dwindling energy sources is to tap the energy in sunlight. The solar energy falling on the earth's surface each year is over 20,000 times the amount presently required by the human race, making for a seemingly inexhaustible supply. For effective utilization of any energy source civilization requires an easily storable, easily transportable form of energy (after all, it is dark at night). This implies that the incoming solar energy should be transformed into electrical energy. In turn this means that we need to utilize photovoltaic (solar cell) conversion of the energy in sunlight.

Photovoltaic effects were initially observed more than a century and a half ago. In 1839 E. Becquerel observed a photovoltage (a voltage depending on the character and intensity of the illuminating light) when sunlight was allowed to shine on one of two electrodes in an electrolytic solution. The first scientific paper on photovoltage using solids was published in 1877 and concerned the semiconductor, selenium. In 1954 research groups at RCA and Bell Telephone Laboratories demonstrated the practical conversion of solar radiation into electrical energy by a silicon pn junction solar cell, and shortly thereafter Chapin, Fuller and Pearson reported on a six percent efficient solar cell (Journal of Applied Physics, Vol. 25, 1954, p. 676).

The modern solar cell is an electronic device, fabricated from semiconducting materials. It converts a fraction of the energy contained in sunlight directly to electrical energy at a voltage and current level determined by the properties of the semiconductor, the solar cell design and construction techniques, and the incident light. To gain an understanding of how solar cells work and to be in a position to design and construct energy conversion systems using solar cells requires a background covering such diverse areas as: the nature of solar radiation; semiconductor physics; quantum mechanics; the techniques of energy storage; optics; heat flow in solids; the nature of elemental, compound, single crystal, polycrystalline and amorphous semiconductors; the technology of semiconductor device fabrication; and the economics of energy flow. It is not physically possible to cover, in depth, all of these areas in a single work. In writing this volume, I have endeavored to create a survey text; a book that explores a number of critical background areas and then outlines the theory of operation of solar cells while considering their design and fabrication. Solar cell performance is treated both in the general sense and for some specific examples. These examples select semiconductor, junction type, optical orientation and fabrication technology and then highlight the problems encountered in solar cell design and illustrate, both in general and specific fashion, areas of promising future research and development. References are provided to facilitate deeper investigations of the various topics of interest--from quantum mechanics to economics.

This is the second edition of this work on solar cells. Historically, this book originated from a series of lectures on energy and solar cells given to engineering students at the University of California at Santa Barbara. These lectures culminated in the first edition of this work, in 1978. Since that time there has been much change in the fields of energy generation and consumption, solar energy and solar cells. Additional lectures at UCSB and at Northern Arizona University, coupled with considerable research into aspects of photovoltaic and solar energy have modified the original work. This, the second edition, is thus the result of more than 20 years of interest in solar energy and solar cells coupled with steady changes in these fields and our understanding of these fields. Since it is virtually impossible to separate design and operating theory, engineering, economics and politics in considering the use of solar cells in addressing the energy problem facing humanity, the systems aspect is present throughout this volume.

The first chapter is a broad (and brief) survey of the elements which make up the "energy crisis". It is devoted to illustrating the limited nature of presently utilized energy sources and to a discussion of the various "non-conventional" energy sources proposed for the future: biological, wind, wave and solar. It has, as its major purpose, three points to make: (1) that our conventional energy sources will be exhausted at some point in the not-too-distant future, (2) that solar energy is capable of supplying the energy requirements of humanity for the foreseeable future, and (3) that photovoltaic energy conversion is a major candidate for supplying mankind with its required energy; perhaps the prime candidate.

The second chapter considers the nature of sunlight, discusses the solar spectrum, the effects of latitude, the earth's rotation and axial tilt, and atmosphere and weather. A brief discussion of optics is included as a background for those individuals interested in this aspect of energy conversion.

The third chapter surveys the nature of semiconductors. Solar cells are theoretically constructed of various semiconductors and their performance is shown to depend upon the properties of these materials. These properties are best understood within the framework of quantum mechanics and solid state physics. Chapter III discusses crystals, quantum mechanics and semiconductor physics with a view towards outlining the principal properties of semiconductors and the manner in which these properties vary with device processing technology, temperature of operation, and the characteristics of the illumination. Because the physics of single crystal semiconductors is best understood (as contrasted with polycrystalline or amorphous structured semiconductors), the emphasis in this chapter is on solar cells constructed from single crystal semiconductors. It is in this chapter that certain specific example semiconductor materials are first introduced.

Chapter IV treats the interaction of light semiconductors including absorption, reflection and transmission. The generation of hole-electron pairs is treated both in the abstract and in detail using the example semiconducting materials introduced in Chapter III. The maximum potential output power density and the optimum output current density for photovoltaic cells are displayed for solar cells fabricated from six sample single crystal semiconductors.

The fifth chapter is devoted to a general discussion of solar cell performance as a function of the junction employed. The current versus voltage characteristics of pn, heterojunctions, mos junctions and Schottky barrier solar cells are considered and a general expression for the output power density as delivered to an optimum external load is obtained. From this expression, the maximum expectable output power density for solar cells, as a function of the energy gap of the semiconductor employed, is derived. This is displayed as a function of the saturation current of the solar cell.

In Chapter VI the six example semiconductors are employed to provide specific values of estimated solar cell performance, based on various technologies of junction fabrication and upon the optical orientation of the solar cells. The solar cell performance levels computed in this chapter, and in later chapters, are not meant as absolute predictions of maximum performance. Rather, they are intended to provide indications of "typical" solar cell performance as structured by technology and materials limitations. It is intended that they will suggest areas in need of research and development.

The seventh chapter considers the effects upon solar cell operation of changes in junction temperature and the use of concentrated sunlight. The power density in natural sunlight is very low (approximately one kw/m^2 at sea level) and hence any sizeable energy requirement implies a large area of solar cells. By utilizing relatively inexpensive mirrors or lenses to concentrate sunlight upon expensive solar cells a significant reduction in cost can be effected. This chapter examines the limits imposed on optical concentration levels by the solar cells and shows that improved solar cell performance is possible using the six example single crystal semiconductors.

Chapter VIII carries the materials of the preceding chapter a step further. In addition to considering the electrical energy output for solar cells operating under concentrated sunlight, the thermal energy available from such a situation is considered. Thus a complete systems approach to producing energy from photovoltaic cells is developed. Later in this chapter various approaches to further improving overall energy output (both electrical and thermal) from photovoltaic systems are considered. Most of these systems involve modifying the spectral characteristics of the light used to illuminate the solar cells. The altered light is a better match for the semiconductors used in fabricating the solar cells and so overall efficiency is improved.

In the ninth chapter the solar cells are constructed using polycrystalline and amorphous semiconductors. The operation of these devices depends strongly on the crystal interfaces and the properties of unsaturated chemical bonds. As a result, the theory of operation of polycrystalline and amorphous material solar cells is not well understood. Thus, this chapter is less theoretical and more empirical in nature than the previous chapters. Numerous examples of polycrystalline and amorphous solar cell operations and materials are provided.

The final chapter, Chapter X, is devoted to a brief survey of such topics as economics, energy storage, and overall systems effects. Potential problems and proposed solutions are noted and briefly discussed. It is intended that the reader treat this chapter as a question mark whose main purpose is to provoke inquiry.

The energy "problem" has not gone away, and will not go away. Without strenuous and continuing efforts on the part of humanity we will see a continuing series of "crises". Fortunately, the field of photovoltaic energy conversion is growing rapidly, both in scope and complexity. Of necessity I have been forced to treat lightly many areas which deserve considerably more intense study. To those readers whose specialty in research or development lies in these areas, my apologies. I can but plead lack of space and time.

In closing I would like to thank my fellow faculty members and my students for many hours of stimulating discussion and my wife, Laura Lou for her encouragement, patience, support and proof reading. In the final analysis, any errors are, of course, my responsibility.

Richard C. Neville
Flagstaff, Arizona 86011
U.S.A.
29 March 1994

TABLE OF CONTENTS

PREFACE   v

**CHAPTER I: ENERGY NEEDS--ENERGY
           SOURCES**   1
Introduction   1
Consumption   3
Conventional Sources of Energy   7
Alternative Energy Sources   16
       Nuclear Fusion   16
       Solar Energy   17
       Temperature Differences   22
       Thermodynamics   24
       Ocean Temperature-Difference Generators   24
       Solar-Thermal   26
       Solar-Electric   29
    References   36

CHAPTER II: THE SUN AND SUNLIGHT   39
Introduction   39
Sunlight   40
Geometrical Effects   44
Weather   54
Light Collection   57
       Lens Systems   57
       Mirrors   60
       Optical Materials   66
       Maximum Optical Concentration   68
    References   69

CHAPTER III: SEMICONDUCTORS   71
Introduction   71
Crystal Structure   72

Quantum Mechanics and Energy Bands   76
Electrons and Holes   83
Currents   91
Recombination and Carrier Lifetime   97
Junctions   108
References   115

CHAPTER IV: LIGHT-SEMICONDUCTOR INTERACTION   119
Introduction   119
Reflection   120
Light Interaction   123
Preliminary Material Selection   129
Absorption   132
Reflection and Absorption   145
References   151

CHAPTER V: BASIC THEORETICAL PERFORMANCE   155
Introduction   155
Local Electric Fields   156
PN Junction Electrical Characteristics   158
Heterojunction Electrical Characteristics   163
Electrical Characteristics of Schottky Junctions   167
Open Circuit Voltage and Short Circuit Current   170
Optimum Power Conditions   175
References   192

CHAPTER VI: SOLAR CELL CONFIGURATION AND PERFORMANCE   197
Introduction   197
Optical Orientation   199
Device Design - Minority Carrier Collection   205
Device Design - Saturation Current   220
Device Design - Series Resistance   222
Solar Cell Performance - Discussion   244
References   253

CHAPTER VII: ADVANCED APPROACHES   257
Introduction   257
Temperature Effects   259
Heat Flow within a Solar Cell   264
Optical Concentration - Photocurrent   269
Performance Under Concentration   272
References   299

CHAPTER VIII: ADVANCED APPROACHES-II   301
Introduction   301
Second Stage Solar Power Systems   302
Third Generation Solar Cell Systems   315
Miscellaneous Approaches   333
References   336

CHAPTER IX: POLYCRYSTALLINE AND AMORPHOUS SOLAR CELLS   339
Introduction   339
Polycrystalline Solar Cells   340
    Cadmium Sulfide/Copper Sulfide   342
    Copper Indium Selenide (CIS)   344
    Polycrystalline Silicon   346
    Thin Film Cadmium Telluride   347
    Other Possibilities for Polycrystalline Solar Cells   347
    Final Comments on Polycrystalline Solar Cells   348
Amorphous Material Based Solar Cells   348
    Amorphous Silicon   349
Concluding Remarks   358
References   359

CHAPTER X: CONCLUDING THOUGHTS   363
Introduction   363
Economics   365
Electrical Energy Storage   382
The System   384
Final Words   387
References   391

xiv

APPENDIX A: CONVERSION FACTORS   397

APPENDIX B: SELECTED PROPERTIES OF SEMICONDUCTORS WITH SOLAR CELL POTENTIAL   403

APPENDIX C: THE SATURATION CURRENT IN PN JUNCTION SOLAR CELLS   407

APPENDIX D: SOME USEFUL PHYSICAL CONSTANTS   411

APPENDIX E: SYMBOLS   411

SUBJECT INDEX   420

CHAPTER I: ENERGY NEEDS--ENERGY SOURCES

Introduction

This work is concerned with the theory, design and operation of solar cells. However, we need to first ask the fundamental question--why consider solar cells at all? The answer to this question involves energy. As a form of life Homo sapiens sapiens requires, as do all other living things, energy in the form of food and energy in the form of heat (ofttimes supplied by food, but sometimes by sunlight or hot water or...). We also use energy for a number of other purposes, such as clothing, shelter, transportation, entertainment, cooling and the construction of tools.

There is a large number of energy sources available to our species and we make use of most of them. In this chapter we will consider, briefly, a number of these energy sources, examine how we utilize them to supply us with energy, how this energy is used to provide us with a lifestyle, how much energy we use, what problems are or may be associated with this use, and describe several scenarios for the future. Note that how much energy a particular human being uses depends on where she/he lives and the nature of her/his lifestyle. Thus the overall quantity of energy used by our species is the net result of a complex interaction between energy sources, energy uses, human interaction (politics), human aspirations and engineering talent. There is no way in which we can obtain complete understanding of this subject in a single chapter. Such a complete understanding involves an exhaustive description of the energy resources available on the planet Earth, a thorough knowledge of the potential ecological interactions and an ability to forecast the numbers and lifestyles of humanity for at least the next thousand years...

What we shall do is to undertake a brief examination of the types of energy we use, how and how much we use them, and the availability of these energy sources. While doing so we will look, very briefly, into politics, ecology and demographics. We will discover that the energy crisis does exist, and, depending on our definitions of such items as satisfactory lifestyle* and our selection of energy sources, serious ecological consequences can develop for a planet whose problems are driven by a large and expanding population.

The sources of energy available to mankind on this planet are commonly divided into two broad categories: (1) energy capital sources, i.e., those sources of energy which, once used, cannot be replaced on any time scale less than millions of years (details to follow); and (2) energy income sources, i.e. those sources of energy which are more or less continuously refreshed (by nature or by man assisting nature) and which may be considered to be available, at potentially their current levels of supply, for millions of years. A listing of energy sources under these two categories would include:

Table I.1

Energy source types

Energy Capital	Energy Income
Fossil fuels (coal, oil and gas)	Biological sources (wood, plants)
Geothermal	Hydropower (dams, tides)
Nuclear fission	Wind energy
Nuclear fusion	Solar energy

The division in energy sources indicated in Table I.1 is not inflexible. For example, if we burn our trees very rapidly, we will outstrip the ability of our forests to grow new trees; making wood a capital energy source. The divisions indicated in Table I.1 are consistent with the way in which we are likely to make use of the energy sources--the capital sources will eventually become exhausted while the income sources will

---

* The reader should understand that, in this text as elsewhere in the literature, it is often implied that an improved lifestyle requires a greater expenditure of energy. Depending on one's viewpoint, this is not necessarily the case.

not. (Note, if our understanding of the physics of the universe is correct, all of the energy sources will eventually fail as the stars use up their fuel and turn dark. This is unlikely to be a problem during the next twenty billion years and will be ignored in this work).

The questions we need to ask are: (1) How long will the capital energy sources last? and (2) How many people can the energy income sources support? To answer these questions we need to consider how much energy the human race uses.

Consumption

In Table I.2 the per capita rate of energy use in the United States is presented for selected years.

Table I.2

Per capita energy use in the United States [1, 2, 3, 4]

Year	Per Capita energy use (kwh/year)	Year	Per Capita energy use (kwh/year)
1800	12,000	1950	66,200
1850	17,000	1960	72,000
1900	32,200	1970	97,000
1925	52,800	1980	96,700
1940	53,000	1990	99,600

The energies in Table I.2 were used for food, transportation, clothing, tools, housing, etc. and the units in which the energy is expressed for each usage varies (see Appendix A for a listing of the various units of energy). The annual consumption of energy for the United States alone is currently in excess of 10^{13} kwh. It is possible to represent this number in Btus, in calories, in barrels of oil equivalent, horsepower-years, or any one of a number of equivalent energy units. For relatively small amounts of energy we will use the kilowatt hour (kwh). For very large amounts of energy, such as the amount annually used in the United States, we shall utilize the Q. The Q is defined by:

$$1 \text{ Q} = 1 \times 10^{18} \text{ Btu} = 2.93 \times 10^{14} \text{ kwh.} \qquad \text{(I.1)}$$

Note that one Q is approximately the amount of energy required to bring Lake Michigan (North America) to a boil.

Over the past two millennia, the total world energy consumption has been approximately 22 Q [5, 6], corresponding to an average annual use of 0.011 Q. However, during the past century and a half (the period of the industrial revolution) some 13 Q of energy has been consumed, corresponding to a rate more than eight times the average annual use above. The world rate of energy consumption has changed from approximately 0.01 Q in 1850 to 0.22 Q in 1970 and to an estimated value of 0.42 Q in 1990 [7]. During this time period, the world's population has increased from approximately one billion to five billion [8]. This implies that the average annual energy consumption for each person in the world rose from 2,930 kwh in 1850 to approximately 24,600 kwh in 1990.

The estimated average energy consumption rate for the world in 1990 is significantly lower than the average energy consumption in the United States (see Table I.2). Indeed, with about five percent of the world's population [8], the United States now consumes an estimated 0.085 Q of energy each year--roughly 20% of the world's usage. What does this imply for the future? If the human race were to remain at its present population of five billion, and if the rest of the world were to "live as well" as the average U. S. citizen, the world's annual energy consumption would increase to 1.7 Q. If we allow for a population increase to 10 billion, then the required energy to live "the good life" rises to an annual value of 3.4 Q. Of course, it is possible to reduce energy consumption by a combination of more efficient energy use and by completely abandoning certain energy-using processes. The name normally given to this type of behavior is conservation. Carried to an extreme limit, we could envisage a world where each year each individual uses no more energy than in 1850 (2,930 kwh). This level of energy consumption is some 12% of the current world usage and is approximately one thirty-fourth of the present usage in the United States.

If the world lived at the average consumption level of 1850, the energy required annually would range from 0.05 Q per year (for a population of five billion) to 0.10 Q/year (for a population of ten billion), much reduced from the several Q values of the preceding paragraph.

Now the energy we consume is spent for transportation, industry and commerce, space heating, electricity, etc. An exact assessment of how much is used in each category depends on geography, lifestyle and weather. As a potential prototype for the future, let us consider the distribution of the energy consumption as averaged over the United States. Figure I.1 presents one view of this problem, dividing energy into comfort heat (heating and cooling residences, stores, factories, hot water, etc.), process heat (in manufacturing) and work (electricity, transportation, etc.).

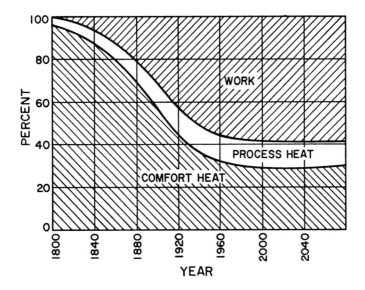

Figure I.1. A projection of the relative proportions of three components of the energy use system, to the year 2050, for the United States. After Putnam [9], with permission.

Note that the present division for energy allocation is predicted (by Figure I.1) to remain constant for several decades. Table I.3 provides a more detailed viewpoint for two selected years. Note the shift in how energy is used, and how the industrial sector is becoming more efficient.

Overall, how efficient is our use of energy? In 1968, the average efficiency appears to have been in the neighborhood of 32% [10] with individual area efficiencies ranging from 60% for heating usage to 15% for the energy efficiency of transportation. In 1990, the average efficiency of energy use was estimated to be 35%. Some additional improvement in efficiencies is possible (for example, replacing 100 watt incandescent light bulbs in homes with 15 watt fluorescent tubes which yield the same amount of light could improve the overall efficiency of energy use by one

Table I.3

Energy usage in the United States

Energy Area of Use	Percent of Total National Energy Use	
	1968 [1, 2, 10]	1990 [4]
Residential	19.2	21.3
--Heating and cooling	12.9	14.5
--Electrical	6.3	6.8
Commercial	14.4	14.6
--Heating and cooling	6.2	6.3
--Electrical	4.9	4.9
--Miscellaneous	3.3	3.4
Industrial	41.2	37.0
--Heat (steam and direct)	28.0	25.0
--Electrical	9.3	8.4
--Miscellaneous	3.9	3.6
Transportation	25.2	27.1
--Fuel	24.9	26.8
--Raw materials	0.3	0.3
National Total	100.0	100.0

to two percent), but it should be noted that most machines are heat engines and their efficiencies are limited to no more than those of the Carnot cycle (see the section on thermodynamics later in this chapter). This implies a realistic upper bound on energy use efficiencies of approximately 40%. This is some improvement in efficiency over current levels, but this improvement is not significant in the sense that the improvement, while reducing energy use, does no more than postpone the day on which our capital energy resources will become exhausted. We will see that new energy sources are still required.

Consider the political and lifestyle aspects of conservation. An example of extreme conservation is the use of energy at the level of 1850

(2,930 kwh per capita) rather than at the 1990 United States level of ~100,000 kwh per capita. For the vast majority of people on this planet, this represents a sharp decrease in the amount of energy consumed, and, therefore, a major reduction in their quality of life. It is clear, from the historical record of our species, that any such abrupt change will be accompanied by major political upheavals. A less violent form of "conservation" might be a reduction in the average U. S. energy consumption figures from ~100,00 kwh per capita per year to the world average of ~25,000 kwh per capita per year. The change in lifestyle dictated by this degree of conservation would affect citizens of the industrialized nations far more than those of the third world. To this extent, the political problems would be simpler than those for the extreme conservation. Clearly, both of these conservation approaches would, of necessity, be accompanied by a change in the lifestyle, of major proportions, for the average U. S. citizen. Perhaps, it is not reasonable to label such approaches as conservation, in that we normally define conservation as a combination of improvements in the efficiency of energy use and decreases in energy requirements such that an overall reduction in energy use of some five to ten percent is achieved. However, whatever the label, we are discussing major changes and the resulting resistance by the human populations involved.

Conventional Sources of Energy

Table I.4 lists the principal energy sources currently being utilized by Homo sapiens sapiens. Referring to Table I.1, the reader will note that both capital and income energy sources are currently being tapped. We now need to briefly examine each of the energy sources in Table I.4, keeping in mind that each source has certain advantage (e.g., petroleum is very portable) and certain disadvantages (burning oil produces carbon dioxide and enhances the greenhouse effect) and that the problems of energy necessarily involve environmental, political and supply factors.

Petroleum, the most widely used energy source, is a fluid and is easily transportable with currently estimated world wide reserves of 12.7 Q [12]. In considering energy reserves it is wise to be cautious, because any estimate of the reserves of any material (be it oil, coal, uranium or breadfruit) depends on two factors: First, how well do we know the geology/geography of our planet and where the substance might be found?

Second, how much is the resource worth? As energy sources become scarce, the price of energy will increase and sources that were once uneconomic to develop will become attractive. To compound this problem

Table I.4

Energy source contributions to energy needs [4, 8, 11]

Energy Source	Current contribution to energy used	
	U.S. % World	
Fossil Fuels	85.5	83.0
--Coal	23.4	25.7
--Oil	41.3	37.6
--Natural gas	23.8	19.7
Nuclear	7.6	5.5
--Fission	7.6	5.5
--Fusion	0.0	0.0
Geothermal	0.1	0.1
Hydropower	3.6	5.9
--Dams	3.6	5.8
--Tidal	0.0	0.1
Biological	3.0	3.4
--Wood	3.0	3.2
--Other	0.0	0.2
Solar	0.2	2.1
--Wind	0.1	0.2
--Thermal	0.1	1.8
--Electrical	0.0	0.1

scarcity can result from depletion of the source material through use or from other human actions (politics, war, etc.). The estimates given here result from attempts to take all of these factors into account; an extremely complex task. As an added complication, petroleum is used for more than an energy source. It is used to make a variety of products, ranging from fertilizers to pharmaceuticals; and some thought needs to be devoted to this fact before all the available petroleum is burned to generate energy.

Besides the classical "oil well", additional sources of petroleum lie in tar sands and oil shale deposits. It is difficult to measure the amount of these additional sources of petroleum , but we shall estimate them to be 9.0 Q. To further complicate matters, the extraction of petroleum from tar sands and oil shale deposits is not, as yet, a practical matter--in other words, considerable engineering needs to be done. All-in-all, we have estimated reserves of oil of some 21.7 Q. If burned to produce energy this amount will yield considerable carbon dioxide and other air pollutants (a 1,000 Mwe power plant burning oil emits over 158 tons of pollutants into the air each year [8]).

Coal is our most plentiful fossil fuel energy source. However, it is a particulate solid and is, therefore, not easy to transport. Additionally, there are environmental problems, both where coal is mined and where it is burned (among these problems are acid rain and the greenhouse effect). Estimated coal reserves are 31.3 Q world-wide [12].

The third fossil fuel is natural gas. This gaseous material is easily transportable, easily stored, but does, like coal and petroleum, lead to carbon dioxide and the greenhouse effect when burned. Edmonds and Riles estimate [12] that approximately 10.8 Q of natural gas are potential-ly available on a world-wide basis.

The sum of known recoverable energy reserves for fossil fuels is some 63.8 Q. Certainly there are undiscovered reserves; perhaps sufficient to double the above number. For our purposes, it is not overly critical what value is chosen for fossil fuel energy reserves, so long as it is clear that they are finite.. For this work, let us take the 63.8 Q reserve value computed above as the practical total of effectively extractable fossil fuel energy.

How many years will these fossil fuel energy reserves last? Let us consider events on a world-wide basis and examine five possible scenarios. These will be: (A) the world population holds at its current five billion with a per capita energy consumption of 3,000 kwh per year (effectively the value for 1850 considered earlier); (B) the world popula-tion and per capita energy use stay at the current levels (five billion people and 25,000 kwh per year); (C) everyone in the world consumes energy at the current rate exhibited by citizens of the United States (five billion people using energy at a rate of 100,00 kwh per person per year); (D) the world population climbs to some 10 billion while each individual uses energy at 100,00 kwh per person per year; and (E) the world population climbs to 10 billion while each individual uses some 150,000 kwh per year.

Clearly, none of the presented scenarios will be the one which is actually followed. However, between them they span the gamut of possibilities from extreme conservation (Scenario A) to an attitude best describe as "if we ignore the problem it will go away" (Scenario E). The intent is not so much as being precise as to future energy use, but to demonstrate that there is a time limit to how long we can use the capital energy resources, and that the time limit is relatively short. The results of scenarios for fossil fuels are provided in Table I.5.

Table I.5

The estimated time to exhaustion for fossil fuel energy reserves, if they are the sole source of energy.

Scenario	A	B	C	D	E
Population (billions)	5	5	5	10	10
Per capita energy consumption rate (kwh/year)	3,000	25,000	100,000	100,000	150,000
World annual energy consumption (Q)	.051	.427	1.71	3.41	5.12
Years until exhaustion of fossil fuel reserves	1,251	149	37.3	37.3	12.7

Historically, a drastic shift in energy consumption has been accompanied by some form of catastrophe (war, pestilence or a climatic change). Scenario A represents such a catastrophic situation, one of extremely violent potential, while giving the human race over a millennium of energy. Scenario B, while maintaining the current world average of energy consumption, yields but a century and a half before we exhaust our fossil fuel reserves and either continues the current imbalance in energy use between industrial and third-world countries or results in major conservation efforts by the industrial nations. Scenarios C, D, and E, run through available fossil fuel resources in extremely short periods of time. Current estimates of the population for the world in the early 21st century fall between six and eight billion. Thus we can expect realistic future energy use to fall somewhere between Scenarios B and E [13]; placing a limit of less than a century on the time to exhaustion of fossil fuels.

What about other conventional energy sources? Another capital energy source is nuclear fission. Estimates of the reserves for this energy source vary widely. A reasonably conservative estimate for the reserves for nuclear fission is 8.5 Q if light-water rector technology (LWR) is employed [12] and perhaps 600 Q [14] for a scenario involving the use of breeder reactor technology*. Returning to the five scenarios, what additional time does the use of nuclear fission provide? To answer this question, consider Table I.6.

The time our energy sources will last is considerable, but not infinite. What price must be paid for this extension in time? Fortunately we do not have additions to acid rain or to the greenhouse effect. However, some of the waste products of nuclear fission are radioactive, with radioactive half-lives extending over tens of thousands of years. The storage of such radioactive isotopes can be accomplished, in glass and underground [15]. There remain potential problems with reactor accidents (such as Three Mile Island or Chernobyl) and the possible use of nuclear materials to make explosive devices. The major problem with nuclear fission is the publicly perceived horror of "things" nuclear.

Geothermal power is the last of the capital energy sources listed in Table I.1 to be in current use. Speaking in human terms, it is possible to treat geothermal energy either as an energy income source or as an energy capital source. The heat energy which is present in the earth's core regions can be pulled quickly from the center of the earth or allowed to slowly trickle out. At present, the normal practice is to locate a geological

---

* In breeder reactor technology, excess neutrons (neutrons not used in producing energy via nuclear fission in the nuclear reactor) from the reactor are employed in converting material (such as thorium) to nuclear fuel.

Table I.6

The time in years to exhaustion of nuclear fission resources when used as the sole energy source

Scenario	A	B	C	D	E
Population (billions)	5	5	5	10	10
Per capita energy consumption rate (kwh/year)	3,000	25,000	100,000	100,000	150,000
World annual energy consumption (Q)	.051	.427	1.71	3.41	5.12
With LWR technology (years)	167	19.9	5.0	2.5	1.7
With breeder reactors (years)	11,760	1,410	351	176	117

site at which heat from the earth's core is close to the surface and to extract heat from this source, either by allowing steam to escape from the heat source or by pumping water down to the underground lava and then recovering the resultant steam. Once steam is available; it can be used to drive an electrical generator, in similar fashion to what is done with the steam produced by burning coal, oil or natural gas; or the steam produced by nuclear fission generators. There are a number of major geothermal "fields" in various locations around the world: Larderello, Italy; Wairaki, New Zealand; The Geysers, California and Cerro Prieto, Mexico; to name a few. The present extraction rate of energy from geothermal sources is close to 0.00042 Q a year. This rate may or may not be sustainable since data from one field, The Geysers, indicates that the heat flow into this geothermal field from within the earth is dropping [4, 16]. Averaged over the surface of the earth, the heat flux of geothermal energy is very low (0.06 watts per square meter) and is at a relatively low temperature. Thus, the field approach indicated above is favored. Note that the majority of these fields are geologically unstable and, hence, are prone to suffer from earthquakes. However, considered as a mined capital source, there are possibilities for geothermal energy; and it is these we study here. An estimate for the total recoverable energy from geothermal sources is 56 Q with a 50 year life span for any given geothermal "field" [17]. Taking the five scenarios considered earlier, the time to exhaustion of our geothermal resources, when used as the sole energy source, is provided in Table I.7.

Table I.7

The time to exhaustion of estimated geothermal energy reserves with geothermal as the sole energy supply

Scenario	A	B	C	D	E
Population (billions)	5	5	5	10	10
Time (years)	1,098	131	32.7	16.4	10.9

We have now covered the principal energy sources which the human race is currently using. Note that they are all capital energy sources and that all of them adversely affect our environment (steam is not the only gas that escapes from a geothermal field, and many of the gases that do escape, for example hydrogen sulfide, are noxious). Adding up the time we have available from these sources, we have the results listed in Table I.8.

Recall that these time scales are based upon using fossil fuels as fuels, not as sources of chemicals or as lubricants. Also consider that, barring a major shift in civilization and the behavior of Homo sapiens sapiens, Scenarios A and B are unlikely and that the futures projected in Scenarios C and D are much more likely to be close to the actual energy demand experienced.

Now consider the energy income sources in Table I.1. We currently tap all of these energy sources, but only two are responsible for any significant energy input. The use of hydropower, primarily from dams but in some instances also from tidal generators, is widespread. Current

Table I.8

The time (in years) until exhaustion of our capital energy resources (fossil fuels, nuclear fissionables--assuming breeder reactor technology--, and geothermal energy)

Scenario	A	B	C	D	E
Population (billions)	5	5	5	10	10
Per capita energy consumption rate (kwh/year)	3,000	25,000	100,000	100,000	150,000
World annual energy consumption (Q)	.051	.427	1.71	3.41	5.12
Exhaustion Time (years)	14,110	1,690	421	216	147

annual energy production is estimated to be 0.025 Q and the long range potential for energy produced from the source is conservatively estimated to be 0.06 Q [16]. The bulk of current energy production from this source is a result of damming rivers and streams. A significant fraction of the eventual potential consists of damming tidal estuaries such as those in Breton and Newfoundland and converting the energy in tidal flows to electrical energy.

Let us return to our five scenarios for the future. With an income resource such as hydropower, we do not ask how long the resource will last, but, what is the carrying capacity of the resource for the human population? Table I.9 provides an answer to this question for each of the five scenarios.

There remains a final conventional source of energy. Mankind has burned wood, farm wastes (straw, animal dung, etc.) and other biological materials for thousands of years. If we are careful to plant new trees, and other flora, so as to replace those burned for fuel, there is a small but steady supply of energy. The estimate provided for wood derived energy provided in Table I.4 is imprecise, but if we make use of it as our basis

Table I.9

Sustained human population level with hydropower as the sole energy source

Scenario	A	B	C&D	E
Annual per capita energy consumption (Q per year)	1.0×10^{-11}	8.5×10^{-11}	3.4×10^{-10}	5.1×10^{-10}
Supported population (billions)	6.0	.706	.176	.118

for discussion, the burning of wood, as an income source of energy is capable of yielding 0.013 Q a year. Note that, if we plant trees and other plants to replace those we burn and maintain a true income energy source, then wood, farm wastes and other biological materials are neutral with respect to the greenhouse effect--the growing plants consume the carbon dioxide produced by burning. In Table I.10 we sum these two energy income sources in present use and determine their carrying capacity in light of our five scenarios.

Table I.10

The sustained population level with hydropower and "wood" as the only energy sources

Scenario	A	B	C&D	E
Annual per capita energy consumption (Q per year)	1.0×10^{-11}	8.5×10^{-11}	3.4×10^{-10}	5.1×10^{-10}
Supported population (billions)	7.3	.859	.214	.143

Note that for the case of Scenario A (the lifestyle of the mid-nineteenth century) hydropower and "wood" are capable of satisfying the energy requirements of the human race. Your author is not at all confident that he, or any of you, the readers, would really enjoy such a lifestyle.

In the case of Scenario B, the energy income sources under consideration are capable of supporting a sizeable portion of the human race and thereby extending the time to exhaustion of the capital energy resources significant.y. The same can be said, on a much more modest basis, for Scenarios C through E. However, in all of these situations, the conventional capital energy sources will eventually become exhausted and the world population will need to undergo a drastic downward shift (in the case of Scenario E this downward shift in population results in a population decrease of over 98%) or the human population will need to change its lifestyle drastically. There has to be a better way, and the following section will consider possibilities for a "kinder, gentler world" as a result of improvements in energy supply.

Alternative Energy Sources

The alternative energy sources considered here are, on a human time scale, both energy capital and energy income. We begin with the remaining capital energy source from Table I.1: Nuclear Fusion.

Nuclear Fusion
If, instead of breaking atoms apart as is done in nuclear fission, we resort to putting them together, nuclear fusion energy can be obtained. If, we use deuterium or tritium (both are forms of heavy hydrogen) as a fuel, several possible fusion reactions are [8]:

$$_1D^2 + {_1}D^2 = {_2}He^3 + \text{a neutron} + 3.27 \text{ Mev},$$

$$_1D^2 + {_1}D^2 = {_1}T^3 + {_1}H^1 + 4.03 \text{ Mev and},$$ (I.2)

$$3\{{_3}Li^6\} + {_1}D^2 = 5\{{_2}He^4\} + 22.4 \text{ Mev}.$$

In the above, H is hydrogen, D is deuterium, T is tritium, He is helium and Li is lithium, and the energy produced is given in millions of electron volts (see Appendix A).

To perform fusion, the fuel is heated to a temperature of approximately 10^8 °K and must be confined, in space, at this temperature, long enough (on the order of 1/4 of a second) to enable the deuterium atoms to collide with each other, realizing more energy than is required to initiate the process*. Work has been proceeding on this source of energy for considerable time [15, 18-20], but various estimates as to the date of achieving practical fusion energy still range from a time 50 years in the future to never. Additionally, there are problems with potential radioactive waste products. The end products of the nuclear reactions taking place (Equation I.2 does not exhaust the total list of possibilities) have considerable kinetic energy. These products are slowed down by collision, producing heat (which, in turn, can be used to produce steam, which is used to turn a turbine, and so generate electricity) and radioactive byproducts, slowed down in the reactor shield/kinetic-to-thermal energy converter surrounding the fusion chamber. These radioactive byproducts must be stored for a sufficient length of time (several thousand years) to "cool down". In the event that mankind is smart enough to solve the puzzles inherent in nuclear fusion (both in the fusion process itself and in disposing of its waste products), there is sufficient deuterium in the oceans of this planet to completely satisfy the energy needs of the human race for several million years.

Solar Energy

The remainder of the energy sources listed in Tables I.1 and I.4 may all be classified as solar energy derived. The length of the casual chain between the nuclear fusion reaction occurring in the sun and our eventual use of some form of energy may very well vary, but all of these energy sources are dependent on the existence of the sun. Each hour the earth receives 173×10^{12} kwh of energy from the sun. Over a year, this corresponds to 5,160 Q, a figure more than 12,000 times the current energy requirements of the human race. Not all of this energy reaches the surface of the earth. A portion is reflected by clouds, by the oceans and by the land. This amounts to some 1,570 Q [21]. An additional 1,120 Q

---

* The fusion process outlined above and discussed in this work is closely allied to that utilized by stars. This work will not address "cold fusion". The mechanisms of cold fusion have not been demonstrated to be fusion, if they exist at all--the nature and reality of this process are not certain.

is employed in evaporating water from the oceans, lakes and rivers. The remainder, 2,490 Q, is available for such purposes as powering photosynthesis, warming the surface of the earth and providing energy for the human race. Utilizing land-based solar energy collector/converters alone, the potential solar energy supply available for use by man is in the neighborhood of 1,100 Q. This value is still over two thousand times the present energy requirements of the human race.

In discussing solar energy we do have the option of considering ground-based solar energy collection systems, as implied in the preceding paragraph, or some type of solar energy collection system that is operated in space. A major advantage of space-based systems is that sunlight is continuously available. The disadvantages inherent in space-based systems are of two kinds. First, a considerable amount of energy must be invested in orbiting the energy collection/conversion system. Second, the energy so collected, must be transported back down to the earth's surface. This transportation system is most likely to be some type of microwave beam and, as such, is likely to cause difficulties with the ozone layer as well as serving as a potential danger should the beam somehow be misdirected from the energy receiver on the ground and strike some nearby population center. For the purposes of this chapter we will consider only ground-based systems and reserve further discussion of the implications of space-based systems for the second chapter.

We have already discussed several solar-based energy schemes. For example, our fossil fuels were once living plants whose growth energy was powered by light from the sun. The hydropower systems (dams on rivers and in tidal estuaries) we discussed depend on the existence of the sun in evaporating water and driving tidal flows. Other possibilities for solar energy are listed in Table I.11.

Consider the possibilities of Table I.11 in broad detail. When considering ground-based energy conversion schemes, it is necessary to remember that the sun is not always shining, due to weather (clouds) or to the earth's rotation. Thus, solar energy conversion occurs, on an average, some 12 hours of each day. Therefore, some method of energy storage must be employed to assure the availability of energy on a 24 hour basis. In some of the following solar energy conversion/collection schemes (such as in the growing of trees for use as firewood) the energy storage system is an integral part of the conversion system; in many other cases (e.g., the generation of electrical energy from sunlight using solar cells) it is quite distinct.

Table I.11

Solar powered energy sources

Immediate source	Remarks
Biological --Trees --Specialized plants --Waste materials	In part, already discussed, this energy source does not utilize the energy in sunlight very efficiently, but it improves the greenhouse situation.
Temperature Differences --Wind --Waves	Primarily windmills. However, it has been estimated that the United Kingdom could use energy from ocean waves as its sole energy source.
Solar-Thermal --Heating & Cooling --Mechanical	The provision of heat for heating and cooling buildings, process heat and hot water.
Solar-Electric --Solar-Thermal-Electric --Thermoelectricity --Photovoltaics	A variety of techniques for converting the energy in sunlight to electrical energy.

Let us begin our discussion of solar energy with the biological utilization of solar energy. In its simplest form, the chemical reaction known as photosynthesis may be written as:

$$H_2O + CO_2 + \text{Light Energy} \longrightarrow CH_2O + O_2, \tag{I.3}$$

where C is carbon and O is oxygen.

Besides forming the carbohydrates $(C_nH_{2n}O_n)$ that frequently provide us with food and fuel energies, this reaction yields oxygen; indeed all of the oxygen in the earth's atmosphere comes from this source.

In actuality, the reaction that takes place in plants is considerably more complicated than that indicated in Equation I.3 [22], and the predicted maximum efficiency attained by plants when converting solar energy into chemical energy (i.e., storing the energy obtained from the sun in assembled carbohydrates) is only about 4.5%[*]. This efficiency is not very high and, moreover, conditions for plant growth are rarely ideal. In desert regions lack of water forces plants to be sparing of leaf and sparsely distributed. Most of the sun's energy, therefore, goes into heating the ground. In other locations, such as salt pans or parts of the oceans' surfaces, the required nutrients may be absent or some harmful material may be present in significant amounts. In winter, deciduous trees must "hibernate" to avoid internal freezing, and so shed their solar energy processing leaves.

In considering the use of biological processes as energy sources, we must consider other requirements or boundary conditions: (1) the world's supply of oxygen and carbon dioxide must be maintained at relatively constant levels; too little oxygen and mankind will be asphyxiated; too much carbon dioxide and there is the problem of the greenhouse effect and a potential major upward shift in the earth's temperature; (2) the human race and other species need to be fed, a process which currently requires that substantial land areas be devoted to food crops[#]. Furthermore, a growing plant requires more than carbon dioxide, water and sunlight. Nitrogen and a host of trace elements are required for the proper functioning of plants. If we grow plants and burn them to provide energy or eat them, these trace nutrients must be returned to the soil if we are to be able to continue to grow plants on the same plot of ground. The energy cost of such fertilizing operations could easily exceed 33% of that

---

[*] Under ideal conditions, the maximum observed efficiency for photosynthesis lies between four and five percent--in excellent agreement.

[#] Field crops average a solar energy conversion efficiency of approximately 0.3% [23]. At 2,000 kilocalories per day per person, in food intake requirements for humanity and with each square meter of ground receiving some 3 kwh of solar energy each day (this number includes adjustments for weather and the earth's rotation and revolution about the sun--see the next chapter), the minimum area of land required per person to produce food is approximately 360 square meters. Considering fertilization of the farming area, crop rotation, summer/winter temperature variations, and pest problems; an amount of land some four to five times this area is more realistic.

represented by the full grown plant [24]. We have already briefly discussed the burning of wood and farm wastes for energy; a practice that has been followed for millennia. We are not restricted to conventional trees in this connection. A number of varieties of fast growing trees have been investigated [24, 25, 26]. If we use one or another variety of these fast growing trees, it has been estimated that approximately one acre of trees would be required to provide the annual electrical energy requirements for an individual [24]. On the average, these fast growing trees require some eight years to reach their full growth. Thus, to supply all of the required electrical energy for an individual, some eight acres of fast growing trees would be needed.

We are not restricted to trees. As a source of energy, large-scale (several hundreds of kilometers square) kelp farms have been proposed [24, 27]. The combination of slightly less than 2% energy conversion efficiency, coupled with the use of the coastal shelf areas of the continents, is a powerful argument for the use of kelp. The kelp grown would be harvested and turned into energy by burning to generate steam. Of course, it requires energy to harvest, transport, and convert the kelp to the energy we require. This, clearly, reduces the overall efficiency.

There are many other plants, ranging from algae to tree sized that can be grown for energy. Corn is grown in the United States and elsewhere for conversion to ethanol, which is then used as a fuel in internal combustion engines (automobiles [26]). There also exist shrubs (members of the genus Euphorbia) which produce significant quantities of a milk-like sap composed of hydrocarbons in water. It has been suggested [28] that this sap could be converted into a gasoline substitute and into other petroleum replacing products. These shrubs grow in arid regions and preliminary estimates suggest that an acre might support enough plants of this type to produce between 10 and 40 barrels of "oil" at $18 to $30 a barrel each year. In such a case, planting an area the size of the state of New Mexico with Euphorbia might go a long way toward supplying the United States with fuel for its automobiles.

On a cellular level, research is being conducted into bacteria such as the purple bacteria [29] that respond well to infrared and visible light. This, in turn, has led to considerable interest in the biology and chemistry of membranes.

Note that all of the preceding biological sources are greenhouse neutral. The carbon dioxide that they release when burned is removed from the atmosphere during the growth of the next generation of plants.

Another biological energy source lies in the waste materials of our sewage and garbage systems (methane gas is a common byproduct of our garbage dumps). Once the inorganic wastes are separated out, the remaining organic material can be burned (or digested by bacteria), yielding heat; methane for use in automobiles, and electrical energy [25]. Since this source utilizes energy rejected by our current lifestyles, the net effect is more efficient operation of civilization. Assuming some five billion people, providing some 1×10^{12} pounds of recoverable wastes per year and that the net recoverable energy is 2 kwh/pound (this is some 1/5 of the energy in soft coal), then this energy source alone would supply 0.003 Q each year.

In summary, on a short-term basis, we have already seen that the energy obtainable from burning plants and animal wastes has an annual value of 0.016 Q. There is promise of considerable additional energy from other biological sources, however, it is too early to estimate which, if any, approach will be a viable future energy source. A conservative estimate for future biological energy is 0.03 Q. It is possible to increase this value, but, considering the low conversion efficiencies inherent in biological processes it would seem that some other energy source may prove to be preferable.

Temperature Differences

The fact that the temperature varies from one point to another on the earth can be utilized to supply us with energy. These temperature differences provide the driving "force" behind winds, waves and ocean currents. An estimate [30] for the amount of solar energy annually going into winds, waves, and currents is 11 Q. A major problem with winds and ocean currents as energy sources is their diffuse nature. The use of ocean currents also presents an environmental problem. For example, if we were to extract energy from the Gulf Stream off the coast of Florida, what would be the effect on the climate of northwestern Europe? Would such an operation result in the possibility of a new ice age?

In the case of ocean currents, energy storage is not a problem as the major ocean currents flow continuously. However, their geographical location does change with the seasons, forcing a mobility on any energy conversion system. Considerable research into wave generated electrical power is also being conducted, with a good deal of optimism. This source is particularly attractive to island nations such as the United Kingdom and to continental sea coasts.

Historically, the use of wind as an energy source dates back to the 13th century in Europe and even earlier in the Near East. The basic procedure, today, is to mount an electrical generator on a tower, put a propeller on the generator and either store the energy produced in batteries or feed the energy into the local utility grid. The storage problem is significant, since we cannot guarantee a constant wind velocity. However, it has been shown that, even without storage, wind generated electric power can produce a significant fraction of the local energy requirements and at a cost competitive with conventional (oil/coal/natural gas fueled) energy systems [26, 31, 32]. Problems encountered in implementing wind powered energy generators include the frequent necessity of erecting the generator propellers on tall towers. These towers: (1) are unsightly, (2) interfere with television transmission, (3) both towers and propellers are susceptible to damage from wind gusts, and (4) the propellers are potentially noisy [33]. Many regions are too calm or too stormy to make wind power practical. However, in areas of reasonably steady wind with relatively constant wind velocity, such as the Midwestern region of the United States, the use of wind power is found to be advantageous. In Figure I.2, the regions of highest available wind power are deliniated on

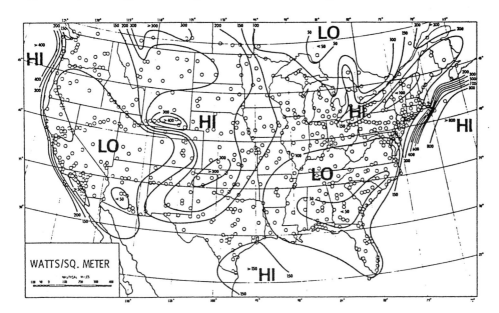

Figure I.2.   Regions of high and low wind power for the continental United States. From Solar Energy, an ERA publication by Easton, with permission.

a map of the continental United States. The highest values of wind energy in the United States are in the Midwest region and on both coasts (where average power levels may rise to in excess of 500 watts per square meter. The potential wind energy in the U. S. is some 0.005 Q [8] and on a world-wide basis we can estimate a value for the energy obtainable from wind of approximately 0.06 Q.

Another energy producing temperature effect is that due to the difference in temperature between the ocean surface (about 15 °C) and the ocean depths (approximately 5 °C). It is possible to construct a generating device which utilizes this temperature difference to generate electrical energy. The generator and support equipment must be several hundred meters in length in order to "reach" regions at both temperatures, and some method is required to transport the generated electrical power from the ocean generating sites to land. A major problem with these devices is due to their low efficiency. A brief aside concerning the topic of thermodynamics is in order.

Thermodynamics

The laws of thermodynamics may be used to set an upper limit to the efficiency with which any heat engine (or pump) can operate. One such type of engine, and the most efficient, is the Carnot cycle engine. The Carnot cycle engine extracts energy from a hot (high temperature) energy reservoir and rejects a portion of this energy to a cold (low temperature) energy reservoir. The net difference in energy is available to do useful work. The efficiency of a Carnot cycle engine, η_c, is given by:

$$\eta_c = [1 - T_c/T_h] \times 100\%. \qquad (I.4)$$

where T_c is the temperature of the cold reservoir and T_h is the temperature of the hot reservoir. The temperatures are in degrees Kelvin (°K), an absolute temperature scale based on the laws of thermodynamics. For any heat-driven, energy conversion system which we can construct, it can be shown, using the laws of thermodynamics, that the efficiency cannot be greater than that of a Carnot cycle engine operating between the same two temperature reservoirs and that this maximum efficiency is given by Equation I.4.

Ocean Temperature-Difference Generators

Returning to the electrical generating system driven by the temperature difference between the surface of the ocean and the ocean depths, the Carnot efficiency for such an OTEC (Ocean Thermal Energy Converter) may now be estimated:

$$\eta_c = [1 - (5 + 273)/(15 + 273)] \, x100 = 3.5\% \qquad (I.5)$$

This is the maximum efficiency obtainable. Realistically, including transport losses and the maintenance for such a deep ocean generator, the operating efficiency will be in the neighborhood of two percent. Approximately 1,740 Q of solar energy falls upon the oceans and is turned to heat, assisting in evaporation [34]. If we were to utilize the entire 1,740 Q as input for OTEC systems, about 35 Q of energy could be obtained. Note that to do this, we would have to fill the oceans with OTEC generators. In turn, this introduces two new problems: (1) the earth's climate would change since we would alter the ocean temperature, its surface reflectivity, stop water evaporation from the oceans, and pump energy from the oceans to the land masses (continents), and (2) in addition to energy, the human race faces a limited set of elemental resources [35]. Filling the oceans with generators would totally exhaust other resources, such as copper for wiring. On a limited basis, and in selected locations which favor such an operation, we can be optimistic about OTEC systems. Combined with wave and ocean current energy, a conservative estimate for the amount of energy available on a yearly basis might well be 0.04 Q.

The alternative energy sources, income energy sources, we have so far considered, and we include hydropower from the conventional source list earlier in this chapter, are all secondary sources. In the final analysis the basic energy source is the sun, and mankind is merely tapping a secondary source when we burn a log, erect a windmill or convert the potential energy of water stored behind a dam. These secondary sources, provide us with a potential annual energy income of 0.06 Q (hydropower), plus 0.03 Q (biological sources), plus 0.09 Q (wind, wave and OTEC), totaling 0.18 Q. Table I.12 indicates the population which this amount of energy can support considering the various proposed scenarios.

None of these secondary energy sources is very efficient and, thus, they do not possess the capacity to even support the world as it currently exists (Scenario B). Let us now investigate some potentially more efficient and greater (in amounts of potential energy) alternate energy sources.

Table I.12

The population supported by secondary solar energy sources (water power,
biological sources and temperature differences)

Scenario	A	B	C&D	E
Annual per capita energy consumption (Q per year)	1.0×10^{-11}	8.5×10^{-11}	3.4×10^{-10}	5.1×10^{-10}
Supported population (billions)	18	2.1	.53	.35

Solar-Thermal

In considering solar energy systems which produce only heat
energy for our use, we are restricting ourselves to approximately 40% of
the energy requirements of modern civilization (See Figure I.1). Additionally, since it is difficult to transport heat energy over any but the shortest
distances, the regions of heat energy production must be close to the areas
of heat energy utilization. We cannot, for example, expect to efficiently
pipe heat from central India to northern Siberia, or from the south Pacific
Ocean to Canada. Thus, we are limited to producing the heat from sources
on land and close to the point of end use.

Of the 5,180 Q in solar energy impinging on the earth each year,
approximately 1,100 Q is available on the land surfaces [13]. This amount
must be further reduced by allowing room for growing foodstuffs, carbon
dioxide absorbing plants, and areas for human and animal activity. Despite
these problems, man has made use of thermal energy from the sun for a
long time. The greenhouse, the medieval solar, solar box cookers, large
scale solar furnaces, hot water heaters and solar heated structures are all
examples of our use of solar energy for heat. There exists an extensive
literature, both in the popular press [36] as well as in the scientific realm
[37-39]. Solar homes and buildings, such as the school shown in Figure
I.3 are prominently featured. Performance calculations for individual
residences [38, 40, 41] generally indicate that in most low latitudes
(within 40° of the equator) sufficient solar energy to supply their heating
requirements impinges on most residential and commercial buildings.

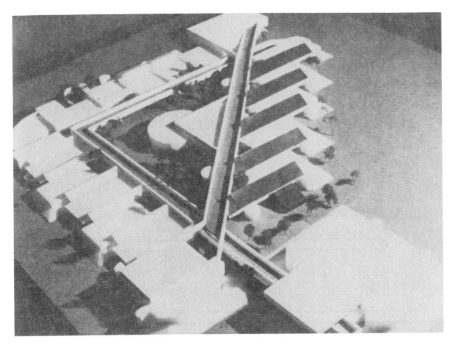

Figure I.3. Corona del Sol High School, Tempe Arizona. The shed like collectors (the black-looking panels on the right side of the model in the photo) collect solar heat and pipe it throughout the 264,000 square foot complex, providing heat and domestic hot water. Predicted savings in fuel costs are in the neighborhood of $70,000/year. Neil Koppes, Arizona Highways, August 1975.

The subject of cost in selecting alternative energy sources has not been mentioned to this point. This is because it is extremely complicated. Considering the assigning of values to such items as depleted resources or to greenhouse effects, cost projections are at best tenuous and at worst totally misleading. This topic will be addressed, briefly, in a later chapter which will consider economic factors in solar cell work. A short word on the economics of solar heating is not, however, out of place here. Tybout and Lof made a number of calculations [42] which show that flat-plate solar energy collectors (see Figure I.4) produce heat energy, and hot water, at prices that are cost competitive with conventional energy systems. Payback periods for domestic water systems in the southern United States range from two to five years; less than half of the lifetime of typical systems. A major problem with many hot water systems is that they require a lot of upkeep, and many home owners seem to be unwilling to perform the required maintenance.

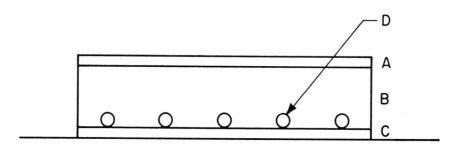

Figure I.4. A flat-plate collector: (A) An optically transparent plate which acts as an infrared trap, (B) A space filled with air (or vacuum), (C) The heat absorber, and (D) Pipes through which air or a fluid is circulated to absorb heat and transport it to the point of use. Instead of pipes, a sheet of fluid is often allowed to flow over (C). Considerably more complex systems have been proposed [43], but, in the final analysis they are essentially as shown here.

How does a solar-thermal system work? The heat from the sun is first trapped within the collector body (see Figure I.4) using some version of the greenhouse effect. This heat is then transported to its use point, if required immediately, or stored (typically as a hot fluid such as water or by raising the temperature of some object such as a bed of pebbles [44-46]. All of these activities (heat trapping, heat transport and heat storage) result in reduced system efficiencies--the net system efficiency is often less than 10%.

Another use of heat energy has been to drive various mechanical devices--such as water pumps for irrigation in Egypt. An excellent early review is provide by Jordan and Ibele [47]. As is the case for solar heating and cooling, the energy obtained for mechanical system usage is utilized locally. Also, as for solar heating and cooling, the net system efficiency is not high and a significantly large number of square miles of solar-thermal collectors is required to supply the necessary energy (see Chapter II).

How much energy could solar heating supply to a future civilization? On a world-wide basis, it is felt that 35% of the total heating and cooling requirements could be met by solar energy. If we assume that heating and cooling require 40% of the total energy load (see Figure I.1), then 14% of the world's energy requirements could be met by various solar-thermal heating and cooling schemes. For our five scenarios we could have the following solar-thermal contributions.

Table I.13

The energy supplied by solar-thermal systems

Scenario	A	B	C	D	E
Population (billions)	5	5	5	10	10
Per capita energy consumption rate (kwh/year)	3,000	25,000	100,000	100,000	150,000
World annual energy consumption (Q)	.051	.427	1.71	3.41	5.12
Energy supplied by Solar-Thermal (Q)	.007	.060	.239	.477	.717

Solar-Electric

There remains one major category from Table I.11--that dealing with techniques of generating electrical energy more or less directly from solar energy. As a form of energy, electricity is extremely convenient. It is easily transportable, can be readily converted into other forms of energy (such as heat in a toaster, or chemical energy in a battery) and can be stored in a number of forms with fair efficiency [48]. For example, we can store electrical energy in electrical form as the charge on the plates of a capacitor, in various chemical forms such as in a rechargeable battery or in hydrogen (obtained by the electrolysis of water*), or in mechanical forms such as in water pumped into a reservoir above a dam, or in a rapidly rotating flywheel.

One method of generating electricity from sunlight which is very successful is the solar-thermal-electric conversion system (STEC). A commonly encountered system of this type is sketched in Figure I.5. In

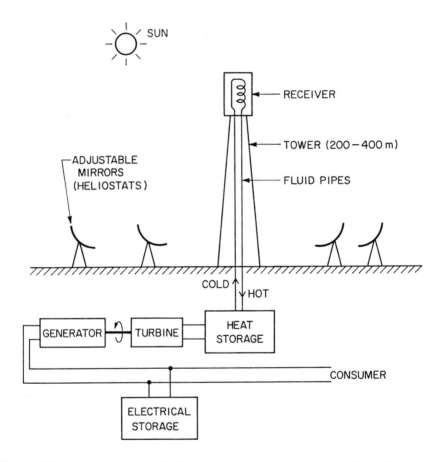

Figure I.5. A solar-thermal-electric-conversion system (STEC).

---

* Study and development of the processes allowing solar energy to lead to the electrolysis of water producing hydrogen continues, but the process cannot be considered to be mature at the present time [49].

this system, sunlight falls on a large number of mirrors which reflect and focus the sunlight onto a central receiver. This is necessary because of the low power density in sunlight. The maximum power density of sunlight at sea level on the earth's surface is a mere 1.07 kw/m^2, corresponding to the sun being directly overhead in a very dry and cloudless atmosphere. When the sun is 70° from the zenith, the power density measured by a collector lying flat upon the ground is approximately 0.25 kw/m^2. If the mirrors in a STEC system are adjustable so that they can always face the sun and reflect the power into the central receiver, then the power density for the situation with the sun 70° from the zenith can be maintained at 0.72 kw/m^2 *.

The sunlight focused on the central receiver acts to raise the temperature of a fluid flowing through the receiver. This fluid is frequently water, which is turned into steam by absorbing the solar energy (other fluids can be and are used, but water/steam is widely used by the electric power industry). The steam is then either sent to some location for heat energy storage or directly to a turbine, which, in turn, provides rotary motion to an electrical generator, which produces electrical power. This electrical energy is normally fed into the local power grid, but can be stored in batteries, by pumping water into a reservoir behind a dam, by electrolyzing hydrogen or in a number of other ways. The heat energy storage is present to assist in smoothing out power production. If, for example, a cloud were to cover the sun, the amount of energy available to the STEC would plummet and the customers of this energy production facility would experience brownouts and other undesirable effects. The heat storage ensures that the system can maintain a constant rate of power production. Typically, the heat storage allows the facility to operate for several hours after sunset.

Overall efficiency of such a system is difficult to estimate. Among other variables is weather (a cloudy day sharply reduces the solar energy density available at the earth's surface). This is a reason for solar-electric power plants being constructed in regions with high solar inputs. One estimate for the output power of a STEC plant is based on a facility located in the deserts of the American Southwest [50] and provides an

---

* Chapter II contains a detailed study of the available power density in sunlight under a wide variety of conditions.

estimated annual maximum electrical generation of approximately 50 kwh/ft^2, which translates into an efficiency of approximately 15%. Since a STEC operates in direct sunlight, some half of the energy provided by such a plant should first be stored (either as electrical or as heat energy) for use at night. Suppose the storage efficiency is 80% and suppose the STEC systems are supplying half of the energy required by Scenario C (0.85Q per year). In such a situation, we would require some 180,000 square miles of mirrored collectors, or approximately 0.3% of the earth's land area.

Aside from the engineering problems involved in energy storage, keeping the mirror structures clean and keeping the mirrors pointing at the sun, there is a potential ecological difficulty. A STEC system requires a substantial, contiguous area of mirrors, all focused on the central receiver (boiler assembly). The sunlight is intercepted and, as estimated earlier, some 15% is converted into electrical energy, which is then shipped to the customers. There is no guarantee that the customers are close to the power generating facility--there could easily be several thousand miles between them, particularly since those areas receiving the most solar energy are not necessarily those which are favored for human residence. A STEC lying in the Sahara could be supplying power to Bern, Switzerland. Ecologically, what this procedure accomplishes is to remove energy that would normally be used in heating the vicinity of the STEC and using it to heat the consuming regions. Thus, there exists a potential for climatic shifts. The areas near the STEC systems would cool, perhaps becoming more cloudy and wet, and there would be a warming trend in the consuming regions. At present, it is impossible to predict the ultimate effects of such STEC induced climactic shifts.

The two remaining energy conversion methods of Table I.11 present opportunities for avoiding this ecological effect. Both convert the energy in sunlight directly to electrical energy. Let us first consider the thermoelectric generator. When one surface of a thermoelectric generator is maintained at a high temperature (e.g., by having sunlight fall on it) and the opposing surface is maintained at a low temperature, the ensuing heat flux results in the production of electrical energy [51]. The efficiency of operation is dependent on both the properties of the materials used in the thermoelectric converters and the Carnot efficiency (Equation I.4). Proper selection of materials and operating temperature, which may require the use of mirrors and or lenses in order to compensate for the relatively weak solar energy density, can yield an efficiency in the neighborhood of 10% [52] for systems employing a high degree of optical concentration.

The area required for thermoelectric systems is larger than that for STEC systems because of the lower energy conversion efficiency of the thermocouples. However, because the operating temperatures do not have to be as high as for a STEC and because no expensive and bulky steam boiler-turbine-generator structure is required, we can disperse the thermoelectric generators over a wide area. It is feasible to consider individual thermoelectric generators placed on separate buildings, supplying individual structures. Not all of the structures would be in dry, high solar energy areas such as the Australian Outback, the deserts of Arabia or the American Southwest. To obtain half the energy of Scenario C, 0.85 Q a year, at an efficiency of 5% (we are now considering dispersed systems, not always sited in areas of favorable solar energy input) with flat horizontal converters would require approximately 540,000 square miles of thermoelectric converter systems. This is close to 0.9% of the land area of the planet. The materials used in thermoelectric devices are generally both rare and expensive. Even using lenses (or mirrors) for optical concentration of sunlight, the use of thermoelectric systems to supply a significant fraction of the energy required by the human race would seem to be profligate of resources and an expensive undertaking. Except in special circumstances, it is difficult to see any use of thermo-electric systems as sunlight powered energy converters[*].

We have been considering solar thermal-electric converters (STECs) and thermoelectric devices, both of which generate electrical energy from sunlight, using heat energy as an intermediate step. Both of these systems utilize solar-tracking mirror (or lens) optical subsystems for concentrating the relatively weak solar energy. We have seen that the STEC system is normally concentrated in a centralized facility while the thermoelectric systems can be either concentrated or dispersed--a procedure with less potential ecological damage [53]. Consider Tables I.14 and I.15. These tables provide an estimate of the amount of effort required to power civilization by various solar-electric means. Note that Scenario A requires no solar-electric input, but that with their increasing population

---

[*] Thermoelectric systems are employed as electrical energy sources in deep space craft. There, the heat from the radioactive decay of some material such as plutonium is used to power the thermoelectric process.

and per capita energy requirements, the other scenarios require significant land areas (the total land surface of the earth is slightly less than 60,000,000 square miles). It is clear that neither the STEC nor the thermoelectric approach is ideal.

Table I.14

The human populations, per capita energy consumption and annual energy consumption for the five scenarios, together with the annual contribution to energy needs supplied by non-capital, non-solar-electric sources

Scenario	A	B	C	D	E
Population (billions)	5	5	5	10	10
Per capita energy consumption rate (Q /year)	1×10^{-11}	8.5×10^{-11}	3.4×10^{-10}	3.4×10^{-10}	3.4×10^{-10}
Annual energy consumption (Q)	.051	.427	1.71	3.41	5.12

Energy supplied by non-capital, non-solar-electric sources:

Annual energy supplied by water, biological and temperature difference sources (Q)

	.18	.18	.18	.18	.18

Annual energy supplied by solar thermal sources (Q)

	.01	.06	.24	.48	.72

The solar cell, the principal subject of this work, is a device constructed from semiconducting material. The solar cell is capable of converting the energy in sunlight directly to electrical energy. It is not a heat engine (unlike the STEC and thermoelectric energy converters) and so it is not limited by the Carnot efficiency factor. Since the bulk of this work deals with solar cells (also known as photovoltaic devices), we will not go into construction techniques or fabrication details at present. What is important to us at the moment, is that the solar cells are potentially more efficient than any of the other solar energy options considered in this chapter. Solar cells made from silicon are routinely capable of generating electricity with efficiencies as high as 18% and those constructed of gallium arsenide have exhibited efficiencies of 27 %. Even

Table I.15.

The energy required of solar-electric sources, together with the area required as a function of scenario

Scenario	A	B	C	D	E
Population (billions)	5	5	5	10	10
Per capita energy consumption rate (Q /year)	1×10^{-11}	8.5×10^{-11}	3.4×10^{-10}	3.4×10^{-10}	3.4×10^{-10}
Annual energy consumption (Q)	.051	.427	1.71	3.41	5.12
Required solar-electric energy contribution (1)					
Energy (Q)	0.00	0.19	1.29	2.75	4.22
Collector area (in thousands of square miles)					
STEC (2)	0	45	305	650	1,000
Thermoelectric (3)	0	305	1,010	2,170	3,330
Solar Cells (4)	0	28	191	406	625

(1) We assume that this energy contribution is met 50% directly and 50% after some form of storage at 80% efficiency.
(2) This assumes that the STEC systems are installed in regions of high solar energy density with an average energy production of 50 kwh/ft^2/year.
(3) We assume that these are dispersed systems with an overall energy production of 15 kwh/ft^2/year.
(4) This assumes that the dispersed solar cells have an annual energy production of 80 kwh/ft^2/year. The details concerning how this is accomplished are covered later in the text.

higher efficiencies have been observed in selected solar cell systems and postulated for more advanced photovoltaic systems; efficiencies as high as 50% [54]. For solar cells made from silicon, gallium arsenide and other materials, solar concentration, using mirrors or lenses, is often employed to improve overall system efficiencies. Here, not only the direct current electrical energy provided by the solar cells, but the excess thermal energy present in the solar cells is commonly extracted from the system as useful energy. With gallium arsenide photovoltaic cells optical concentrations close to 1,000 have been used, whereas with silicon the limit of optical concentration has been found to be approximately 300.

With solar cells, the prime output is direct current electrical energy, so that the system may be dispersed, with sets of solar cells attached to the roofs of buildings, or at least sited close to the locations where the energy is required. In systems utilizing optical concentration, the annual output energies (combined heat and electrical) can exceed 100 kwh/ft^2/year. Study of Table I.15 makes it clear that photovoltaic devices have a high potential when it comes to supplying the energy required by civilization. Let us now move to the next chapter and commence our detailed investigation of solar cells.

References

1 U.S. Energy - A Summary Review, U. S. Dept. of Interior, 1972.
2 Project Independence Report, U. S. Federal Energy Administration, Nov. 1974.
3 R. C. Dorf, The Energy Fact Book, McGraw-Hill, New York, 1981, p. 13.
4 1990 annual Energy Review, Energy Information Administration, Washington, D. C.
5 Information Please Almanac, Houghton Mifflin, Co., New York, 1972, p. 617.
6 D. Hedly, World Energy, The Facts and The Future, Euromonitor Publications Limited, Detroit, Michigan, 1984, p. 26.
7 D. B. Hughes, et. al., Energy in the Global Arena, Duke University Press, North Carolina, 1985, p. 136.
8 R. Mills and A. Toke, Energy Economics and the Environment, Prentice Hall, Englewood Cliffs, New Jersey, 1985, p. 5.
9 P. C. Putnam, Energy in the Future, Rowman & Littlefield, Lanham,

Maryland, 1972.

10 Patterns of Energy Consumption in the United States, Office of Science and Technology, Washington D. C., 1972.

11 W. Eager, Renewable Energy: A Global Option for Global Warming: in Solar Today, Nov.-Dec. 1989, p. 17.

12 J. Edmonds and J. M. Riley, Global Energy--Assessing the Future, Oxford University Press, Great Britan, 1985.

13 S. W. Angrist, Direct Energy Conversion, 3rd ed., Allyn and Bacon, New Jersey, 1976, Fig. 1.2.

14 A National Plan for Energy Research, Development and Demonstration Creating Energy Choices for the Future, U. S. Energy Research and Development Agency, Washington D. C., Vol. I, 1976.

15 D. J. Rose, Learning About Energy, Plenum Press, New York, 1988.

16 J. T. McMullin, R. Morgan and R. B. Murray, Energy Resources and Supply, John Wiley and Sons, New York, 1976.

17 J. Byrne and D. Rich, Planning for Changing Conditions, Energy Policy Studies - Vol. 4, Transaction Books, New Bruswick, New Jersey, 1988, p. 72.

18. W. O. Metz, in Science, Vol. 192, 1976, p. 1320.

19 J. D. Lindl, R. L. McCroy and E. M. Cambell, in Physics Today, Sept. 1992, p. 32.

20. W. J. Hagan, R. Bangerter and G. L. Kulcinski, in Physics Today, Sept. 1992, p. 42.

21 E. H. Thorndike, Energy and Environment: A Primer for Scientists and Engineers, Addison-Wesley, Reading, Massachusetts, 1976, p. 20.

22 Reference 21, Chap. 3.

23 A. M. Zarem and D. D. Erway, Introduction to the Utilization of Solar Energy, McGraw-Hill, New York, 1963, Chap. 9.

24 E. E. Robertson, Invited paper # J.1.3, 1977 Annual meeting of the American Section of the International Solar Energy Society, Orlando, Florida.

25 D. A. Andrejko, editor, Assessment of Solar Energies, American Solar Energy Society, 1989.

26 C. J. Weinberg, and R. H. Williams, Energy from the Sun: in Scientific American, Sept. 1990, p. 146.

27 M. Calvin, in American Scientist, Vol. 64, 1976, p. 270.

28 Science, Vol. 194, 1976, p. 46.

29 W. Stoeckenius, in Scientific American, Vol. 234, 1976, p. 38.

30 Reference 21, p. 51.

31 B. Sorenson, in Science, Vol. 194, 1976, p. 935.
32 L. C. Evans, Wind Energy in Europe, in Solar Today, May-June 1991, p. 32.
33 J. G. Holmes, et. al., 1977 Annual meeting of the American Section of the International Solar Energy Society.
34 Reference 21, p. 20.
35 B. Christiansen and T. H. Clack, Jr., in Science, Vol. 194, 1976, p. 578.
36 For an early example, see Arizona Highways, August 1975.
37 T. N. Veziroglu, editor, Alternate Energy Sources III, Hemisphere Publishing Co., New York, 1989.
38 A, M. Zarem and D. D. Erway, Introduction to the Utilization of Solar Energy, Chaps. 1 and 5.
39 Solar Today, a publication of the International Solar Energy Society.
40 J. A. Duffie and W. A. Beckman, in Science, Vol. 191, 1976, p. 143.
41 R. C. Neville, in Solar Energy, Vol. 19, 1977, p. 539.
42 R. A. Tybout and G. Lof: in Natural Resource Journal, Vol. 10, 1970, p. 268; G. Lof and R. A. Tybout: in Solar energy, Vol.14, 1973: ibid, Vol. 16, 1974, p. 9.
43 A. Goetzberger, et. al., in Solar Energy, Vol. 49, 1992, p. 403.
44 Reference 13, Chap. 2.
45 See the various proceedings of the Annual Meetings of the American Section of the International Solar Energy Society.
46 J. M. Cruido, M. Macias and S. Macias-Machin, in Solar Energy, Vol. 49, 1992, p. 83.
47 Jordan and Ibele, Solar Energy--Mechanical Energy Systems, in the Proceedings of the 1955 World Symposium on Applied Solar Energy.
48 Reference 13, p. 52.
49 N. Serpone, D. Lawless and R. Terzian, in Solar Energy, Vol 49, 1992, p. 221.
50 W. G. Poland, in American Scientist, Vol. 64, 1976, p. 424.
51 S. M. Sze, Physics of Semiconductor Devices, 2nd Edition, Wiley-Interscience, New York, 1981, p. 42.
52 Reference 13, Ch. 4.
53 For further discussion of STEC and Thermoelectric systems, see the Annual Proceedings of the International Solar Energy Society.
54 R. C. Neville, Gallium Arsenide Thermophotovoltaic Systems and Converter Temperature, 9th Miami International Congress on Energy and the Environment, Miami Beach, Florida, Dec. 1989.

CHAPTER II: THE SUN AND SUNLIGHT

Introduction

Any discussion of solar energy and solar (photovoltaic) cells should begin with an examination of the energy source, the sun. Our sun is a dG2 star, classified as a yellow dwarf of the fifth magnitude. The sun has a mass of approximately 10^{24} tons, a diameter of 865,000 miles, and radiates energy at a rate of some 3.8×10^{20} megawatts. Present theories predict that this output will continue, essentially unchanged, for several billion years. It is necessary to say essentially, because the sun's energy output may fluctuate by a few percent from time to time. For our purposes we will consider the solar energy output to be a constant.

Between the sun and the earth there exists a hard vacuum and a distance which varies from 92 to 95 million miles. This distance variation implies, using the inverse square law, that the light energy reaching the earth in June (when the earth is at its maximum distance from the sun) is approximately 94% of the light energy reaching the earth in December [1]. In this work, we will follow the common practice of assigning an average value to the light energy density available to a collector positioned just outside the earth's atmosphere. This quantity is known as the solar constant and has been measured variously as lying between 0.1338 and 0.1418 W/cm^2 [2]. Four our purposes, we will take for the solar constant [3]:

Solar Constant = 0.1353 W/cm^2. (II.2)

This power density is available, on the sunlit side of the earth, 24 hours a day, each day of the year, yielding an annual energy flux to the earth of 1,186 kwh/cm^2.

Once the sunlight has reached the earth's atmosphere a number of additional effects play a part. Those effects resulting from weather, and photon absorption by water vapor, ozone, and other atmospheric constituents will be considered in detail later in this chapter. They have the general overall effect of reducing the energy density in sunlight at the earth's surface. Atmospheric effects are at a minimum on a dry, cloudless day with the sun directly overhead (and at the zenith). Under these conditions (known collectively as air-mass-one or AM1) the power flux in sunlight at the earth's surface is [2, 4]:

AM1 power flux $= 0.107$ W/cm^2. (II.2)

Two other factors influencing the availability of solar energy are geometric in nature. The first we have already considered, in part. The earth rotates on its axis with a period of approximately 24 hours, hence sunlight is available for only an average of 12 hours a day. Next, the earth's axis of rotation is tilted approximately 23.5° to the normal of its plane of revolution about the sun. Together, these two effects act to produce a shift in the number of hours of daylight and a geometrical situation in which the sun is almost never directly overhead, thereby enhancing light losses due to various atmospheric phenomena.

Sunlight

The observed solar light output (spectrum) is not constant with respect to wavelength or to time. Time variations depend on solar flares, sunspot activity, and other phenomena. Overall, the variation in solar output with wavelength corresponds quite closely to that expected of a 5,800 °K black-body radiator. However, there are departures due to Fraunhofer absorption lines and a number of emission lines from the solar corona and elsewhere close to the surface of the sun [5, 6]. For our purposes, the solar spectral irradiance, defined as the power density per unit wavelength in sunlight, may be presented as a time and wavelength smoothed envelope. This is done in Figure II.1 [4-11]. The outer curve depicts the solar spectral irradiance above the earth's atmosphere; the condition known as air-mass-zero (AMO). The inner curve in Figure II.1 is the solar spectral irradiance at sea level with the sun at the zenith on a clear, dry day--the air-mass-one condition. Note that the inner curve shows the effects of ozone on the ultraviolet portion of the spectrum and the effects of water vapor, smog, carbon dioxide and other atmospheric gases and pollutants on the long wavelength portions of the spectrum.

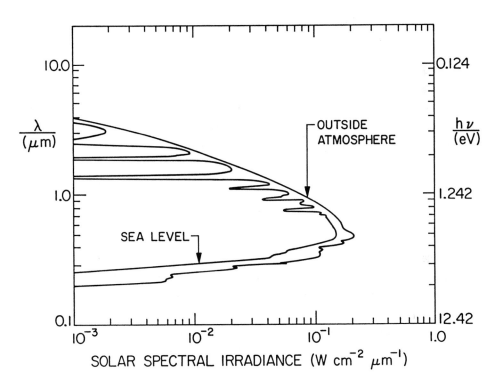

Figure II.1. The solar spectral irradiance as a function of wavelength and photon energy, with atmospheric location as a parameter.

As indicated earlier, the air-mass-zero condition corresponds to a solar power flux density of 0.1353 watts per square centimeter and a potential of 24 hours of sunlight each day. The air-mass-one condition corresponds to a maximum power flux density of 0.107 watts per square centimeter and a potential average of approximately 12 hours a day of sunlight.

Under air-mass-zero (AMO) conditions the annual energy flux density is 1.188 kwh/cm^2, but under air-mass-one (AM1) conditions the maximum energy density is reduced to 0.469 kwh/cm^2 per year. In actuality, the earth's rotation, combined with weather effects, reduces this value still further. It is this difference in potential energy density that prompts discussion of energy collection stations in orbit about the earth. There are, naturally a number of problems to be considered in the construction of extra-terrestrial energy stations. Not the least of these are the effort involved in boosting the station components to orbit, the construction of the station and the maintenance of the station. (The potential energy difference between the earth's surface and a space station orbiting over a fixed point on the earth's surface is 5.3×10^8 joules per kilogram of mass. To supply a net of 4 Q of energy to the earth, an orbiting solar energy collection/conversion facility, with a 20% operating efficiency, must have an area of approximately a million square kilometers--implying a considerable mass.) A potential ecological problem with space power lies in the fact that, as a result of the energy collection, a net input of additional energy is made to the earth (in the case of the preceding example, slightly less than 0.1% of the naturally occurring solar energy flux). Whether this additional energy would have any environmental effect, we cannot answer at present. The economics and technical design aspects of solar space stations are discussed in the literature [12]. We will not go into additional detail here. However, in connection with the question of space- or surface-based solar energy converter/collectors, note that more energy exists in the ultraviolet under AMO conditions. This implies that the design of a photovoltaic (solar) cell for AMO conditions must be different from the design of a solar cell for AM1 conditions. This topic will be discussed in a later chapter.

In Figure II.1 there are two ordinate scales. In one, the wavelength of the solar radiation is portrayed. However, in working with solar cells it is frequently more convenient to discuss the effects in terms of the energies of the individual photons--thus leading to the second scale. The conversion from one scale to the other is easily made using the following:

Photon energy (in ev) = 1.243/λ(in μm)   (II.3)

The basic quantity is, of course, the energy available in sunlight. Integrating over Figure II.1, we arrive at the wavelength versus energy distribution of Table II.1. Note that a significant fraction of the solar spectral irradiance lies at wavelengths greater than 1.15 μm. This region of the solar spectrum is often termed the far infrared, as opposed to the near infrared (which lies between 0.7 and 1.15 μm). In later chapters it will be demonstrated that this far infrared region is not available to solar cells as an energy source. This is a major limitation on the efficiency of energy producing photovoltaic devices.

Table II.1

Solar spectral irradiance under air-mass-zero and air-mass-one conditions as a function of wavelength

Wavelength (μm)	Photon Energy (ev)	Solar Spectral Irradiance (milliwatts/cm^2)	
		AMO	AM1
∞ - 1.15	0.00 - 1.08	31.77	25.24
1.15 - 1.00	1.09 - 1.24	9.51	8.41
1.00 - 0.90	1.25 - 1.38	8.29	6.02
0.90 - 0.80	1.39 - 1.55	9.93	8.35
0.80 - 0.70	1.56 - 1.78	12.37	8.05
0.70 - 0.60	1.79 - 2.07	15.15	13.25
0.60 - 0.50	2.08 - 2.49	17.70	14.30
0.50 - 0.40	2.50 - 3.11	18.77	15.10
0.40 - 0.30	3.12 - 4.14	10.17	7.91
0.30 - 0.20	4.15 - 6.22	1.63	0.37
0.20 - 0.00	6.23 - ∞	00.01	00.00
	Total	135.30	107.00

We will return to the data of this table in later chapters as the selection of semiconducting materials for and the design of solar cells is considered in detail. Let us now turn our attention to the effects of geometrical situations on solar energy.

Geometrical Effects

Perhaps the simplest geometrical effect is that due to altitude. Simply moving up a tall mountain reduces the atmospheric losses, a fact evidenced by the darker sky in regions of high elevation. Observations by Laue [13] indicate an increase of approximately 18% in solar power flux density for an increase in altitude from sea level to three kilometers, when the sun is at the zenith. This enhancement in solar energy, while significant in the sense that there are many areas of sun and high altitude on our planet, will not be considered in the rest of this work. Thus, what follows may be considered to be a conservative estimate of the available solar energy, as all locations are treated as being at sea level.

As discussed in the introduction to this chapter, the earth's rotation and axial tilt introduce a variation in the length of day, both as a function of the time of year and or latitude. While this has no effect on an orbiting solar energy collection/conversion system, it strongly affects collectors on the earth's surface.

The solar input power under air-mass-one is 1.07 kw/m^2. This condition implies a vertical orientation for the sun and solar collectors that are at right angles to the direction to the sun. The earth rotates with an axis angle of $23.5°$ and a period of approximately 24 hours. Hence, the sun is seldom, if ever, directly overhead (as depicted in Figure II.2). Should the sun make an angle, ψ, with the normal of a detector lying flat upon the ground (again, as shown in Figure II.2), the available solar power density, P, is given by:

$$\text{Solar Power Density} = P = \alpha(\psi)\cos\{\psi\}, \tag{II.4}$$

where $\alpha(o) = 1.07$ kw/m^2 under AM1 conditions. In general, the factor alpha is a function of the angle, ψ. As light traverses the atmosphere, it is partially absorbed and partially scattered by the gases in the air and any contaminants in the atmosphere. The greater the angle, ψ, the further the sun is from the normal and the greater the thickness of atmosphere which must be traversed and the more photons are scattered and/or absorbed. Thus, the value of alpha decreases with increasing psi. Table II.2 lists values of alpha as a function of the angle, psi, [14].

The angle, ψ, varies daily from sunrise to sunset, and also on a monthly basis, as the sun angle at noon (noon being defined as that time of day when the sun angle, ψ, is at a minimum) varies from L - 23.5° in

June to L + 23.5° in December, where L, the latitude is in degrees[*].

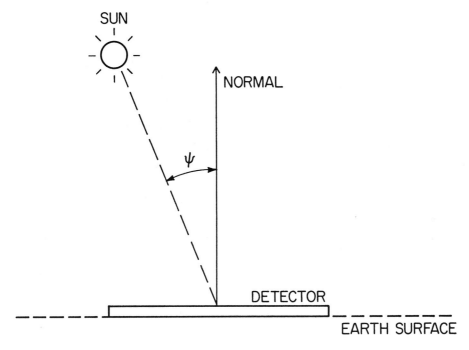

Figure II.2. Sun and detector orientations for typical daylight conditions [11].

Table II.2

Atmospheric effects on solar power input as a function of the solar angle, ψ (see Figure II.2)

ψ (°)	0	10	20	30	40	45	50	55
$\alpha(kw/m^2)$	1.07	1.06	1.05	1.03	0.993	0.968	0.939	0.903

ψ (°)	60	70	75	80	85	87	88
$\alpha(kw/m^2)$	0.855	0.717	0.603	0.451	0.233	0.142	0.095

---

[*] This assumes a Northern Hemisphere location for the solar collector. In the Southern Hemisphere, the signs should be reversed.

It is possible to eliminate the geometric portion, $\cos\{\psi\}$, of Equation II.4 by tilting the detector until the sun is aligned with the collector normal. Figure II.3 indicates the angular positions for such a tiltable collector (not yet perfectly aligned) when positioned in the Northern Hemisphere. However, such an action will not reduce the length of the atmospheric path, and, hence, the effect on $\alpha(\psi)$.

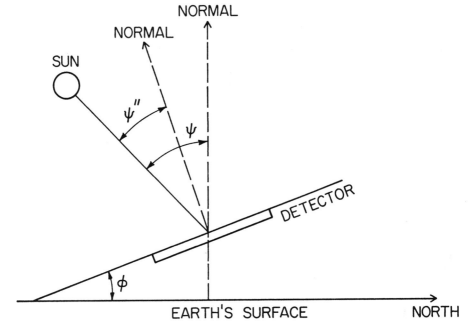

Figure II.3.   Detector, sun and earth orientations for a tiltable solar detector/collector [11].

The average year is 365 and a fraction days long, during which time the sun traverses approximately 47° from north to south and back again. Therefore, the change in solar angle measured on a north-to-south basis is approximately eight degrees per month, a relatively small number. Each day, however, the sun rises in the east and sets in the west, covering about 180 degrees of arc in an average period of 12 hours. In computing the total solar insolation (TSI), defined as the energy received each day directly from the sun, a reasonable value can be obtained using the following expression:

$$TSI = 2 \ \int^{D/2} \alpha(\psi)\cos\{\psi''\}dt, \qquad\qquad (II.5)$$

where D is the amount of daylight for a given day, α is determined by computing the solar angle ψ (see Figure II.3) and using Table II.2, and ψ'' is defined in Figure II.3.

There are several solar energy collector configurations of interest. (A) We can consider the collector to be a device lying, unmoving, flat upon the ground. In this case, the angle of the collector with respect to the ground, ϕ in Figure II.3, is zero. (B) The collector can be mounted upon some mechanism that insures that the collector is always pointed at the sun with the angle, ψ'', in Figure II.3 zero at all times. (C) A collector can be mounted upon the ground, unmoving, but at some fixed angle, ϕ, to the horizontal. This angle will be southward facing in the Northern Hemisphere, as shown in Figure II.3 and will be, clearly, northward facing in the Southern Hemisphere. A final class of collector, (D), is fixed at some angle, ϕ, in the north-to-south direction, but is capable of following the sun in its east-to-west motion[*].

For a horizontally mounted, unmoving solar energy collector:

$$\tan^2(\psi) = \tan^2(\psi') + \tan^2(\theta), \qquad\qquad (II.6)$$

where ψ' is the average north-south angle, for a given day, between the sun and the zenith, and θ is the east-west angle between the sun and the surface normal as the sun traverses the sky from sunrise to sunset. As a function of time, we may write for θ:

$$\theta = (1 - 2t/D)\pi/2 \text{ for } 0 < t \leq D/2, \qquad\qquad (II.7)$$

where t is the time (a value of zero corresponds to sunrise).

For an unmoving, but angled collector:

$$\tan^2(\psi'') = \tan^2(\psi' - \phi) + \tan^2(\theta). \qquad\qquad (II.8)$$

---

[*] This type of collector is known as a polar-axis tracking collector or single-axis polar tracking collector, as opposed to type B which is a two-axis tracking collector.

For a collector that tracks the sun only from east-to-west (a single-axis collector):

$$\psi'' = \psi - \phi. \tag{II.9}$$

For a full (two-axis) tracking collector:

$$\psi'' = 0. \tag{II.10}$$

In Figure II.4 the solar insolation (TSI), in kwh/m^2, per year is presented for latitudes from zero (the equator) to 70° with collector orientation (the angle ϕ in Figure II.3) and tracking mode as parameters. As before, the No Tracking mode implies a collector which is fixed at some north-to-south angle; East-West Tracking implies a mechanism that moves the collector only in the east-to-west direction, while the north-to-south angle is fixed at the given angle (this omits corrections for changes in the seasonal north-to-south solar orientation); and Ideal Tracking signifies a two-axis tracking mode that maintains the collector normal pointed at the sun at all times that the sun is above the horizon.

Study of Figure II.4 shows that the use of an east-west tracking mode system yields an insolation which is close to the maximum value provided by two-axis tracking whenever the optimum north-to-south angle is selected. This optimum angle is clearly a function of the latitude. It is also evident from Figure II.4 that fixed collectors are much less effective as energy collectors than either the single-axis or two-axis tracking collectors.

Later in this work we will encounter other and even more powerful arguments for the use of tracking mechanisms. However, the enhanced efficiency evidenced in Figure II.4 is more than sufficient to make serious consideration of solar collectors with tracking mechanisms attractive.

In Figure II.5, the optimum collector angle, ϕ, and the accompanying maximum insolation are plotted as a function of latitude, with the tracking mode as a parameter. This figure serves to emphasize, once more, the desirability of some form of tracking. Note that the optimum collector angle (on a north-to-south basis) for east-west tracking systems is approximately equal to the latitude.

Figures II.4 and II.5 do not convey the total picture. For some locations and some climates, the maximization of insolation over an entire year may not be as desirable as the maximization of insolation for a

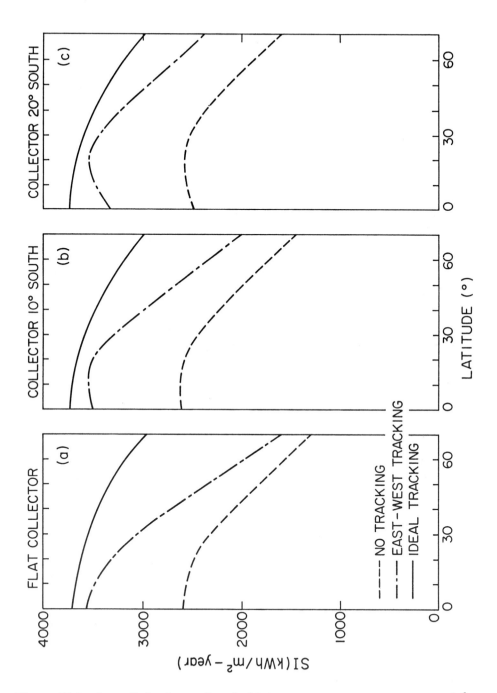

Figure II.4 a,b,c. Solar input (insolation) per square meter per year at the earth's surface as a function of latitude and collector tilt, with the tracking mode as a parameter [11].

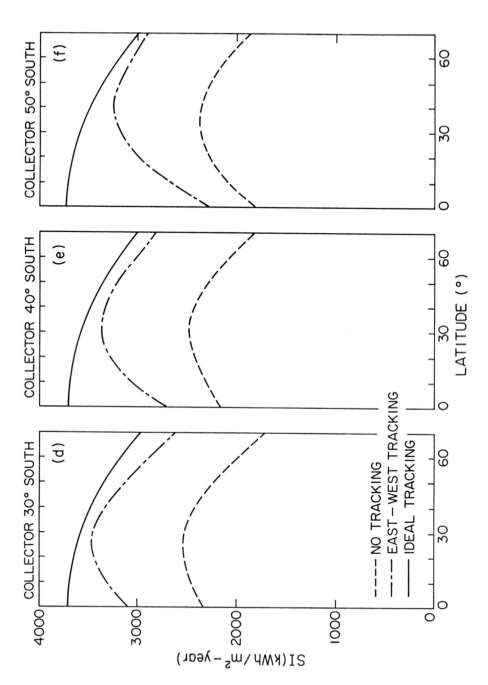

Figure II.4 d,e,f. Solar input (insolation) per square meter per year at the earth's surface as a function of latitude and collector tilt, with the tracking mode as a parameter [11].

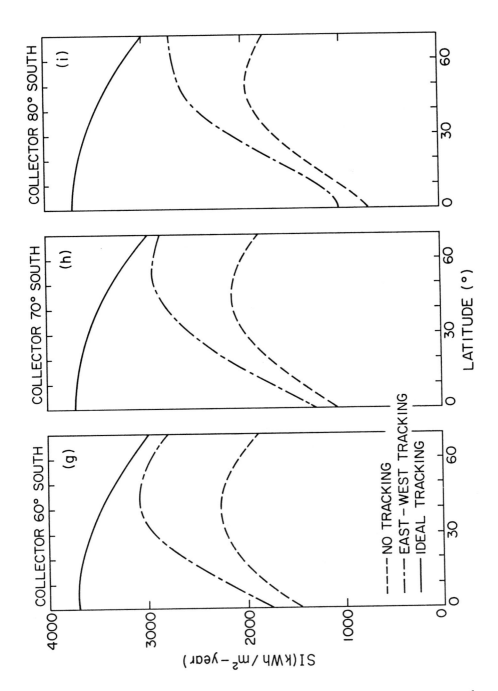

Figure II.4 g,h,i. Solar input (insolation) per square meter per year at the earth's surface as a function of latitude and collector tilt, with tracking mode as a parameter [11].

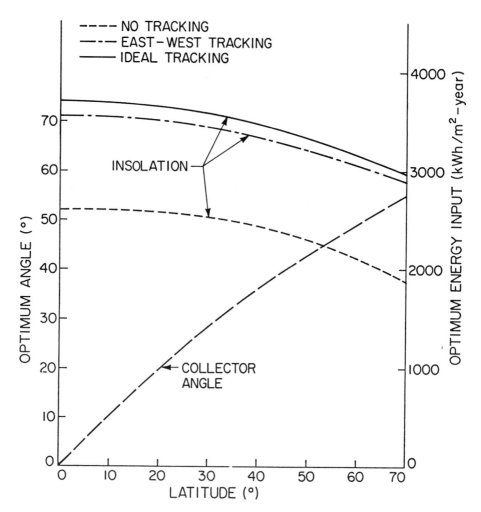

Figure II.5. The optimum collector angle and maximum energy input as a function of latitude for No Tracking (fixed), East-West Tracking (single-axis) and Ideal (two-axis) Tracking [11].

particular season, such as summer in Tucson, Arizona. Table II.3 presents the insolation, by season for the three tracking modes as a function of collector orientation angle, φ, with latitude as a parameter. Clearly, the collector orientation angle, φ, is important only for fixed position and single-axis tracking systems. For the purposes of Table II.3, winter is defined as that third of the year that has as its midpoint, the winter solstice; summer is defined as that third of the year that has as its midpoint, the summer solstice.

Table II.3

Seasonal available insolation (kwh/m^2) as a function of latitude and tracking mode with north-to-south tracker angle, ϕ, (in degrees), as a parameter

ϕ	0	10	20	30	40	50	60
			Latitude of zero degrees				
Winter							
--Two-axis	1220	1220	1220	1220	1220	1220	1220
--One-axis	1160	1210	1210	1200	1150	1050	930
--Fixed	850	880	890	880	850	800	730
Summer							
--Two-axis	1220	1220	1220	1220	1220	1220	1220
--One-axis	1160	1070	950	800	620	430	220
--Fixed	850	820	740	650	530	390	220
			Latitude of ten degrees				
Winter							
--Two-axis	1170	1170	1170	1170	1170	1170	1170
--One-axis	1020	1100	1150	1160	1150	1090	1000
--Fixed	770	810	840	850	840	810	760
Summer							
--Two-axis	1280	1280	1280	1280	1280	1280	1280
--One-axis	1260	1200	1110	990	830	650	440
--Fixed	920	890	840	770	680	550	400
			Latitude of twenty degrees				
Winter							
--Two-axis	1090	1090	1090	1090	1090	1090	1090
--One-axis	840	950	1030	1070	1080	1070	1020
--Fixed	650	710	760	780	790	780	750
Summer							
--Two-axis	1330	1330	1330	1330	1330	1330	1330
--One-axis	1320	1310	1280	1190	1060	880	680
--Fixed	970	960	930	870	800	700	570
			Latitude of thirty degrees				
Winter							
--Two-axis	990	990	990	990	990	990	990
--One-axis	650	770	860	940	980	980	980
--Fixed	520	590	640	680	700	710	700

Table II.3, continued

φ	0	10	20	30	40	50	60
Summer							
--Two-axis	1390	1390	1390	1390	1390	1390	1390
--One-axis	1360	1380	1370	1310	1200	1010	900
--Fixed	1000	1000	990	960	910	830	730

<u>Latitude of forty degrees</u>

Winter							
--Two-axis	850	850	850	850	850	850	850
--One-axis	430	550	660	740	830	840	840
--Fixed	360	440	500	550	580	600	600
Summer							
--Two-axis	1460	1460	1460	1460	1460	1460	1460
--One-axis	1370	1440	1450	1440	1380	1270	1130
--Fixed	1010	1040	1050	1040	1000	950	870

<u>Latitude of fifty degrees</u>

Winter							
--Two-axis	650	650	650	650	650	650	650
--One-axis	230	340	430	510	570	620	640
--Fixed	200	280	340	380	420	440	450
Summer							
--Two-axis	1540	1540	1540	1540	1540	1540	1540
--One-axis	1330	1450	1520	1520	1520	1450	1340
--Fixed	1010	1040	1070	1080	1070	1040	980

<u>Latitude of sixty degrees</u>

Winter							
--Two-axis	340	340	340	340	340	340	340
--One-axis	70	130	180	230	270	300	320
--Fixed	60	100	140	170	200	220	230
Summer							
--Two-axis	1710	1710	1710	1710	1710	1710	1710
--One-axis	1300	1470	1600	1680	1700	1680	1610
--Fixed	1010	1040	1080	1100	1130	1160	1120

<u>Weather</u>

There remains a major variant that must be considered in

determining the insolation available to any collector, the climate of the particular region in which the collector is operating.

This volume is concerned with solar (photovoltaic) cells. Such devices require direct solar radiation for efficient operation. Weather is a highly variable factor, both in the short and long term (15-20), so variable that we are forced to use statistical approaches and average values in order to assess its effects.

Consider, for example, Table II.4. In this table we present, for twenty cities, the maximum insolation obtainable by a two-axis tracking collector and, with statistically averaged weather effects included, the insolation available to a tracking collector and to a horizontally mounted unmoving solar collector. This last quantity is the amount of solar energy that is normally available at the earth's surface at any given location. The variation observable in Table II.4 explains the known datum that the polar regions are cold and the equatorial regions of the earth are hot!

Since weather data, particularly cloud cover, are extremely location dependent, any specific solar collector design must be based on local weather conditions. General pictures such as those furnished in Table II.4 or the literature [12] may rule out an area such as Alaska or Terra del Fuego, or suggest regions for solar cell collectors (the Sahara, Arizona, etc.), but local information is required for actual system and device design.

Other factors that must be considered are altitude and the nature of the local horizon. Perhaps the most important effect in urban areas is shadowing from local structures. Consider, for example, the fact that the net sea level solar energy input to the earth is 3,770 Q each year (ignoring clouds). If we return to Figure II.5 and compute the insolation predicted for an earth completely covered with ideal two-axis tracking collectors, the calculated value for insolation is 6,160 Q annually. The difference between this value and the 3,770 Q actually annually available lies in the fact that in the case of an earth totally covered with collectors, the collectors would, in part, shadow each other, reducing the insolation to the actually available value. The use of tracking collectors cannot increase the total amount of insolation available, but they do allow an increase in the energy density of the collected energy. For example, the 3,770 Q that does reach the earth's surface each year corresponds to an average energy density of 2,120 kwh/m^2 per year for unshadowed collectors. Thus, the net effect of tracking collectors is to reduce the total collector area required to provide a given amount of energy.

Table II.4

Effective insolation for selected cities with and without weather effects and for tracking two-axis and horizontal, fixed position collectors

Location	Lat (°)	Insolation (kwh/m²)		
		Maximum Two-axis Collector	Including Weather Two-axis Collector	Fixed Collector
In the United States [11]				
City/State				
Honolulu, HI	21.3	3,610	2,340	1,590
Miami, FL	25.8	3,570	2,320	1,550
New Orleans, LA	30.0	3,550	2,120	1,370
Phoenix, AZ	33.4	3,520	3,020	1,890
Santa Barbara, CA	34.4	3,510	2,530	1,560
Flagstaff, AZ	35.2	3,490	2,510	1,550
Fresno, CA	36.7	3,480	2,850	1,740
Washington, DC	38.9	3,460	2,040	1,210
New York, NY	40.8	3,440	2,060	1,220
Boston, MA	42.2	3,420	1,980	1,150
Burlington, VT	44.5	3,370	1,680	970
Seattle, WA	47.6	3,350	1,570	880
Fairbanks, AK	64.8	3,030	1,450	670
Elsewhere in the world [21-26]				
City/Country				
Athens, Greece	38.0	3,470	2,770	1,920
Mexico City, Mexico	19.2	3,620	2,700	1,870
Port Harcourt, Nigeria	4.9	3,700	2,137	1,444
Jakarta, Indonesia	7.5*	3,670	2,090	1,453
Rio de Janero, Brazil	22.5*	3,600	2,420	1,682
Perth, Australia	32.2*	3,530	2,690	1,905
Wellington, New Zealand	42.0*	3,420	2,740	1,922

* South of the equator

Light Collection

Because of the high degree of purity required of the semiconducting materials used in solar cells, the materials are expensive. The fabrication processes involved in converting the raw semiconductor into solar cells is labor intensive and costly (see Chapters VIII-X). As a result, current solar cell systems cost in excess of $4,000 per peak kilowatt of generating capacity, as opposed to a conventional power plant which costs approximately $1,200 per kilowatt of installed generating capacity. An attractive method of reducing the cost of a solar cell energy system is to substitute less expensive mirrors (or lenses) for solar cells and to use these mirrors (or lenses) to focus sunlight on a relatively small area of solar cells--as depicted in Figure II.6. Besides reducing the overall cost, such a concentration has an additional advantage. Any solar cell has a certain number of ways in which energy can be lost--internal loss mechanisms. In a light concentrating system the higher solar flux saturates many of these loss mechanisms (a topic to be discussed in detail in a later chapter), yielding a higher operating efficiency for the solar cells and for the entire collecting system. However, it is important to note that the performance of any large light concentrating system will depend strongly on the difference in direction between the system normal and the sun. This requirement, implies that the light collecting/concentrating system must be of the tracking type. Thus, any net reduction in overall system cost will depend on both the relative cost of the lenses (or mirrors) and of the solar cells, as well as the expense incurred in providing the tracking mechanism. This section will discuss, briefly, the important optical aspects of light collecting/concentrating systems; leaving until later any interactions between the light and the semiconducting materials.

In general, two types of optical systems are utilized in solar energy collectors with solar cells; lens and mirror systems. Both types of system appear in cylindrical and circular symmetries. The shapes and sizes of the optical components in these solar energy concentrating/collecting systems are largely a result of the twin requirements of minimum cost and of large aperture-to-focal-length ratio (i.e., the requirement for a high degree of light concentration).

Lens Systems

Two types of lenses are commonly employed in solar cell concentrating systems, the so-called standard lens, as depicted in Figure

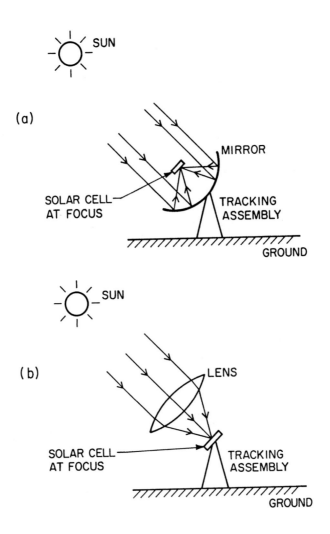

Figure II.6. Techniques for concentrating sunlight on a solar cell.

II.7a, and the Fresnel lens, as shown in Figure II.7b. The Fresnel lens is generally favored because it is thinner and uses far less material, resulting in reduced weight and cost.

Optically, the Fresnel lens is a close equivalent to the thin, conventional lens exemplified by Figure II.7a. (A thin lens is any lens in which the thickness of the lens is small compared to the focal length, F, of the system.) The focal length of a thin lens, with one plane surface is given by:

$$F = R/(n - 1), \tag{II.9}$$

where n is the index of refraction of the material from which the lens is constructed and R is the radius of curvature of the curved surface.

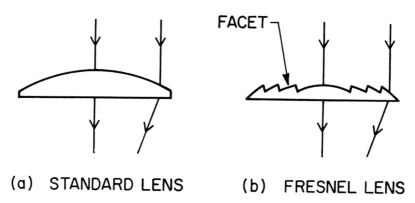

(a) STANDARD LENS   (b) FRESNEL LENS

Figure II.7. A standard plano-convex lens and the equivalent Fresnel lens. At the same radial displacement, both lens types induce the same deviation of the incoming solar rays.

As evident in Figure II.7b, in a Fresnel lens, the smooth, even curvature of the standard lens surface is replaced by a number of thin, concentric rings, or facets, with flat surfaces. Each facet is set at the correct angle to refract incoming light rays to the focus[*]. The discrete facet steps introduce some error into the precision of the focus and some light is lost upon scattering from the vertical faces of the facet steps. Even though the "picture" at the focus is highly distorted by a Fresnel lens, it is still important that any system with a Fresnel lens as the focusing device, track the sun. Light rays entering the system, not parallel to the optic axis have different focal lengths from those entering parallel to the optic axis. The result of failure to accurately track the sun is an increasingly hazy image and a rapid decrease in the collectable optical energy density. Exact corrections for the focal location may be found in the literature [27, 28].

---

[*] Such a lens, whether it is in a circular or in a cylindrical symmetry, as in Figure II.8, has a focal length as given by Equation II.9.

Mirrors

The simplest form of mirror system is a flat, reflecting surface with a light path as shown in Figure II.9a. A light-collecting system using a plane mirror implies that the mirror has the same size of area as the solar cell, since no focusing effect is present. Light concentration can be achieved by using several mirrors aligned at specified angles as shown in Figure II.9b. Tracking devices for each mirror are clearly required in order to follow the sun during its daily motion.

A system utilizing two plane mirrors to focus light on a photo-

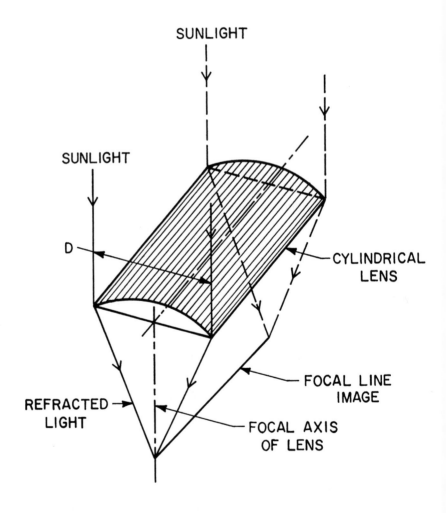

Figure II.8. A diagram of a cylindrical configuration of a Fresnel lens.

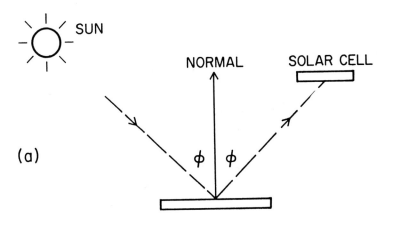

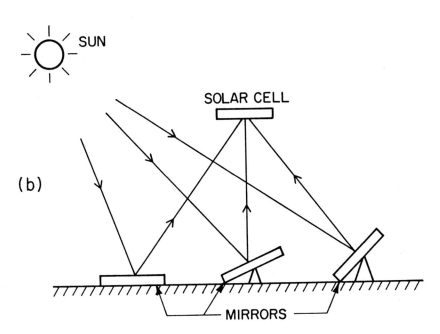

Figure II.9. Simple mirror systems for use in connection with solar cells:
(a) a single mirror and (b) a system of mirrors.

voltaic energy converter (solar cell) is depicted in Figure II.10. Note that
such a system does not trap all of the incoming light, and that it requires
a very deep and, therefore, large area mirror. This system could be

converted to conical symmetry by rotation about the central axis. Combinations of plane mirrors with Fresnel lenses, to provide a focusing effect, are also possible [29].

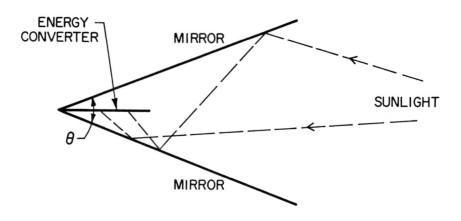

Figure II.10. A deep two mirror focusing system.

The mirrors employed do not have to be of the plane type. An optical diagram for a curved spherical symmetry mirror is displayed in Figure II.11. In this figure, consider an incoming ray of light, at a height, h, above the center of curvature. This ray of light reflects off the mirror

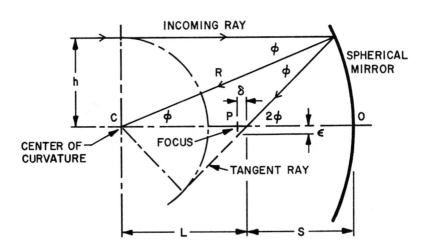

Figure II.11. A basic spherical mirror. P denotes the paraxial focus position. Note that L is always greater than S.

surface and passes close to the focus point, P. The nominal focal length, S, is given by:

$$S = R(2 - 1/\cos\phi), \tag{II.10}$$

where:

$$\phi = \text{arc } \sin\{h/R\}. \tag{II.11}$$

The intercept of the reflected ray, S, with the center line, CPO, is the small distance, δ, from the point of focus, P. This distance is called the longitudinal spherical aberration. The lateral spherical aberration is ε; h is the vertical distance from P to the light ray at a distance, S, from the mirror center, O. Clearly, both the longitudinal and lateral spherical aberrations are a function of the ray height, h. Thus, the use of a spherical mirror leads to a "fuzzy" focus. This requires that the energy converting device, which is to be placed at the focus of the mirror, must be carefully shaped to maximize the collection of light and the conversion of the light energy.

In the upper portion of Figure II.12, the outline of a parabolic mirror is displayed. The mathematical description of the mirror surface is:

$$x = (1/4S)y^2. \tag{II.12}$$

The focus of a parabolic mirror is sharp for all light rays entering parallel to the optic axis. However, the focus point will shift if the incoming light rays are not parallel to the optic axis. The end result is an ellipsoidal focal volume, as depicted in the lower portion of Figure II.12. This is a powerful argument for tracking mirrors because tracking will keep the incoming light rays as close to parallel to the optic axis as possible and so maintain a sharp focus.

In considering solar cell energy light converting/collecting systems, there are three basic configurations of parabolic mirrors [30]. These are displayed in Figure II.13. The initial case, (a), is a parabolic mirror facing the sun and focused on a photovoltaic energy converter which has its back towards the sun. The light concentration for this example is nine to one, since the back of the converter is considered to be nonconverting. Case (b) in Figure II.13 folds the energy converter, so that both sides are capable of converting sunlight to electrical energy, and places the

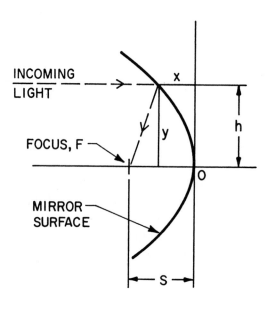

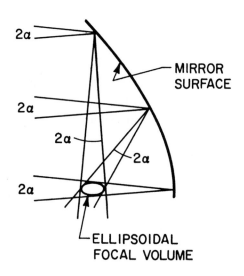

Figure II.12. Parabolic mirror configurations. The lower figure demonstrates the change in focus with mirror-to-light source orientation.

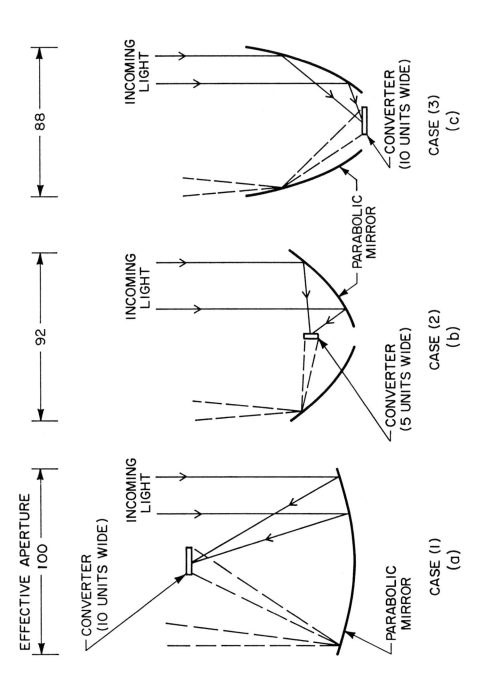

Figure II.13. Three types of parabolic focusing mirror systems for use with solar energy converters [after 30].

converter parallel to the original light ray direction. Here, the optical concentration is also approximately nine to one, owing to some non-reflecting areas along the central axis. The final case, (c), has both the converter and a very deep mirror facing the sun. This configuration utilizes more of its mirrors, but the effective optical concentration is smaller than the other two options. The principal advantage of this variety of mirror/solar cell lies in the fact that the solar cells are close to the ground and mechanical cooling is more easily effected.

There are many other types of lens and mirror energy collection systems which have been proposed over the years, both for fixed and for tracking systems. For a more detailed presentation, the reader is referred to Meinel [31].

Optical Materials

Before leaving light-collecting systems, we should discuss one additional feature of these systems; the nature of the materials through which the light must pass (in a lens system) or off which it must reflect (in a mirrored system). In all cases, we need to strive for minimum cost and maximum performance, while maintaining both initial surface accuracy and maximizing the life expectancy of the optical surfaces involved. The life expectancy is a vital factor as the solar collector is mounted out-of-doors and will be subjected to weather, temperature variations and sunlight for many years.

Consider possible materials for constructing lenses. Generally, we are limited to quartz, glass or plastics. Quartz passes light of a wide range of wavelengths, becoming opaque only for wavelengths less than 0.12 microns. From Figure II.1, we see that the solar output for wavelengths less than this number is effectively zero. However, quartz is expensive and, therefore, is not an economically wise choice. Glass is also transparent over a wide range of wavelengths, being opaque only for ultraviolet wavelengths less than 0.38 microns and for selected spectral regions in the infrared. Study of Figure II.1 indicates that the amount of energy contained in these regions of the solar spectrum is small at the earth's surface (approximately 8% of the total). It will be shown in subsequent chapters that, owing to the nature of the materials normally used in solar cells, infrared light of wavelengths longer than 1.2 microns is relatively unimportant. Thus glass, which is inexpensive, easily cast into optically desirable shapes is an excellent possibility even though it is somewhat heavy.

There is a wide variety of plastic materials from which lenses (standard or Fresnel) can be constructed, including vinyl, polyethylene, etc. Plastics can be tailored to provide efficient light transmission in desired wavelengths with low reflection. The desirable property which is most difficult to obtain in plastics is durability when exposed to sunlight and various weathering elements. However, like glass, plastics can be cast into desired shapes, and, unlike glass, plastics are lightweight.

Besides the transmission of light through a lens, we must consider the phenomenon of reflection of light from the surfaces of the lens. Consider the case for light traveling through a medium with an index of refraction, n_1, and an extinction coefficient, k_1. The photons, which make up this light, impact a lens material with an index of refraction, n_2, and extinction coefficient, k_2, at an angle, θ, to the lens normal. The fraction, f, of the light which is reflected from the lens/medium interface is given by:

$$f = [(n_2 - n_1)^2 + (k_2 + k_1)^2]/[(n_2 + n_1)^2 + (k_2 + k_1)^2], \qquad (II.13)$$

where the index of refraction and the extinction coefficient of a material are related to the permittivity of the material, ϵ, the conductivity of the material, σ, and the frequency, υ, of the photons by:

$$\epsilon = n^2 - k^2 ,$$

and   (II.14)

$$2nk = \sigma/(2\pi\upsilon\epsilon).$$

Note that the index of refraction for air (or for a vacuum) is unity and that the extinction coefficient for both vacuum and air is zero. Using this fact, together with Equation II.13, it is possible to determine the fraction of energy reflected as a function of the index of refraction of a lens material with a close-to-zero extinction coefficient. The results of such a calculation are presented in Table II.5.

Clearly, a low index of refraction material is desirable in a lens system. We can compensate, to some extent, by the use of antireflection coatings--at the cost of added system complexity and expense. (Antireflection coatings of as many as seven layers are employed in high precision camera lenses. Such a complex process is too expensive to be used in solar cell systems.)

Table II.5

The fraction of light energy reflected as a function of the index of refraction for lenses constructed of low extinction coefficient materials

Index of refraction	1.0	1.25	1.5	2.0	2.5	3.0
Fraction reflected	0.0	0.012	0.04	0.111	0.184	0.444

In the construction of mirrors it is possible to use, as the reflecting surface, a metal such as aluminum or some alloy (e.g., the commercial alloy, Alzak). We can use a sheet of metal, or evaporate a thin reflective surface over some, adequately smooth substrate. This latter technique produces front-surface mirrors and has the disadvantage that the reflecting surface is subject to abrasion and chemical weathering processes; influences that may degrade the reflecting surface within a relatively short time--days, in some instances. To circumvent this problem, we can make use of second-surface mirrors. Here, the reflective layer is placed behind a transmissive material (glass, quartz or some plastic) that acts as the mirror substrate as well as protecting the reflective layer. The disadvantage in this situation arises from transmission losses as light passes through the protecting layer[*].

Maximum Optical Concentration

After considering aberration effects and the problems inherent in the construction of large lens (or mirror) light collecting systems, an estimate can be formed of the maximum flux concentration that may reasonably be expected from commercially viable lens and mirror systems. These values are provided in the following table, Table II.6.

As we will observe later, the values for optical concentration in Table II.6 are more than adequate for work with solar cells. A smaller concentration would underuse the cells, a greater would overstress them.

---

[*] Note that the light must pass through the transparent layer twice!

Table II.6

Estimated maximum commercially viable light flux concentrating ratios [32]

Type of collecting system	Fresnel Lens	Mirror
Parabolic optics	10 - 30	40 - 60
Circular optics	100 - 400	1400 - 3200

References

1 A. J. Duffle and W. A. Beckman, Solar Energy Thermal Processes, John Wiley and Sons, New York, 1974.
2 M. P. Thakaekara, in Solar Energy, Vol. 14, 1973, p. 3.
3 C. E. Backus, Solar Cells, IEEE Press, New York, 1976, p. 3.
4 M. P. Thakaekara and A . J. Drummond, in Nature of Physical Science, Vol. 229, 1971, p. 6.
5 E. G. Gibson, The Quitet Sun, NASA, Washington, D. C., 1973, Appendix.
6 N. Robinson, Solar Radiation, Elsevier, Amsterdam, 1976.
7 Reference 5, p. 14.
8 E. Petit, in Astrophysics, J. A. Hynek, editor, McGraw-Hill, New York, 1951.
9 The International Symposium on Solar Radiation Simulation, Institute of Enviornmental Sciences, 1965.
10 S. I. Rasool, Physics of the Solar System, NASA, Washington, D. C., 1972.
11 Figures 11.1 - 11.5 and portions of Table II.3 originally appeared in an article by the author entitled, "Solar Energy Collector Orientation and the Tracking Mode", in Solar Energy, Vol. 19, 1971. Table II.4 is derived from an article by the author entitled, "Solar Energy and the Residence: Some Systems Aspects", in Solar Energy, Vol. 19, 1977, p. 589.
12 P. E. Glaser, in Physics Today, Feb. 1977, p. 30.
13 E. G. Laue, in Solar Energy, Vol. 13, 1970, p. 43.

14 A. M. Zarem and D. D. Erway, editors, <u>Introduction to the Utiliz-ation of Solar Energy</u>, McGraw-Hill, New York, 1963.

15 J. A. Eddy, in Scientific American, May 1977, p. 80.

16 J. D. Hays, J. Imorie and N. J. Shakleton, in Science, Vol. 194, 1976, p. 1121.

17 J. Ruffner and F. E. Bair, <u>The Weather Almanac</u>, Gale Research, New Jersey, 1974.

18 <u>Climates of the States I and II</u>, U. S. National Oceanic and Atmospheric Administration, Washington, D.C., 1974.

19 F. A. Berry Jr., E. Bullay, N. R. Beers, <u>Handbook of Meterology</u>, McGraw-Hill, New York, 1975, p. 988-995.

20 H. McKinley and C. Liston, <u>The Weather Handbook</u>, Conway Research, 1974.

21 G. L. Morrison and Sudjito, in Solar Energy, Vol. 49, 1992, p. 65.

22 A. Kuye and S. S. Jagtap, in Solar Energy, Vol. 49, 1992, p. 139.

23 M. Nunez, in Solar Energy, Vol. 44, 1990, p. 343.

24 B. E. Western, in Solar Energy, Vol. 45, 1990, p. 14.

25 E. S. C. Cavalcanti, in Solar Energy, Vol. 47, 1991, p. 231.

26 <u>World Weather Records</u>, U. S. National Oceanic and Atmospheric Administration, Washington, D. C., 1987, Vol. I and II.

27 A. B. Meinel and M. F. Meinel, <u>Applied Solar Energy, An Inroduc-tion</u>, Addison-Wesley, Reading, Massachussets, 1976, p. 146-147.

28 Reference # 27, p. 144-145 and 148-149.

29 R. R. Aoarisi, Y. G. Kolos and N. I. Shatof, <u>Semiconductor Solar En-ergy Converters</u>, V. A. Barn, editor, Translation Consultant's Bureau, New York, 1989, p. 10.

30 V. B. Veinberg, <u>Optics in Equipment for Utilization of Solar Energy</u>, State Publishing House, Moscow, 1959, p. 100.

31 Reference # 27, Chap. 27.

32 Reference # 27, p. 172.

CHAPTER III: SEMICONDUCTORS

Introduction

In this chapter we will introduce a number of concepts and terms that will be most useful in considering aspects of the design, construction and operation of solar (photovoltaic) cells. For the reader who wishes a more in-depth presentation of semiconductor theory there are a number of recommended references [1-8], both for physical insight and for study of the technologies involved in device construction.

Solar cells are constructed from materials commonly known as semiconductors. Electrically, we can divide all materials into three classifications, relating to the ability of a specific material to carry electrical current. Conductors are those materials that contain many free electrons; electrons that are not bound to specific sites within the material and, hence, are free to move, and thereby constitute a current. An insulator, on the other hand, has not mobile electrons and, therefore, is unable to sustain an electrical current. Semiconductors are those materials whose ability to carry an electrical current lies between these two extremes. We can use Ohm's law to differentiate between these materials. Ohm's law states:

$$J = \sigma \mathscr{E}, \qquad\qquad\qquad (III.1)$$

where J is the current density (typically in amperes per square centimeter), $\mathscr{E}$ is the applied electric field in volts per centimeter, and σ is the conductivity of the material in mhos per centimeter. A metal will possess a conductivity in the neighborhood of 10^6 mho/cm, an insulator will have a conductivity in the region of 10^{-11} mho/cm, while a semiconductor will exhibit a conductivity lying in the range from 10^{-6} to 10^4 mho/cm.

Semiconductors possess two properties that considerably enhance their usefulness. The first property is that, unlike insulators and conductors whose conductivity is essentially fixed, it is possible to alter the conductivity of a given sample of semiconductor. The alteration can be by several orders of magnitude and can be accomplished by relatively simple methods. For example, the conductivity of the commonly used semiconductor, silicon (Si), can be easily varied from 0.001 to 1000 mho/cm. This capability for alteration permits a great deal of flexibility in the design and construction of solar cells. The second advantageous property is that, in a semiconductor, the electrical current may be carried by either negatively charged electrons or by positively charged particles known as holes, or by both. It is this property that leads to our ability to construct solar cells, transistors and integrated circuits.

Crystal Structure

The crystal structure, or atomic arrangement, of any material has a great deal to do with its electrical properties. When analyzing the physical structure of any material, we can speak of its existing in a crystalline, a polycrystalline or an amorphous form. In its crystalline form a material is characterized by an ordered array of component atoms. This array is repetitive with displacement through the material sample. Where a polycrystalline material is concerned, the object is composed of a number of sub-sections, each of which is crystalline in form. These sub-sections, however, are independently oriented so that at their interfaces the atomic order and regularity undergo sharp discontinuities. The final category, the amorphous material, displays no atomic regularity of arrangement on any macroscopic scale (scales with dimensions greater than 10 to 20Å).

Semiconductors exhibit all three forms of crystal arrangement. Amorphous semiconductor devices have been studied intensively and fabricated into a number of useful devices [9-13] exhibiting both diode- and transistor-like phenomena, in addition to switching characteristics and photoconductivity. As we shall learn later in this work, some 25-30% of all current solar cell fabrication is conducted using amorphous semiconductors. We will discuss solar cells made from amorphous materials in a later chapter. A number of polycrystalline solar cells have also been constructed   [13-17]. Photovoltaic cells made from polycrystalline materials are less expensive to construct per unit area than single crystal solar cells (both in terms of finance and in terms of energy). However, they are less efficient and often more sensitive to changes in ambient conditions--both undesirable attributes. The use of optical concentration, as considered in subsequent chapters, favors the more efficient and less ambient sensitive single crystal solar cells. For these reasons, coupled with a lack of space, we will not discuss polycrystalline solar cells in much detail at present.. We will, however, spend some time in a subsequent chapter on polycrystalline solar cells even though their performance is generally dictated by poorly understood interactions at the crystal boundaries within the solar cells [17].

Many of the properties that we find to be useful in semiconductors and solar cells arise from the basic crystal structure of the semiconductor and are dependent on the direction in the crystal toward which the observer is facing. It is, therefore, useful to consider some of the physical properties and descriptive terms for crystals. An ideal crystal is built up by the regular repetition in physical space of a number of identical structural units (parallelepipeds) called unit cells. In the simplest crystal, e.g., gold, platinum, the alkali metals, etc., the basic structural unit contains but a single atom. In complex organic crystals, such as DNA, the basic structural unit often exceeds several million atoms. The crystal structure of a unit cell is described in terms of a single mathematical, periodic structure of discrete points, known as the lattice. Attached to each lattice point is an atom (or group of atoms) known as the basis. We can state:

Lattice plus basis equals crystal structure.   (III.1)

Suppose we are located in a crystal lattice at a point R_1. Then, to reach any other lattice point, R_2, we must perform the mathematical operation:

$$R_2 = R_1 + n\underline{a} + m\underline{b} + l\underline{c} ,$$ (III.3)

where n, m and l are integers and $\underline{a}$, $\underline{b}$ and $\underline{c}$ are the three sides of the parallelepiped that makes up the basic structural unit of the crystal, the unit cell. The three sides of the unit cell are parallel to the crystal axes, but are not necessarily of equal lengths. Nor are the crystal axes necessarily at right angles to each other. Let the angles between $\underline{a}$, $\underline{b}$ and $\underline{c}$ be α, β and γ. In three dimensions there are 14 types of lattices, summarized in Table III.1. Figure III.1 illustrates the three-dimensional aspects of the conventional unit cells for the 14 types of lattices.

Table III.1

The 14 conventional lattice types [18]

System	Number of lattices in system	Lattice symbols	Unit cell axes and angles
Triclinic	one	P	$a \neq b \neq c$ $\alpha \neq \beta \neq \gamma$
Monoclinic	two	P,C	$a \neq b \neq c$ $\alpha = \gamma = 90° \neq \beta$
Orthorhombic	four	P,C,I,F	$a \neq b \neq c$ $\alpha = \beta = \gamma = 90°$
Tetragonal	two	P,I	$a = b \neq c$ $\alpha = \beta = \gamma = 90°$
Cubic	three	P,I,F	$a = b = c$ $\alpha = \beta = \gamma = 90°$
Trigonal	one	R	$a = b = c$ $\alpha = \beta = \gamma < 120°, \neq 90°$
Hexagonal	one	P	$a = b \neq c$ $\alpha = \beta = 90°, \gamma = 120°$

Study of Figure III.1 discloses that, in any crystal, there must be a large number of crystal planes, each determined by three non-collinear points. There exists a conventional scheme for defining a crystal plane in terms of numbers known as Miller indices. To determine the Miller indices for a given plane, first select a lattice point as an arbitrary origin; next determine the intercepts, K_i, at which the plane crosses the extended axes (e.g., K_a may be 4a, K_b may be 2b and K_c may be 1c). Next the reciprocals of these numbers (e.g., 1/4, 1/2 and 1/1) are multiplied by the smallest integer necessary to create three integers (e.g., 1, 2, 4). The result, enclosed in parenthesis, (1,2,4), is the Miller index description of

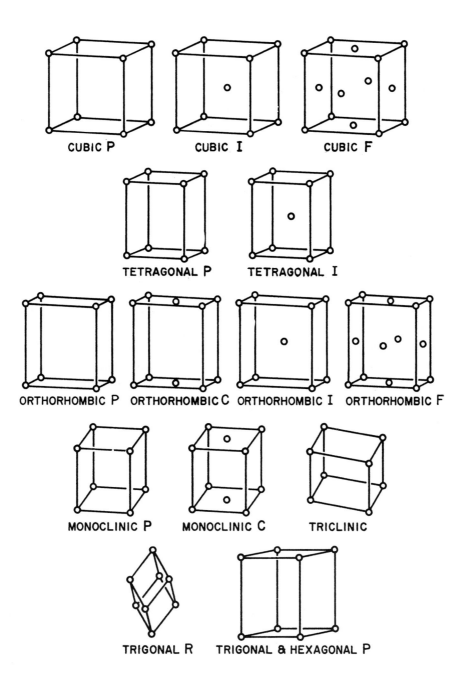

Figure III.1. The 14 conventional three-dimensional lattices shown as unit cells. After Kittel [4], with permission.

a crystal plane. For example a (100) plane crosses only the a axis, and in a cubic crystal describes one face of the cube that is the unit cell.

When real atoms are added to the lattice to make up the physical crystal, they may or may not be added to the crystal at the lattice points. For example, many semiconductors are in the so-called diamond structure[*]. The space lattice for diamond is face-centered cubic (F, in Figure III.1) and the atoms of the unit cell are located at 0,0,0 and 1/4, 1/4, 1/4; where 0,0,0 is one corner of the unit cell cube. Another structure typical of many semiconductors is the zinc blende structure. This, too, is face-centered cubic. For example, in zinc sulfide, there are four molecules of ZnS per unit cell with zinc atoms at 0,0,0; 0,1/2,1/2; 1/2,0,1/2; and 1/2, 1/2, 0; and sulfur atoms at 1/4,1/4,1/4; 1/4,3/4,1/4; 3/4,1/4,3/4; and 3/4,3/4,1/4.

Figure III.2 contains four views of a diamond lattice material (e.g., silicon or germanium) observed along various axes; displaying the crystal order, and the variation in appearance of the crystal when seen in different directions. These differences contribute to the electrical and other physical differences in the properties of the crystals, which, in turn, fundamentally influence the performance of solar cells when made from these materials. When a solar cell, or other electronic device, is made from silicon (or some other semiconductor) we speak of the starting crystal as being (100), (110) or (111) material--a shorthand description for the crystal plane upon which the processing necessary for fabrication of the solar cell is to be performed.

Quantum Mechanics and Energy Bands

That branch of physics known as quantum mechanics is complex, intricate and basically mathematical. It is not our purpose here to affect a complete and exhaustive review of the subject. However, we do need to review certain of the fundamental concepts of this subject, and so provide a basis for what will be said later about semiconductors, their properties and how they perform as solar cells. When considering an object,

---

[*] Including, of course, crystalline carbon or diamond as well as the most used semiconducting material, silicon.

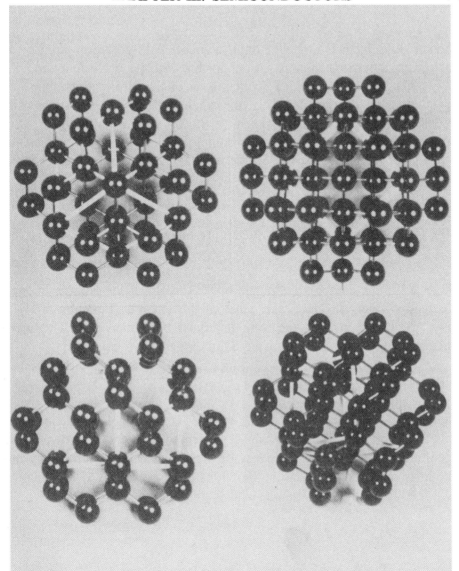

Figure III.2. Four views of a diamond or silicon semiconductor crystal. Clockwise from the top left-hand corner, these views are along the [111], [100], a few degrees off the [110] and the [101] axes*.

---

* Note that while the (100) plane is enclosed in curvilinear brackets, the [100] direction is enclosed in squared brackets. In a cubic structure the [100] direction is the normal to the (100) plane. Also note that the commas, separating the various Miller indices are customarily omitted.

whether this object is a freight train, a soccer ball, an argon atom or an electron, quantum mechanics assigns to this object a wave function, Ψ.

The wave function, Ψ, contains all of the information about the object that is observable (i.e., measurable), but has no physically observable property in and of itself. If, however, we multiply the wave function, Ψ, for some object by its complex conjugate and integrate this over a given volume of space, V, the resultant value of the integral is the probability that the object exists in the volume of space, V. Thus:

$$\int_0^V \Psi\Psi^*dv = \text{probability of existence in volume, V,} \tag{III.4}$$

where Ψ^* is the complex conjugate of Ψ. Clearly, if the volume integral is extended over all space, the value of the integral (Equation III.4) is unity, for the train, the electron, etc. must be somewhere. The wave function is also the solution to the Schrödinger wave equation:

$$\mathcal{H}\Psi_n = (\hbar/i)(\partial[\Psi_n]/\partial t) = E_n \ , \tag{III.5}$$

where i is the square root of minus one; Ψ_n is a particular version of the wave function describing the object under consideration, known as an eigenfunction; and E_n is the energy of the object when described by this particular eigenfunction, and is known as the eigenenergy. The quantity, $\mathcal{H}$, is known as the Hamiltonian and is given by:

$$\mathcal{H} = -(\hbar^2/2m)\nabla^2 + V_p(x_j), \tag{III.6}$$

where m is the mass of the object, $\hbar$ is Planck's constant divided by 2π ($\hbar = 1.054 \times 10^{-34}$ joule-seconds), ∇^2 is a mathematical operation[*] and $V_p(x_j)$ is the potential energy of the object as a function of location.

---

[*] The exact nature of this operation depends on the nature of the space in which it is being conducted. For example, in a cartesian coordinate space (x, y, z), this function, when applied to the vector function F, where F has components F_x, F_y and F_z in the x, y and z directions, provides the following mathematical operation: $\nabla^2 F = (\partial^2 F_x/\partial x^2) + (\partial^2 F_y/\partial y^2) + (\partial^2 F_z/\partial z^2)$.

The general form of a wave function solution to Equation III.5 is provided by Equation III.7:

$$\Psi_n = A_n\psi_n(x,y,z)\exp(-E_nt/\hbar). \qquad (III.7)$$

For an electron traveling in a region of constant potential energy, where the potential energy, V_p, is less than the total energy of the electron, E_T, the solution to Schrödinger's equation further simplifies to:

$$\Psi = A\exp(ik_xx + ik_yy + ik_zz)\exp(-i\omega t), \qquad (III.8)$$

where k_x, k_y and k_z are constants of motion (real) of the general form $1/[\hbar\sqrt{(2m\{E_T - V_p\})}]$, $\omega = E_T/\hbar$, and A is a normalizing constant.

Once the wave function for an object has been determined by solving Schrödinger's equation, it is possible to derive the expected value of various physical observables, the object's momentum, P, for example, by a simple mathematical operation:

$$<P> = (\hbar/i)\int_V \Psi^*\nabla\Psi dv, \qquad (III.9)$$

where Ψ^* is the complex conjugate of Ψ, V is the volume of space being considered and $<P>$ is the expected value of the momentum[*]. Using the traveling-wave solution for a free electron (Equation III.8), the momentum in each of the x, y and z directions:

$$P_x = \hbar k_x, \ P_y = \hbar k_y \text{ and } P_z = \hbar k_z, \qquad (III.10)$$

where $\hbar k_i$ is termed the crystal momentum in the i direction.

For objects that are large (such as a pencil, a button or a human being) the values of the various properties which we determine in a manner analogous to Equation III.9 are indistinguishable from the predictions made by classical (Newtonian) physics. However, when we enter the world of an electron moving through a crystal, or of electrons

---

[*] The expected value of some observable is the value which we are most likely to obtain when making some physical measurement.

moving in their orbits around an atom, then the solutions of quantum mechanics may differ, and do so dramatically, from the predictions of classical mechanics. For example, the uncertainty principle requires that the precision with which the position of a particle and the momentum of that particle may be simultaneously determined is finite, e.g.:

$$\Delta P_x \Delta x \geq \hbar/2, \tag{III.11}$$

where ΔP_x and Δx are the potential variability (the uncertainty) of the x-directed momentum and the location in the x-direction. As a result of this fact, and the permissible superposition of wave functions, a particle will, in general, be described by a wave function, Ψ_T, made up of a number of individual wave functions, Ψ_n:

$$\Psi_T = \int A_n(E) \psi_n(x,y,z) \exp(-iE_n t/\hbar) dE, \tag{III.12}$$

where $A_n(E)$ is a normalizing function dependent on particle energy. In many important situations, such as within any finite crystal, there exist boundary conditions which limit the permissible wave functions so that we may replace the integral of Equation III.12 by a summation:

$$\Psi_n = \Sigma A_n(E) \psi_n(x,y,z) \exp(-iE_n t/\hbar). \tag{III.13}$$

As an example, consider a particle, confined in a box of dimensions, L, M and N. The potential energy within the box is taken to be zero and is assumed to be infinite without the box. The solution to the Schrödinger equation (Equation III.5) in this case is of the form:

$$\Psi_T = \Sigma A_{l,m,n} \sin\{k_l x\} \sin\{k_m y\} \sin\{k_n z\} \exp(-iEt/\hbar), \tag{III.14}$$

where

$$k_l = l\pi/L, \quad k_m = m\pi/M, \quad k_n = n\pi/N, \tag{III.15}$$

and l, m, n are non-zero integers. The momentum of this particle in the x-direction is (from Equation III.9) equal to $\hbar\pi l/L$. Thus, the particle has a large number of possible values for its x-directed momentum--but not an infinite number of values. The periodic nature of the wave function (induced by the boundary conditions) requires that Ψ_T vanish at x = 0 and L; at y = 0 and M and at z equals 0 and N (the edges of the box). It can further be shown that the allowed energies for this particle are:

$$E = [(\hbar k_l)^2 + (\hbar k_m)^2 + (\hbar k_n)^2]/2m = [P_x^2 + P_y^2 + P_z^2]/2m, \qquad (III.16)$$

where m is the mass of the particle and P_i is the momentum of the particle in the i-direction. Note that the particle in a box (which serves as a useful approximation to an electron confined within the dimensions of an atom) is restricted to certain specified energies and their accompanying wave functions. When a measurement of the energy of the particle is made, the result of any single measurement is one of the possible energies (e.g., that for $k_l = 2$, $k_m = 4$ and $k_n = 5$). If a second measurement of the energies is made, the result may well be another value of energy. In any realistic physical measurement system, the energy we obtain will be an average of many individual measurements, weighted by statistical probabilities.

The motion of electrons through a semiconductor crystal is of major interest to us. The potential energy of the region through which the electron moves is not constant (as assumed by Equation III.8), but is a periodic function (since the crystal is an ordered periodic array of atoms), which leads to a wave function which, in one dimension, is the sum of a number of terms of the form:

$$\Psi(x) = U(x)\exp(ikx)\exp(-iEt/\hbar), \qquad (III.17)$$

where U(x) is periodic in x with a period equal to that of the crystal.

When Schrödinger's equation (Equation III.5) is solved for a crystal using the wave function type of Equation III.17, subject to the periodicity of the lattice, it is found that only certain energies are permitted for electrons in the crystal [19]! In figure III.3 we display the allowed energies, as a function of k, divided by the spatial period, a, for a one-dimensional crystal. The dashed line represents the energy versus k relationship for a free electron moving in a constant potential field-- clearly, a continuum of energies.

When translating Figure III.3 to a three-dimensional crystal, the most useful approach is to use directions within the crystal as dictated by the crystal structure itself. In Figure III.4 the energy-band structures for the semiconductors gallium arsenide, germanium and silicon are presented for k in the [100] and [111] directions [20].

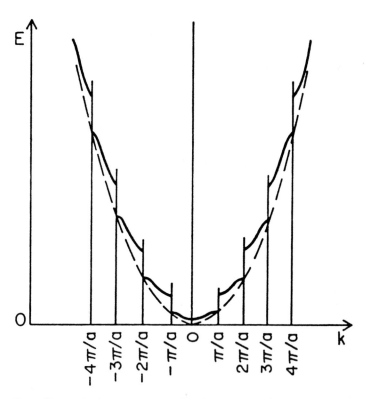

Figure III.3. The allowed electron energies in a one-dimensional crystal. For comparison, the dashed line indicates the allowed continuum of energies for a free electron.

In Figures III.3 and III.4 electrons are permitted only energy and k ("momentum") values that place them on the solid lines. In Figure III.4 the difference in energy between the highest point in the lower set of allowed energy bands and the lowest point in the upper set of allowed bands is defined as the energy gap (often called the band gap) and is delineated as E_g.

In a semiconductor, at a temperature of zero Kelvin, there are just sufficient electrons to fill the lower energy versus momentum curves, and there are no electrons in the upper curves ($E > E_g$). The lower curves are known as the valence band and the electrons therein are responsible for the chemical properties of the semiconductor. The upper set of curves are collectively known as the conduction band. Note that when the valence band (or any other band) is full of electrons it is incapable of conducting current. This is so because when we apply an electric field to some object to carry current, the electrons in the object must accelerate in the direction dictated by the field. If an electron accelerates, it must change energy by some, small, amount ($\Delta E \ll E_g$). If the electron is to change energy, it must shift to another location on the energy versus momentum diagram. However, in a full energy band, this location is already occupied by another electron. The Pauli Exclusion Principle states that, in a crystal, no two electrons may occupy the same energy-momentum position! Therefore, the electron cannot change position and cannot accelerate or change energy. Thus, at a temperature of absolute zero, with a full valence band and an empty conduction band, the semiconductor is unable to carry current and acts as an insulator.

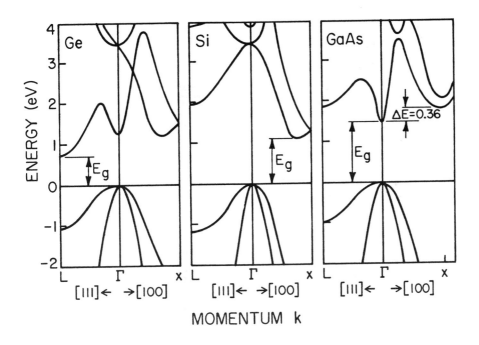

Figure III.4. The energy band versus momentum diagram for germanium, Ge; silicon, Si; and gallium arsenide, GaAS; in the [111] and [100] directions [20]. E_g is the energy gap and x, L and Γ are crystal points.

Electrons and Holes

Let the temperature of a semiconductor increase. The band gap energy in the semiconductors employed for solar cells falls between 0.8 and 2.4 electron volts. As the temperature of a semiconductor crystal increases, some of the electrons will receive enough thermal energy (via collisions with the lattice atoms) to reach the conduction band. When they do so, they enter the now almost empty conduction band, and can play a part in current flow. Moreover, as each electron exits the valence band, it leaves behind it a vacancy or missing electron in the valence band. This missing electron is known as a hole. The valence band is no longer full and, therefore, can now serve as a vehicle for carrying current. From a consideration of the crystal structure of semiconductors and from applying the rules of quantum mechanics, we may treat the current due to the large number of electrons in the almost full valence band as if it were the result of the motion of a few particles with a positive charge, the holes.

In a pure (intrinsic) semiconductor, at any temperature above absolute zero, the number of holes in the valence band is equal to the number of electrons in the conduction band. The actual number can be determined from the statistical laws of quantum mechanics, the application of the Pauli Exclusion Principle and an accurate description of the shape of the energy versus k diagram (e.g., Figure III.4) for a given semiconductor. For hole and electron densities (given by p and n, respectively) that are small when compared to the potential number of places for electrons in the conduction band, N_c, and the number of places for electrons in the valence band, N_v, we can write:

$$pn = N_c N_v exp(-E_g/kT) = n_i^2, \qquad\qquad (III.18)$$

where p is the number of holes per unit volume in the semiconductor, n, is the per unit volume number of electrons, k is Boltzmann's constant, T is the absolute temperature, and n_i is known as the intrinsic carrier concentration. The quantities, N_c and N_v are known as the effective densities of states in the conduction and valence bands. The conduction band density of states, N_c, is given by:

$$N_c = (\sqrt{\{\pi/2\}/\{\pi^2 \hbar^3\}})(m_{ce}k)^{3/2}T^{3/2}, \qquad\qquad (III.19)$$

where m_{ce} is the effective mass for electrons in the conduction band. This effective mass is generally not equal to that of a free electron and is dependent on the shape of the E-k diagram. The properties of a number of semiconductors are listed in Appendix B. Included in this list are the effective mass for electrons in the conduction band, for holes in the valence band, the effective density of states for both the conduction and valence bands, the energy gap and a number of other useful properties. The effective mass for holes in the valence band is used to compute the valence band effective density of states, using an expression similar to Equation III.19, replacing m_{ce} by the hole effective mass in the valence band, m_{ve}.

Examination of Equations III.18 and III.19 demonstrates that the intrinsic carrier concentration is a strong function of the temperature and the energy gap. Table III.2 presents values of n_i^2, for a number of semiconductors, as a function of temperature, including the effects caused by changes in energy gap width with temperature (see also Appendix B).

Table III.2

The square of the intrinsic carrier concentration, n_i^2, for eight semiconductors with potential solar cell applications, as a function of temperature

Semi-conductor	Ge	Si	InP	GaAS	CdTe	AlSb	CdSe	GaP
				----at 300°K----				
n_i^2 (cm^{-6})	5.76 x10^{26}	2.10 x10^{20}	6.49 x10^{14}	4.84 x10^{12}	3.17 x10^{11}	1.33 x10^{10}	2.20 x10^{8}	0.58
				----at 350°K----				
n_i^2 (cm^{-6})	6.06 x10^{28}	2.22 x10^{23}	2.73 x10^{18}	3.77 x10^{16}	2.97 x10^{15}	3.28 x10^{14}	1.10 x10^{13}	5.27 x10^{5}
				----at 400°K----				
n_i^2 (cm^{-6})	2.28 x10^{30}	3.88 x10^{25}	1.50 x10^{21}	3.29 x10^{19}	2.98 x10^{18}	6.85 x10^{16}	3.92 x10^{16}	1.64 x10^{10}
				----at 500°K----				
n_i^2 (cm^{-6})	4.14 x10^{32}	7.49 x10^{28}	1.55 x10^{25}	4.82 x10^{23}	5.26 x10^{22}	3.38 x10^{22}	4.08 x10^{21}	3.55 x10^{16}

The most probable distribution of electrons in the conduction band and holes in the valence band of a semiconductor is that which minimizes the energy of the crystal. Therefore, the electrons in the conduction band first fill the conduction band minimum and the holes in the valence band first fill the valence band maximum (the minimum energy position for holes).

We have been discussing pure (intrinsic) semiconductors. While we can alter the carrier concentrations (both electrons and holes) and, hence, the conductivity of such semiconductor crystals by altering their temperatures, the overall usefulness of an intrinsic semiconductor is small. A property of semiconducting materials which is most useful is that, by simple chemical and or physical means, we can introduce certain classes of impurity atoms, in very small amounts. These impurities, in turn, yield either electrons or holes (depending upon the impurity used), thus influencing both the type and amount of the charge carrier and the conductivity of the semiconductor.

If the added impurity results in additional electrons, the impurity is known as a donor, and if the net effect is the addition of holes, the impurity is known as an acceptor. This nomenclature arises from the action of the impurity. For a donor atom, the energy of the electron (which initially orbits the donor atom, but which is soon added to the conduction band) is slightly below the conduction band edge (see Figure III.5*). The difference in energy, $E_c - E_D$, between the location of the electron in the conduction band and when orbiting its impurity atom is on the order of the available thermal energy (kT) and, therefore, the electron can easily acquire sufficient thermal energy to enable its excitation into the conduction band, where it acts as a charge carrier. The fixed donor atoms thus acquire a net positive charge of +1q (1.6x10^{-19} Coulombs) becoming ions. In an acceptor, the energy level of an electron orbiting an acceptor atom, E_A, is approximately kT above the valence band maximum, E_v. In this situation, electrons in the valence band can be easily thermally excited onto the acceptor atoms; each excited electron leaving behind, in the valence band, a mobile hole and producing a fixed, negatively charged acceptor ion.

---

* While Figure III.4 is very useful in determining the electrical properties of a semiconductor, it is more convenient, when describing a semiconductor device (solar cell, diode, or integrated circuit) to consider a picture that displays the conduction and valence bands as a function of the physical position (as done in Figure III.5) rather than as functions of the momentum, $\hbar k$.

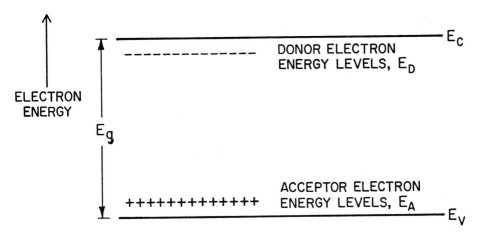

Figure III.5. An energy band diagram showing the lower conduction band edge, E_c; the upper edge of the valence band, E_v; the energy gap, E_g; and the locations of the donor electron energy levels, E_D, and acceptor electron energy levels, E_A.

It can be shown [1,2] that the product of the numbers of holes and electrons per unit volume is a constant for any semiconductor in thermal equilibrium. This product is:

$$pn = n_i^2 , \qquad\qquad (III.20)$$

where n_i^2 is provided by Equation III.18 and selected values are presented in Table III.2. Note that, should donors be added to a semiconducting crystal, the electron concentration will increase above the intrinsic value, while the hole concentration will be depressed.

Clearly, the addition of donor or acceptor impurities may not alter the overall macroscopic charge neutrality of a semiconducting crystal. Besides satisfying Equation III.20, the holes and electrons in a semiconductor must satisfy a charge neutrality condition given by:

$$(p + N_D^+) - (n + N_A^-) = 0, \qquad\qquad (III.21)$$

where N_D^+ is the concentration of ionized donor atoms (atoms that have donated an electron to the conduction band) and N_A^- is the concentration of ionized acceptor atoms (atoms that have accepted an electron from the valence band).

As indicated previously, at room temperature (approximately 300°K) it is "easy" for a donor to "lose" an electron to the conduction band and for an acceptor to "gain" an electron from the valence band. This procedure may be deemed to be "easy", but it is by no means a certainty. From quantum mechanics we can derive, for an electron, a mathematical expression for the probability, $f(E)$, that an electron has a particular energy:

$$f(E) = (1 + \exp\{[E - E_f]/kT\})^{-1}, \tag{III.22}$$

where E_f is a reference energy termed the Fermi energy.

In determining the Fermi energy (also known as the Fermi level) we solve Equation III.21 subject to the conditions of Equations III.20 and III.22. This yields:

$$n = N_c\exp\{[E_f - E_c]/kT\},$$

$$p = N_v\exp\{[E_v - E_f]/kT\},$$

$$N_D^+ = N_D\exp\{[E_D - E_f]/kT\}/(1 + \exp\{[E_D - E_f]/kT\}),$$

and (III.23)

$$N_A^- = N_A/(1 + \exp\{[E_A - E_f]/kT\}),$$

where N_D is the per unit volume concentration of donor impurities, N_A is the concentration of acceptor impurities, E_D is the energy of an electron on a donor atom and E_A is the energy of an electron on an acceptor atom (see Figure III.5).

Examination of Equations III.21 and III.22 for semiconductors of practical use in solar cells and for the temperatures commonly experienced on the earth's surface, indicates that, for N_D and N_A less than one-tenth of the appropriate density of states and for these quantities larger than 10 times the intrinsic carrier concentration, we can state the following:

$$n \approx N_D = N_D^+, \text{ if } N_A = 0,$$

and (III.24)

$p \simeq N_A = N_A^-$, if $N_D = 0$.

Hence, from Equations III.20 and III.24, if we add to a semiconductor crystal a number of donor atoms per cubic centimeter equal to $120n_i$, the number of electrons per cubic centimeter in the altered semiconductor will be $120n_i$ and the hole volume concentration will be $n_i/120$. Such a semiconductor is known as an n-type semiconductor, referring to the fact that the primary current carriers are electrons, with their negative charge. If, on the other hand, $650n_i$ of acceptors are added to each cubic centimeter of a pure semiconductor, the hole volume concentration will now be $650n_i$ and the electron concentration will fall to $n_i/650$, producing a p-type semiconductor, with the principal charge carrier possessing a positive charge. If both donors and acceptors are present in a semiconductor crystal in moderate concentrations we can assume $N_D = N_D^+$ and $N_A = N_A^-$ and state:

$$n + N_A = n_i^2/n + N_D. \tag{III.25}$$

For donor or acceptor concentrations at or above the effective densities of states, N_c and N_v, Equations III.23 through III.25 do not hold, owing to the simplifying assumptions made in the derivation of the densities of states (see Equation III.19). In this situation, one of very large impurity concentrations, the position of the Fermi level is within the conduction band ($N_D > N_c \gg N_A$) or within the valence band ($N_A > N_v \gg N_D$). The method to be employed in determining the exact location of the Fermi level for this case may be found in the literature [21].

In general, impurity atoms enter the semiconductor and become effective donors or acceptors when they replace host semiconductor atoms on their lattice sites within the crystal. For a Group IV semiconductor (refer to a periodic chart of the elements) such as silicon or germanium, each host atom has four valence band electrons which it shares with four neighboring atoms (see Figure III.2 and reference [19]) in a form of atomic interaction known as covalent bonding. Addition of impurity atoms from Group V[*], such as arsenic or phosphorus, to a group IV semicon-

---

[*] A Group V element has five valence band electrons and a Group III element has three valence band electrons.

ductor results in an excess electron. This electron is loosely attached to its Group V atoms and is easily detached from the atom, making such an element a donor. Introduction of a Group III element, such as boron or indium, results in an electron being missing from the crystal. This missing electron acts as a hole. In Table III.2 we considered a number of semiconductors. Germanium (Ge) and silicon (Si) are Group IV semiconductors and the donors and acceptors mentioned above apply to them. Indium phosphide (InP), gallium arsenide (GaAs), aluminum antimonide (AlSb) and gallium phosphide (GaP) are III-V semiconductors (so-called from the columns of the periodic chart in which their constituent atoms may be found). These compounds are bound together in crystals by both the covalent sharing of electrons evidenced by Group IV semiconductors and by electron exchange between the constituent atoms (ionic bonding) [14,22]. In III-V semiconductors a Group VI element such as selenium (Se) or tellurium (Te) acts as a donor and a Group II element such as zinc (Zn) acts as an acceptor. However, the situation in these materials is more complex than this. A Group IV element such as Si can act either as a donor or an acceptor in a III-V semiconductor. Such a Group IV element can either be a donor or acceptor depending on whether it is located in the crystal on a III-atom site (in which case it acts as an acceptor) or a V-atom site (in which case it acts as a donor). For details the reader is referred to the literature [23]. To complete our survey of Table III.2, there are two II-VI compounds, cadmium telluride (CdTe) and cadmium selenide (CdSe). Two-six compounds exhibit primarily ionic bonding. In these materials not only are there donor and acceptor impurities, but the absence of a crystal constituent often acts as a donor or an acceptor. For example, cadmium vacancies act as donors in these two semiconductors.

There are a number of methods for introducing impurities into any particular semiconductor. We can insert the impurities into the crystal when the crystal is initially formed. This initial construction is normally performed by pulling the semiconductor crystal from a molten bath, using a seed crystal to define the nature and orientation of the lattice. Construction of a crystal can also be by growth of the semiconductor on a substrate which is either composed of the semiconducting material or of another material with a similar crystal orientation. In this case we have liquid-phase epitaxy if the substrate is submerged in a liquid containing the required atomic ingredients of the semiconductor; gas-phase epitaxy if the materials required for growth of the semiconducting crystal are gases; or molecular beam epitaxy, if the crystal is essentially assembled atom by atom and molecule by molecule in a vacuum chamber using electric and magnetic fields to place the constituent atoms (both the host semiconductor atoms and the impurity atoms) in their proper locations.

We can utilize high temperatures to move impurity atoms into the semiconductors in a process known as diffusion. The result of this process is an impurity concentration in a semiconductor which is high at the surface and which decreases with displacement into the crystal. Another method, known as ion implantation, involves the acceleration, to high velocities, of ions of the required impurity species. These ions are then impacted on the semiconductor crystal and penetrate, to a distance which is dependent on ion, semiconductor and accelerating voltage. The exact nature of any of these technologies is not important at this point. What is important to note is that each technology has unique effects on the performance of the photovoltaic cells made using that technique. These effects will be discussed subsequently in this work.

We have reviewed the origin of holes and electrons and the effect on hole and electron concentrations of changes in temperature and the additions of trace amounts of donors and acceptors[*]. We have seen how it is possible to vary the electron and hole concentrations in a semiconductor over many orders of magnitude and have briefly reviewed techniques for accomplishing this. Let us now consider the motion of the charge carriers (holes and electrons) and the resulting currents.

Currents

Consider a semiconductor with a uniform impurity carrier concentration. Ohm's law (Equation III.1) provides the current for a given applied electric field. The conductivity of a given substance is a function of the number of charge carriers and can be written:

$$\sigma = q(\mu_n n + \mu_p p), \qquad\qquad (III.26)$$

---

[*] By trace amounts we mean amounts less than 0.1% of the host atomic concentration--often amounts less than a billionth of the host concentration.

where μ_n and μ_p are the electron and hole mobilities. These quantities are defined by:

$$\mu = v/\mathscr{E}, \tag{III.27}$$

where v is the effective net velocity of the charge carrier and $\mathscr{E}$ is the electric field strength.

In a semiconductor, or in any material medium, an electron (or a hole) cannot accelerate indefinitely, but suffers collisions with the atoms of the host crystal, with other charge carriers, and with the impurity atoms in the crystal. This complex situation has received extensive study in the literature [24-28]. For small electric fields, such as those normally encountered in the bulk regions of a solar cell, the carrier velocity is linearly dependent on the electric field as indicated by Equation III.27. The mobility is dependent on the temperature, the impurity concentration and upon the crystalline direction in which the electron is moving. The mobility values provided in Appendix B and in this chapter are essentially averaged values taking into account all of these factors[*] and are under small electric field conditions (less than 1000 volts/centimeter]. In all semiconductors, at large electrical fields, a number of complex interactions are encountered, leading to such phenomena as the Gunn effect in gallium arsenide and the maximum carrier drift velocity in a semiconductor. These conditions are of small, if any, importance in solar cells.

The mobilities of both holes and electrons for a number of potentially useful (in solar cells) semiconductors are provided in Appendix B. The values listed are for low impurity concentrations (the number of impurity atoms = the number of charge carriers = 10^{14} to 10^{16}/cm^3) at a temperature of 300°K. Note that the electron mobility is generally higher than that for a hole. This is a general result of the fact that the hole is, in essence, the net electrical effect of the motion of many electrons in the

---

[*] The mobility of the charge carriers in a polycrystalline semiconductor is a complex function, controlled by the grain boundaries between the internal crystallites, making the carrier transport dependent on processing and difficult to discuss theoretically. The mobility of carriers in amorphous materials will be discussed in a later chapter.

valence band and is subject to the quantum mechanical restraints imposed on the crystal by its finite size and the large number of electrons in the valence band. Note that AlSb is an exception to this generalization.

No mobility is given for holes in CdSe. This semiconductor, like many other II-VI semiconductors, is subject to severe stoichiometric restrictions. The addition of acceptors to cadmium selenide during crystal growth is a thermal process in which the crystal approaches thermodynamic equilibrium. The processes involved result in the production of cadmium vacancies during the introduction of the acceptors. These cadmium vacancies act as donors [29,30]. The net outcome of this phenomenon is that it has not been practical to produce p-type CdSe. Hence, it is difficult to measure the hole mobility in this material [31]. It is theoretically possible that the use of ion implantation, which does not require a crystal in thermodynamic equilibrium, will allow fabrication of p-type CdSe.

For non-polar semiconductors, such as Ge and Si, the major elements affecting the mobility are scattering of the charge carriers from acoustic phonons (lattice vibrations) and charge carrier scattering from the ionized impurity atoms. The temperature behavior of mobility as a result of lattice vibrations is:

$$\mu_L \sim (m^*)^{-5/2} T^{3/2}. \tag{III.28}$$

The temperature dependence of mobility as affected by scattering from impurity atoms (known as impurity scattering) is:

$$\mu_I \sim (m^*)^{-1/2} T^{3/2}/N_I, \tag{III.29}$$

where N_I is the density of ionized impurities.

The combined mobility (i.e., the effective mobility of an electron considering scattering by both acoustic phonons and impurity scattering), μ, is given by:

$$1/\mu = 1/\mu_L + 1/\mu_I. \tag{III.30}$$

For a polar semiconductor (compound semiconductors are generally polar) the presence of optical phonons (high energy lattice vibrations) is also important. The combined mobility for such polar semiconductors is [32]:

$$\mu = (m^*)^{-3/2} T^{-1/2}. \tag{III.31}$$

The potential temperature range of operation for earth-based solar cells is approximately 250°K to 400°K. The lower limit is established by ambient conditions during the daylight hours and the upper limit by a consideration of the loss mechanisms within the solar cells (see subsequent chapters). The temperature dependence of mobility, in the vicinity of 300°K, is indicated for a number of semiconductors in Appendix B. Specific examples of electron mobility in selected intrinsic semiconductors are presented in Table III.3.

Table III.3

Electron mobility, μ_n, in cm^2/volt-second, for selected intrinsic semiconductors as a function of temperature (after [27, 28, 33 and 34])

Semiconductor	Si	GaAS	CdTe	CdSe	GaP
T = 300°K	1700	8500	1050	800	110
T = 400°K	920	3800	750	500	100
T = 500°K	400	3200	650	350	80

We have been examining the carrier mobilities in relatively pure semiconductors (impurity concentrations between 10^{12} and 10^{15} per cm^3). While such impurity concentrations are met in the bulk regions of solar cells, it will be seen, later in this work, that heavier impurity concentrations are frequently encountered. Tables III.4 and III.5 present mobility data for several semiconductors as a function of impurity concentration.

In practice, the mobility exhibited by a charge carrier in a semiconductor depends on the processing history of the semiconductor. If we have a solar cell, was it constructed by diffusion, by ion implantation, or by some epitaxial process? Was the original crystal, from which the device was made, grown by pulling from a melt or by epitaxy? Exact details of the processing play a major role. The mobility values furnished in Tables III.3 through III.5 and in Appendix B are the normally maximum values that are exhibited by holes and electrons--smaller mobility values are often encountered in practice. It is worthwhile noting that, while germanium, silicon and gallium arsenide have been under intense study for several decades and their properties are well documented, the other semiconductors listed are much less well understood and improvements in processing and crystal preparation techniques may increase the mobility values listed here.

Table III.4

Electron mobility, μ_n, in cm^2/volt-second, for selected semiconductors at 300°K, as a function of impurity concentration (after [27, 34-38])

Semiconductor	Ge	Si	GaAs	CdTe	CdSe
Impurity concentration (cm^{-3})					
10^{14}	3600	1900	7000	1000	620
10^{16}	3700	1200	6000	900	580
10^{18}	2000	320	3000	800	500

Table III.5

Estimated hole mobilities, μ_p, in cm^2/volt-second, for selected semiconductors at 300°K, as a function of impurity concentration (after [27, 34-37])

Semiconductor	Ge	Si	GaAs	CdTe
Impurity concentration (cm^{-3})				
10^{14}	1900	600	400	100
10^{16}	1300	460	320	80
10^{18}	360	200	160	60

There is a another current mode possible in semiconductor devices, a mode which is generally present in active solar cells. Consider a semiconductor in which the electron (or hole) concentration is not uniform with distance. Let us consider those situations in which:

$$n = n(x,y,z) \text{ and } p = p(x,y,z). \tag{III.32}$$

At any temperature above absolute zero, the charge carriers are in random, thermally driven motion, and the distribution of the carriers dictates that they will, on the average, move from regions of high concentration to regions of low concentration. It can be shown that the current due to this is [40]:

$$J_n = qD_n \text{grad}\{n(x,y,z)\},$$

and   (III.33)

$$J_p = -qD_p \text{grad}\{p(x,y,z)\},$$

where J_n and J_p are the electron and hole current densities and D_n and D_p are the electron and hole diffusion constants. The diffusion constants for electrons and holes are related to the mobilities for these carriers by the Einstein relationship [41]:

$$D_{n,p} = (kT/q)\mu_{n,p}. \tag{III.34}$$

Combining Equations III.1, III.26 and III.33, the overall expression for current density in a semiconductor is given by:

$$J = J_n + J_p, \tag{III.35}$$

where

$$J_n = q\mu_n n + qD_n \nabla n,$$

and   (III.36)

$$J_p = q\mu_p p - qD_p \nabla p.$$

In subsequent chapters we will utilize these expressions for current density while determining the efficiency of solar cells.

Recombination and Carrier Lifetime

In Chapter IV we will consider the interaction of light with semiconductors. We will see that light of sufficiently short wavelengths will stimulate electrons from the valence band and "boost" them into the conduction band. This process creates electron-hole pairs. In subsequent chapters we will consider means for separating these electron-hole pairs into their individual components and for delivering the energy the pairs contain to a load external to the semiconductor, thus forming a solar cell and circuit. Not all of these electron-hole pairs can be collected and separated; a fraction will recombine within the semiconductor before the electrons and holes can be physically separated. There are two principal regions within any semiconductor in which recombination of holes with electrons can occur. First, there is the bulk or internal region of the crystal, characterized by the regular periodic structure of the crystal lattice with its attached host semiconductor atoms, a small percentage of acceptors (and, or donors), and a small number of defects[*].

The other principal recombination region is the surface of the crystal, characterized by the abrupt ending of the semiconductor crystal structure and the presence of foreign atoms on or in the surface; a region some 20 to 100 Å thick. Let us first consider the effects of recombination in the bulk region. The approach to be followed is based on the well-known Shockley-Read-Hall theory of recombination for holes and electrons in semiconductors [42-44].

Consider a semiconductor with a density, N_t, of traps (defects which facilitate electron-hole recombination); each trap with a single energy level, E_t, all under equilibrium conditions. The probability of an electron occupying the energy level of any trap is given by:

$$f_t = 1/(1 + \exp[\{E_t - E_f/kT\}]),\hspace{3cm} \text{(III.37)}$$

---

[*] A defect in a crystal is any irregularity in the periodicity of the crystal, such as a foreign impurity atom not on a normal lattice site, an atom interstitially sited (squeezed in between normally located atoms), a missing atom (known as a vacancy), or some change in the crystal order such as might be occasioned by a tendency to polycrystallinity.

and the probability that an electron is not in the trap is:

$$f_{t-} = 1 - f_t. \tag{III.38}$$

Considering each of the four possible trapping scenarios as outlined in Figure III.6, and assuming that each trap can hold but a single electron, the rate, r_i, at which each process occurs is given by:

(a) the rate for electron capture:

$$r_{ec} = v_{th}\sigma_n n N_t f_{t-}, \tag{III.39}$$

where σ_n is the trap capture cross section for electrons (a number close to 10^{-15} cm^2 in silicon and in many other semiconductors, v_{th} is the average thermal velocity of the carriers ($v_{th} = \sqrt{\{3kT/m^*\}} \approx 10^7$ cm/sec.), n is the number of electrons per cubic centimeter in the conduction band, and $N_t f_{t-}$ is the number of empty traps.

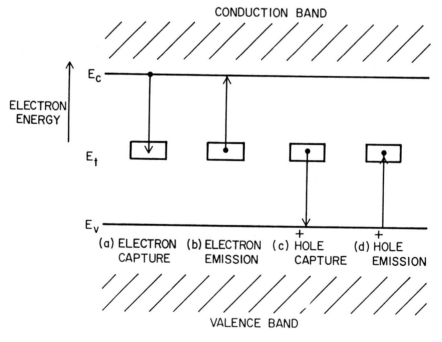

Figure III.6. The basic processes involved in electron-hole recombination.

(b) the rate for electron emission to the conduction band is:

$$r_{ee} = e_n N_t f_t, \tag{III.40}$$

where $N_t f_t$ is the number of traps filled with electrons and e_n is the probability of electron emission from a trap. This probability is a function of the specific nature of the trap, the electron density in the conduction band and the degree of degeneracy. Assuming a non-degenerate situation, implying that e_n is independent of n, we have:

(c) the rate of hole capture is:

$$r_{hc} = v_{th} \sigma_p p N_t f_t, \tag{III.41}$$

where σ_p is the capture cross section for holes, and,

(d) the rate of hole emission is:

$$r_{he} = e_p N_t f_{t-} , \tag{III.42}$$

where e_p is the probability of hole emission from the trap (i.e., the capture of a valence band electron by the trap). As in the case of emission of electrons to the conduction band, hole emission to the valence band is independent of hole concentration in the valence band, p, so long as p is much less than N_v (i.e., the material is non-degenerate).

In thermal equilibrium, the hole and electron concentrations are constant and the capture and emission rates must be equal ($r_{ec} = r_{ee}$ for electrons and $r_{hc} = r_{he}$ for holes). Here, it is possible to solve for the electron and hole emission probabilities. Suppose we now illuminate the bulk semiconductor with light of sufficient energy to generate electron-hole pairs. The electron and hole concentrations now increase above their equilibrium levels. Assuming an n-type semiconductor, the minority carrier (in this case, the holes) concentration is given, as a function of time, by:

$$\partial p_n / \partial t = G_L - U, \tag{III.43}$$

where G_L is rate of generation of holes as a result of the illumination of the semiconductor and U is the net recombination rate for holes ($r_{hc} - r_{he}$).

We define the hole carrier lifetime (the minority carrier lifetime in this situation) by:

$$\tau_p = \delta p / U, \tag{III.44}$$

where $\delta p = p_n - p_{no}$ is the difference between the existing hole concentration (under illumination), p_n, and the thermal equilibrium hole concentration, $p_{no} = n_i^2/n_{no} = n_i^2/N_D$. The solution to Equation III.43, when the light is extinguished (G_L turns to zero at $t = 0$) is:

$$p_n(t) = p_{no} + (p_L - p_{no})\exp(-t/\tau_p), \tag{III.45}$$

where p_L is the equilibrium hole concentration under illumination. While the semiconductor is under equilibrium under illumination we may write for the majority carrier density:

$$\partial n_n/\partial t = 0 = G_L - (r_{ec} - r_{ee}), \tag{III.46}$$

and for the minority carrier holes:

$$\partial p_n/\partial t = 0 = G_L - (r_{hc} - r_{he}). \tag{III.47}$$

Assuming that the probabilities for emission (e_n and e_p) are independent of the illumination level (this implies low level illumination with the majority electron concentration under illumination effectively equal to the majority electron concentration in the dark) the net recombination rate can be written:

$$U = \frac{\sigma_p \sigma_n v_{th} N_t (p_n n_n - n_i^2)}{\sigma_n(n_n + n_i\exp[\{E_t - E_i\}/kT]) + \sigma_p(p_n + n_i\exp[\{E_i - E_t\}/kT])}, \tag{III.48}$$

where E_i is the intrinsic Fermi level.

Clearly, this expression may be applied to both n-type and p-type semiconductors. When the hole and electron trap capture cross sections, σ_p and σ_n, are equal, this expression simplifies to:

$$U = \frac{\sigma v_{th} N_t (p_n n_n - n_i^2)}{n_n + p_n + 2n_i\cosh[\{E_t - E_i\}/kT]}. \tag{III.49}$$

For an n-type semiconductor under low-level optical illumination, we may take $n_n \gg p_n$, and for trap energy at the intrinsic Fermi level (this corresponds to maximum recombination):

$$U = \sigma v_{th} N_t (p_n - p_{no}). \tag{III.50}$$

Similarly, for a p-type semiconductor under low-level optical illumination, $p_p \gg n_p$ and for E_t equal to E_i:

$$U = \sigma v_{th} N_t (n_p - n_{po}). \tag{III.51}$$

From the definition of lifetime, we have for the hole lifetime in an n-type semiconductor, and for the electron lifetime in a p-type semiconductor:

$$\tau_{po} = 1/(\sigma v_{th} N_t),$$

and $$\tag{III.52}$$

$$\tau_{no} = 1/(\sigma v_{th} N_t).$$

For high levels of carrier injection (owing to intense optical illumination, the minority carrier lifetime monotonically approaches [42]:

$$\tau_{\infty} = \tau_{po} + \tau_{no}. \tag{III.53}$$

Should the trap have multiple energy levels, E_{t1}, E_{t2}, etc., the effective lifetime and net recombination rates exhibit the same overall features as do single level traps. For a more detailed examination of single and multi-level traps, the reader is referred to Moll [45].

When the prime mechanism of recombination is via traps, the lifetime is dependent on the impurity concentration, N_i. The general form of dependency is [46]:

$$\tau_i = \tau_{io}/(1 + N_i/N_{io}). \tag{III.54}$$

Here, τ_{io} is the observed lifetime at impurity concentration, N_{io} (either donor or acceptor) and τ_i is the lifetime for an impurity concentration, N_i. In n-type silicon, for example, $\tau_{po} \approx 4 \times 10^{-4}$ seconds, where $N_{io} = 7 \times 10^{15}/cm^3$.

To this point we have considered recombination as it occurs via defect levels (traps) sited in the forbidden gap. Recombination can also occur directly from the conduction band to the valence band; the energy released either yielding a photon (this is known as radiative recombination) or an energetic charge carrier (Auger recombination). Radiative recombination is of particular importance in direct-gap semiconductors. In this type of semiconductor, when an electron transfers from the conduction band minimum to the valence band maximum, only a change in energy is required, and the momentum remains the same (see Figure III.4). In indirect-gap semiconductors a momentum change is required of the electron dropping from conduction to valence band--as well as the energy shift. In the following chapter we will consider the absorption of light by semiconductors as a means of creating electron-hole pairs. The momentum of a photon is, in general, insufficient to supply the required momentum change for an electron switching bands in an indirect-gap semiconductor. Therefore, to effect an interband transmission, an electron must acquire (or shed) momentum by collision with a phonon[*]. This makes for qualitative and quantitative differences in light absorption (and emission, but we are primarily interested in absorption in this work) between direct- and indirect-gap semiconductors; differences which will be considered in the following chapters. The required momentum shift makes radiative recombination unlikely for indirect-gap semiconductors, while in direct-gap semiconductors, radiative recombination normally dominates the recombination phenomena. The minority carrier lifetime for radiative recombination has the following impurity concentration dependency [47]:

$$\tau_{p,n}^{R} \propto 1/N_{D,A}. \tag{III.55}$$

For example, in n-type gallium arsenide, $1/\tau_p = 0.8 \times 10^{-8} N_D$ and in p-type GaAs, $1/\tau_n = 0.47 \times 10^{-9} N_A$ [47]. This dependence of lifetime on doping level is consistent with the direct gap of GaAs and indicates that radiative recombination is the principal mechanism.

Auger recombination also depends on the impurity densities, N_A

---

[*] A phonon is the particle representation of the energy and momentum inherent in lattice vibrations--normally due to the non-zero temperature of the material.

and N_D, and can be shown to have the following impurity concentration dependency [42]:

$$\tau_{p,n}^A \propto 1/N_{D,A}^2. \tag{III.56}$$

Obviously, in the construction of a solar cell it is important to maintain the lifetime at as high a value as possible in order to preserve the optically generated electron-hole pairs. In Appendix B, we see that the trap dominated minority carrier lifetimes for the indirect-gap semiconductors silicon and germanium are much longer than those for direct-gap materials. Even in gallium phosphide, an indirect-gap material, radiative transitions play a major role in reducing the lifetime of the material. With the exceptions of Si and Ge, the semiconductors listed in Appendix B are compound semiconductors with relatively primitive fabrication technologies. This accounts, in part, for the poor lifetimes exhibited and allows for some future improvement. Table III.6, lists the author's estimates of the effective carrier lifetimes, at low optical injection levels, that can be reasonably expected in the near (a decade) future.

Table III.6

Estimated minority carrier bulk lifetimes (in seconds) in solar cells at low carrier concentration levels and for low to moderate optical illumination

Semiconductor	Electron Lifetime	Hole Lifetime
Ge	0.5×10^{-4}	0.5×10^{-4}
Si	0.8×10^{-4}	0.8×10^{-4}
InP	6.0×10^{-8}	3.0×10^{-8}
GaAS	6.0×10^{-8}	3.0×10^{-8}
CdTe	2.0×10^{-6}	1.0×10^{-7}
AlSb	1.0×10^{-7}	0.9×10^{-7}
CdSe	1.5×10^{-8}	1.5×10^{-9}
GaP	7.0×10^{-8}	4.0×10^{-8}

Let us turn to surface recombination. Consider an n-type semiconductor uniformly illuminated with surface recombiantion occurring solely at one edge, as depicted in Figure III.7a. Because of the surface recombination, there is a reduction in minority carrier concentration at the surface as $x = 0$. There will be a flow of holes from the bulk regions of the semiconductor to the outside surface at $x = 0$. There must also be a flow of electrons in the same direction to maintain charge neutrality. The net current flow from bulk to surface is therefore zero, and the hole distribution, p_n, is given as a function of space and time by:

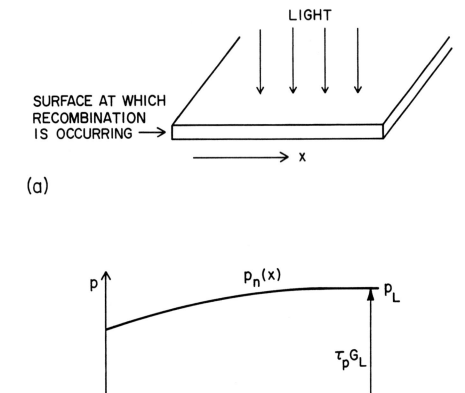

Figure III.7. (a) Uniform single-surface illumination of a semiconductor with recombination occurring at one edge. (b) The minority carrier concentration with single-surface recombination.

$$\partial p_n / \partial t = -[D_p\{\partial p_n / \partial x\} + \mu_p \mathscr{E} p_n] + G_L - U, \qquad \text{(III.57)}$$

where the quantity in brackets is the whole flux, dependent upon both the diffusion constant, D_p, and the electric field, $\mathscr{E}$. From the requirements that hole and electron flows towards the recombining surface must be equal, and for overall charge neutrality, we can solve for the electric field. For a low level of illumination, $p_n \ll n_n$, the drift component of Equation III.57 may be written:

$$\mu_p \mathscr{E} p_n = \frac{-(D_n - D_p)(\partial p_n / \partial x)}{\mu_p p_n + \mu_n n_n} - \mu_p p_n \tilde{\_} - D_p\{\partial p_n / \partial x\}\{1 - D_p / D_n\}\{p_n / n_n\}, \text{ (III.58)}$$

where the hole and electron mobilities are assumed to be approximately equal. Comparison of Equations III.57 and III.58 shows that, under the low-level assumption, the drift component of hole flow toward the surface is much less than the diffusion current component. Therefore, we can solve for the steady-state hole minority hole concentration utilizing only the diffusion component of Equation III.57. For boundary conditions, we have the hole concentration in the bulk regions and the diffusion current at the surface:

$$p_n(\infty) \tilde{\_} p_L = p_{no} + \tau_p G_L \, , \, p_{no} = n_i^2 / n_{no},$$

and   (III.59)

$$D_n\{\partial p_n / \partial_x\}_{x=0} = s_p\{p_L - p_{no}\},$$

where s_p is the surface recombination velocity for holes. Under these conditions, the hole concentration is:

$$p_n(x) = p_L - \frac{(p_L - p_{no})s_p\{\tau_p / L_p\}}{1 + s_p \tau_p L_p} \exp\{-x / L_p\}. \qquad \text{(III.60}$$

In Equation III.60, L_p is the diffusion length for holes, given by $L_p = \sqrt{\{D_p \tau_p\}}$.

In Figure III.8, the minority carrier density is displayed as a function of the surface recombination velocity. Note that an infinite value of surface recombination velocity results in a surface carrier concentration equal to the thermal equilibrium value. This is the condition that is obtained at an ideal ohmic contact.

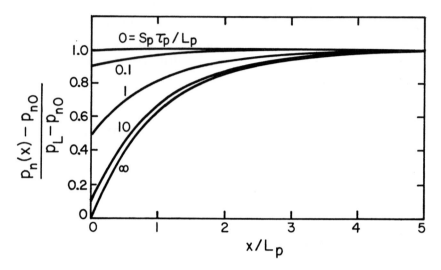

Figure III.8. The steady-state minority charge carrier distribution as a function of surface recombination velocity. Drawn for holes in an n-type semiconductor and after Grove [48], with permission.

For practical semiconductor surfaces, not considering ohmic contacts, the value of s is non-zero and has a magnitude that is very dependent on crystal surface orientation and upon the processing technologies employed in forming the surface. We can write for the surface recombination velocity:

$$s = s_s = \sigma v_{th} N_{st}, \tag{III.61}$$

where N_{st} is the concentration of surface traps. Note that this concentration is higher for [111] silicon surfaces than for [100] silicon surfaces [48].

The surface recombination is also dependent on factors outside of the semiconductor crystal. For example, a typical solar cell process with silicon as the semiconductor involves the oxidation of the surface, producing silica (SiO_2). Charged impurities are frequently trapped within this oxide. These impurities give rise to an electric field that acts upon the holes and electrons near to the semiconductor surface. With the electric field present, space-charge neutrality is no longer required. If the electron and hole capture cross sections (σ_n and σ_p) are equal, the rate of recombination at a semiconductor surface may be written [5]:

$$U = s \; \frac{p_s n_s - n_i^2}{n_s + p_s + 2n_i}, \qquad (III.62)$$

where s is the surface recombination rate with no electric field present, as provided in Equation III.61, and n_s and p_s are the hole and electron concentrations at the surface.

Later, we will examine the various configurations of solar cells and the mechanisms by which light enters them. At that time we will note that the net photocurrent of a solar cell depends on the physical configuration and upon the processing technology employed; a major portion of the dependency being recombination effects. For the moment, it is sufficient that we have introduced the nomenclature and have observed that both the bulk lifetime and surface recombination depend strongly on the individual semiconductor material and the processing technology employed in the manufacture of the semiconductor crystal and the devices made from it [51,52].

While we have not specifically considered temperature in discussing bulk lifetime and surface recombination velocity, it should be clear that both are dependent on temperature, since both are a function of the thermal velocity of the charge carriers.

The preceding consideration of surface recombination rate has been conducted under the general assumption of a low illumination level. In this case, the minority carrier concentration is much less than the thermal equilibrium majority carrier concentration, which is approximately equal to its non-illuminated value ($p_n \ll n_n \tilde{} n_{no}$ for n-type semiconductors and $n_p \ll p_p \tilde{} p_{po}$ for p-type semiconductors). In a subsequent chapter, we will consider the effects of optical concentration on solar cell performance. For sufficient optical concentration, the level of illumination is high enough to yield effectively equal numbers of holes and electrons ($n = p$). Study of Equations III.57 and III.58 indicates that under these conditions the drift component of the minority current is of the same order of magnitude as the diffusion current. In general, under these conditions, when considering the hole and electron densities within a bulk piece of semiconductor, we must consider not only the drift, but also the diffusion current components for both charge carriers. This leads to the following expressions for electron and hole densities within a semiconductor:

$$\partial n/\partial t = G_L - U + \{1/q\}\nabla J_n,$$

and   (III.63)

$$\partial p/\partial t = G_L - U - \{1/q\}\nabla J_p,$$

where J_n and J_p are provided by Equation III.36; the bulk recombination rate, U, is dependent on carrier lifetime; the impurity concentrations, n and p, are functions of the temperature; the generation rate, G_L, depends on the intensity and wavelength of illumination (see the following chapter); and the boundary conditions at the surfaces of the crystal involve the surface recombination velocity.

Junctions

The purpose of a solar cell is to convert the energy in sunlight to electrical energy and to deliver that electrical energy to an external load. This implies that a current must flow within the semiconductor; a current composed of the positively charged holes and negatively charged electrons which the incoming light generates within the semiconductor (see Chapter IV). In the preceding section we observed that these holes and electrons will recombine unless separated from one another. We could separate the holes and electrons by applying a battery to the semiconductor and using the resultant electric field. This would defeat the purpose of the solar cell, since the energy delivered to the external load in this case would come from the battery and not from the photons[*]. We need to generate an electric field within our "piece" of semiconductor, an electric field which is "built into" the semiconductor and is not related to any external energy source. There will always be some recombination in the bulk of the semi-

---

[*] Semiconductor devices biased in such a way, with an externally (battery) powered reverse bias, are used to detect light (of a multiplicity of frequencies, not just visible light). Light detection devices, needless to say are a completely different subject and will not be considered here.

conductor and at its surface, but a majority of the optically generated electron-hole pairs will be separated by the electric field and the individual charge carriers can then be used to carry a current.

We will consider various methods of generating internal electric fields in a semiconductor in Chapter V. The methods most generally employed involve the presence of a junction within the semiconductor. A junction may be the result of a single semiconducting crystal being part p-type and part n-type, or being composed of two different semiconductors, or by applying a metal to the extremely clean surface of a semiconductor. The current versus voltage characteristics of such devices are discussed in Chapter V and will not be covered in any detail here. However, the pn junction device is worth noting as a typical means of separating the optically generated holes and electrons. Consider, as an example, a single crystal of semiconductor, one side of which is p-type and the other, n-type. In this physical situation, the portion of the semiconductor close to the junction[*] is depleted of mobile charge carriers (the electrons and holes). Here, the charge neutrality condition (Equation III.21) is no longer operative, and a powerful electric field is generated[#], satisfying Poisson's equation:

$$\nabla \mathscr{E} = \rho/\epsilon = \{q/\epsilon\}\{N_D - N_A\}, \qquad\qquad\qquad\text{(III.64)}$$

where ρ is the charge density which is equal to the difference between the donor and acceptor densities at any given point and ϵ is the permittivity of the semiconductor.

In Figure III.9 we present the energy versus displacement diagram for an n-type semiconductor and for a p-type semiconductor. Included are the locations of: the Fermi level (as determined by solutions of Equations III.21 and III.23), the energy levels of electrons when on the donor

---

[*] The plane within the semiconductor at which the n- and p-type regions meet is defined as the junction.

[#] The electric field arises on the positively charged donor atoms of the n-side of the junction and ends on the negatively charged acceptor atoms on the p-side of the junction. The electric charge on these atoms is no longer masked by the mobile charge carriers.

and acceptor atoms, the lower edge of the conduction band, and the upper edge of the valence band.

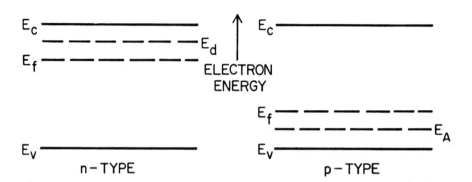

Figure III.9. The energy of the conduction band lower edge, the valence band upper edge, the donor and acceptor levels and the Fermi level for n- and p-type semicondcutors.

When these regions are joined into a single semiconductor crystal, the Fermi levels of the n- and p-type regions must be aligned (at the same energy level), leading to the configuration of Figure III.10.

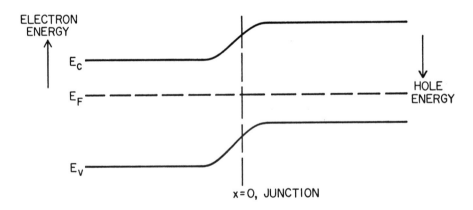

Figure III.10. The energy versus displacement diagram for a pn junction in thermal equilibrium.

Note the bending of the conduction and valence band edges in Figure III.10. The pn junction device of this figure is generally known as a diode and is assumed to be in thermal equilibrium.. It is electrically neutral and electric field free, except in the vicinity of the junction itself. A strong electric field develops at the junction and results in the bending of the energy bands. This field sweeps all of the electrons and holes away from the junction leaving behind the fixed-position, ionized donor and acceptor atoms. This creates a charge dipole as seen in Figure III.11.

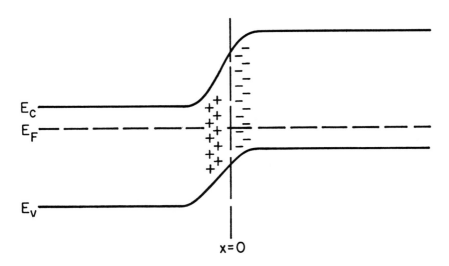

Figure III.11.   The charge dipole produced at the pn junction.

Under the conditions of thermal equilibrium, no net current flows in the dipole. The "tendency" for a large diffusion current to flow from right to left (induced by the huge differences in concentration for the holes and electrons on the two sides of the junction) is counterbalanced by the drift current driven by the electric field across the junction. In thermal equilibrium the energy versus distance configuration of Figure III.10 implies that any electron in the conduction band on the p-side, within a diffusion length (L_n, the diffusion length for electrons, equals the square root of the product of D_n and τ_n), will tend to move toward the n-type region because it has a lower energy when in the n-region. In similar fashion, holes in the n-region and near to the junction will move toward the p-region, the region of lower hole energy. Because p_n and n_p, the minority carrier concentrations of the two regions, are much less than the intrinsic carrier concentration, n_i; these currents are small. Now, let us place a battery across this pn junction diode as shown in Figure III.12.

Motion of electrons from the left to the right and of holes from the right to the left is further impeded by the increased barrier energy when the diode is reversed biased. The charge dipole around the junction will expand in width, and the electric field that arises from this dipole will increase. Such current as does flow in the pn junction will be small and is composed of the scarce holes originating on the n-side and electrons coming from the p-side.

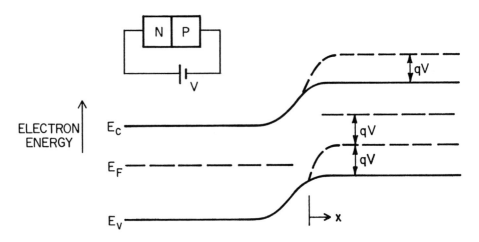

Figure III.12. The reverse biased circuit for a pn junction and the effect of that reverse bias on the energy versus distance diagram.

When the pn junction is forward biased, as depicted in Figure III.13, the diode internal energy barrier is reduced, enhancing electron diffusion across the barrier from left to right and hole diffusion from right to left. With many electrons available on the left-hand (the n-region) side and many holes available on the right-hand side, the diode current flow will increase sharply. A thorough analysis of the flow of current in a pn junction diode [24] leads to the quantitative current versus voltage characteristics outlined in Chapter V. The qualitative characteristics exhibited by forward biased pn junctions are those displayed in Figure III.14. Note that the current and voltage scales are very different in the forward and reverse biased directions.

The current versus voltage characteristics of the other types of junction mentioned earlier are qualitatively similar to that of a pn junction. The quantitative differences will be discussed later, as required. What makes diodes useful in solar cells is the existence of the internal energy barrier. This barrier produces the internal electric field which is used to separate the electron-hole pairs. For an abrupt pn junction this electric

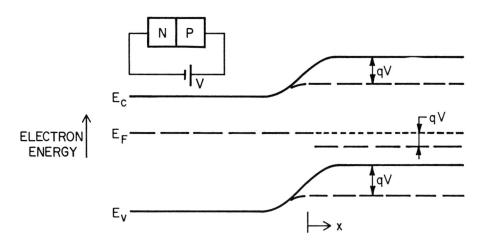

Figure III.13. The forward biased circuit configuration for a pn junction (a diode) and the effect of forward bias on the energy versus displacement diagram.

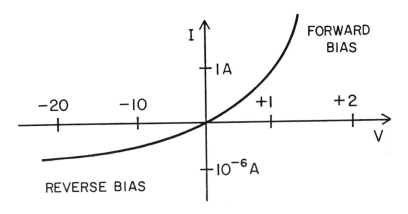

Figure III.14. The qualitative current versus voltage characteristic of a typical pn junction diode.

field extends into both the n- and p-type regions for the following distances [53]:

$$x_n = \sqrt{\{2\epsilon V_D/qN_D\}}\sqrt{\{1/[1 + N_D/N_A]\}}, \qquad (III.65)$$

and

$$x_p = x_n\{N_D/N_A\}.$$

Here, an abrupt junction is one in which the n- and p-type regions merge within a few lattice constants, a situation typical of solar cells. In Equation III.65, x_n is the distance, from the junction, in the n-type region, to which the electric fields extend; and x_p is the distance, from the junction, in the p-type region, to which the electric fields extend. V_D is the effective voltage across the diode junction, given by:

$$V_D = V_B - V_A, \qquad\qquad (III.66)$$

where V_A is the externally applied voltage (positive for forward bias), and V_B is the internally generated barrier voltage. It can be shown that this barrier voltage is [54]:

$$V_B = \{kT/q\}\ln\{n_n p_p/n_i^2\} = \{kT/q\}\ln\{N_A N_D/n_i^2\}. \qquad (III.67)$$

We will return to Equations III.65 through III.67 in later chapters as we consider techniques for the construction of solar cells. For the moment, it is enough if we realize that the junction creates an electric field, that this electric field allows us to separate the electron-hole pairs generated when the incoming light (photons) interact with the semiconductor, and that this electric field also makes possible the collection of individual holes and electrons. The electrons are collected in the n-type semiconductor region and the holes in the p-type region. This "separation/collection" of the individual components of the electron-hole pairs results in more electrons on the n-type side of junction and more holes on the p-type side of the junction--more than are evidenced in thermal equilibrium with no light shining on the diode (solar cell). This uneven distribution of charge carriers generates a forward bias on the diode. Hence, under illumination a solar cell will have the current versus voltage characteristic depicted in Figure III.15.

In Figure III.15, note that the diode voltage and current are present in the first, third and fourth quadrants. Normal diode operation occurs in the first quadrant where the voltage and current are both positive and their positive product denotes power dissipation. Light detecting diodes operate in the third quadrant wherein the voltage and current are both negative. Again the product of diode voltage and current is positive, denoting power dissipation within the diode. Solar cells operate in the fourth quadrant of Figure III.15. Here the voltage across the diode is positive, but the current through the diode is negative when the diode (solar cell) is illuminated.

The negative product of voltage and current is indicative of the generation of electrical power--power which has as its source the photons which make up the light.

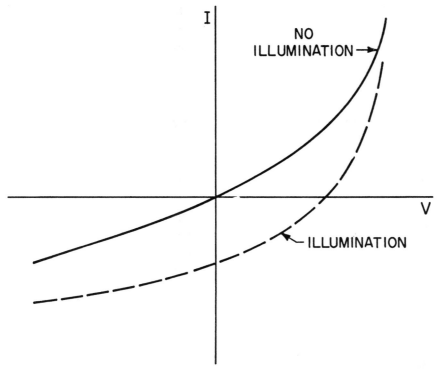

Figure III.15. The current versus voltage characteristic for a pn junction solar cell both in the dark and under illumination.

References

1 A. S. Grove, Physics and Technology of Semiconductor Devices, John Wiley & Sons, New York, 1967.
2 J. L Moll, Physics of Semiconductors, McGraw-Hill, New York, 1964.
3 S. M. Sze, Physics of Semiconductor Devices, 2nd Edition, John Wiley and Sons, New York, 1981.
4 C. Kittel, Introduction to Solid State Physics, 4th edition, John Wiley and Sons, New York, 1966.
5 R. A. Smith, Semiconductors, Cambridge University Press, Cambridge, 1959.

6   B. G. Streetman, <u>Solid State Electronic Devices</u>, 2nd edition, Prentice-Hall, Englewood Cliffs, NJ, 1980.

7   R. A. Smith, <u>Wave Mechanics of Crystalline Solids</u>, Chapman and Hall, London, 1963.

8   R. F. Pierret, editor, <u>Modular Series on Solid State Devices</u>, a multi-volume series from Addison-Wesley, Reading, MA., 1983--.

9   J. Tauc, in Physics Today, Oct. 1976, p. 23.

10 G. Lapidus, in IEEE Spectrum, Jan. 1973, p. 44.

11 S. H. Wemple, in Physics Review B., Vol. 7, 1973, p. 3767.

12 G. Gross, R. B. Stephens and J. Turnbull, in Journal of Applied Physics, Vol. 48, 1977, p. 1139.

13 D. Adler, in Scientific American, Vol. 236, May 1977, p. 36.

14 J. duBow, in Electronics, 11 Nov. 1976, p. 86.

15 L. C. Burton and H. M. Windari, in Journal of Applied Physics, Vol. 47, 1976, p. 4621.

16 L. R. Shiozawa, G. L. Sullivan and F. Augustine, in Conference Record, Seventh IEEE Photovoltaic Specialists Conference, Nov. 1968, p. 39.

17 There is an extensive literature on many polycrystalline based photovoltaic cells (e. g., $CdS:Cu_2S$) in the records of the various annual reports of the IEEE Photovoltaic Specialists Conferences.

18 Reference [4], p. 17.

19 Reference [7], p. 148.

20 M. L. Cohen and T. K. Aberstresser, in Physics Review, Vol. 141, 1966. p. 789.

21 Reference [5], p. 79.

22 S. Rames, <u>The Wave Mechanics of Electrons in Crystals</u>, North Holland Publishing Co., Amsterdam, 1963.

23 The list of semiconductors in Table III.2 is very brief. A much more extensive, but still not all-inclusive, list may be found in J. I. Pankove, <u>Optical Processes in Semiconductors</u>, Prentice-Hall, Englewood Cliffs, N J, 1971, page 412, or in Reference [3], Chap. 14. The question of what makes a satisfactory donor or acceptor for a given semiconductor depends, in part, on how far the impurity energy level is from the nearest band edge (E_c or E_v), upon quantum mechanical considerations concerning the ease with which an electron can be excited, and upon a number of technological considerations such as the relative size of the impurity and host atoms in the semiconducting crystal. The reader is referred to the general scientific

literature (e. g., The Journal of the Electrochemical Society, The Journal of Applied Physics, Semiconductor Products, The IEEE Transactions on Electron Devices, etc.) for details on specific semiconductors and particular donors and acceptors.

24 W. Shockley, Electrons and Holes in Semiconductors, Van-Nostrand-Reinhold, Princeton, NJ, 1950.

25 Reference [2], Chap. 4.

26 Reference [5], Chap. 5.

27 M. Aven and J. S. Presner, editors, Physics and Chemistry of II-VI Compounds, S. S. Devlin, John Wiley and Sons, New York, 1967, Chap. 11.

28 R. K. Willardson and A. C. Beer, editors, Semiconductors and Semi-Metals, Vol. 10, Transport Phenomena, Academic Press, New York, 1975.

29 J. D. Joseph and R. C. Neville, in Journal of Applied Physics, Vol. 48, 1977, p. 1941.

30 Reference [27], A. Abers, Chap. 4.

31 Itakara and H. Toyada, in Japanese Journal of Applied Physics, Vol. 4, 1965, p. 560. They report on the measurement of hole mobility using a pulse-field technique, obtaining μ_p = 50 cm^2/volt-sec.

33 W. Gartner, Transistors: Principles, Design and Applications, Van-Nostrand, Princeton, NJ, 1960.

34 M. R. Lorenz and R. E. Halstead, in Physics Review, Vol. 129, 1963, p. 2471.

35 S. M. Sze and J. C. Irwin, in Solid State Electronics, Vol. 11, 1968, p. 599.

36 M. B. Prince, in Physics Review, Vol. 92, p. 1953.

37 K. B. Wolstirn, In Bell System Technical Journal, Vol. 39, 1960, p. 205.

38 B. Segal, et. al., in Physics Review, Vol. 129, 1963, p. 247.

39 S. J. Yomada, in Journal of Physical Society of Japan, Vol. 15, 1960, p. 1940.

40 See, for example, Reference [2], Chaps. 6 and 7.

41 R. D. Middlebrook, An Introduction to Junction Transistor Theory, John Wiley and Sons, New York, 1957, p. 99.

42 W. Shockley and G. T.Read, Jr., in Physics Review, Vol. 87, 1952, p. 835.

43 R. N. Hall, in Physics Review, Vol. 83, 1951, p. 228.

44 R. N. Hall, in Physics Review, Vol. 87, 1952, p. 387.

45 Reference [2], Chap. 6.
46 J. G. Fossom, in Solid State Electronics, Vol. 19, 1976, p. 229.
47 J. I. Pankove, Optical Processes in Semiconductors, Prentice-Hall, Englewood Cliffs, NJ, 1971.
48 R. M. Das and R. C. Neville, in Journal of Applied Physics, Vol. 48, 1977, p. 3185.
49 Reference [1], p. 125.
50 Reference [1], p. 136.
51 Among the various texts which describe processing technology, several interesting ones are: Silicon Processing, D. C. Gupta, editor, American Society for Testing and Materials, Baltimore, MD; Semiconductor Integrated Circuit Processing Technology, W. R. Runyan and K. E. Bean, Addison-Wesley, Reading, MA, 1990; and Semiconductor Devices, Physics and Technology, S. M. Sze, John Wiley and Sons, New York, 1985.
52 There are a number of journals which devote considerable space to aspects of semiconductor fabrication technology, device performance, surface phenomena and other factors which we have been discussing. The reader is invited to browse in: Surface Science, The IEEE Transactions on Electron Devices, The Journal of the Electrochemical Society, The Journal of Applied Physics and Semiconductor Manufacturing.
53 Reference [6], Chap. 5.
54 Reference [2], p. 126.

CHAPTER IV: LIGHT-SEMICONDUCTOR INTERACTION

Introduction

Consider a semiconductor illuminated as demonstrated in Figure IV.1. The incoming light, composed of photons in a wide range of wavelengths, is normally incident on the semiconductor surface. When the photons strike the surface of the semiconductor a fraction of the photons is reflected from the surface, and the remaining photons enter the semi-

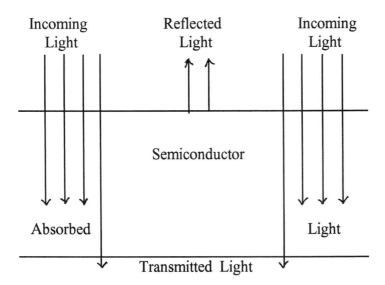

Figure IV.1. A vertically illuminated semiconductor illustrating the incoming light, the reflected light, the light absorbed in the semiconductor and the light passing through the semiconductor.

conductor. Some of these photons are absorbed within the semiconductor, and some pass entirely through the semiconductor and are lost. Most of those photons that are absorbed in the semiconductor give rise to electron-hole pairs[*]. It is these electron-hole pairs that are the source of the photocurrent and it is their separation into component parts and collection that gives rise to the photovoltage exhibited by solar cells. The photovoltage and photocurrent are the visible evidence of the production of electrical power by the solar cells. Notice that, in a solar cell, the fundamental conversion of light energy into electrical energy is performed when the incoming photons collide with valence band electrons, boosting these electrons into the conduction band, leaving behind valence band holes, and thus creating an electron-hole pair.

In the design of solar cells, the first consideration must be the interaction of light with the semiconductor. Study of Figure IV.1 indicates that this interaction begins with the reflection of light from the front (the illuminated) surface.

Reflection

Consider that light is incident upon some material at normal incidence after traveling through some medium with an index of refraction, n_{c1}, where:

$$n_{c1} = n_1 - ik_1, \qquad \text{(IV.1)}$$

where n_1 is the real index of refraction, k_1 is termed the extinction coefficient and i is the square root of minus one. Since the incident media which we will most frequently encounter are air and vacuum, we may take k_1 to be zero for the incident medium. Now, the index of refraction of the material upon which the light is falling (the semiconductor in our case of solar cells) is given by n_{c2}, where:

$$n_{c2} = n_2 - ik_2. \qquad \text{(IV.2)}$$

The fraction of light, f, reflected from the surface of the illumin-

---

[*] For reasons that will be discussed later in this chapter we will assume that all of the absorbed photons give rise to electron-hole pairs.

ated material is given by [1]:

$$f = \frac{(n_2 - n_1)^2 + k_2^2}{(n_2 + n_1)^2 + k_2^2}, \tag{IV.3}$$

where f is also known as the reflection coefficient and, for solar cells in which air (or a vacuum) is the incident medium, n_1 is unity.

It is common practice to apply antireflection coatings to the optically absorbing surface of solar cells as a means of reducing energy losses due to reflection of photons. The literature on antireflection coatings is extensive [2-7], but no single technology has emerged, to date, as preferable. The problem in applying any antireflection coating is multifaceted. First, the coating must not damage the semiconductor when applied. Any damage is often observed to result in an increase in surface recombination velocity, producing smaller photocurrent. A second problem with antireflection coating is a consequence of the broad range of optical wavelengths that are present in sunlight. As will be made clear later in this chapter, the wavelengths that are of interest in solar cell work lie between 2,000 Å and 13,000 Å. Since the thickness and index of refraction required of an antireflection coating are a function of the wavelength of the light, the situation is one in which no single-layer coating can be considered to be ideal. A third problem with antireflection coating is one of aging. The antireflection coating is exposed to both ambient atmospheric and direct (frequently concentrated) sunlight. Chemicals in the air must not degrade the coating in any time period short of several years. Sunlight, too, is capable of inducing chemical changes in an antireflection coating, especially in those instances where mirrors or lenses are employed to concentrate the sunlight which falls on the solar cells.

Antireflection coatings on solar cells range from simple single layers to upwards of seven-layer combinations. The exact choice depends on the semiconductor involved and the economics of the overall system. For those readers desiring more information on such coatings, References [4, 6 and 7] are suggested initial reading.

Before leaving the subject of reflection, a short example is in order. Eight semiconductors with potential usefulness in solar cells are: germanium, silicon, indium phosphide, gallium arsenide, cadmium telluride, aluminum antimonide, cadmium selenide and gallium phosphide. From data provided in Appendix B on semiconductor properties, the reflection coefficient for these materials, with air as the incident medium, is provided in Table IV.1.

Table IV.1

The reflection coefficient, f, of uncoated semiconductor surfaces for eight example semiconductors

Semiconductor	Ge	Si	InP	GaAs	CdTe	AlSb	CdSe	GaP
f	0.36	0.30	0.262	0.286	0.207	0.272	0.184	0.294

A simple antireflection coating that may easily be applied to many semiconductors is silicon dioxide (also known as silica). This material has an index of refraction of approximately 1.5, and is particularly easy to apply to silicon[*]. If a single layer of silica is applied, with optimum thickness, to the semiconductors listed in Table IV.1, the reflection coefficient is modified to the values provided in Table IV.2.

Table IV.2

The reflection coefficient, f, of SiO_2 coated semiconductor surfaces for eight example semiconductors

Semiconductor	Ge	Si	InP	GaAs	CdTe	AlSb	CdSe	GaP
f	0.21	0.15	0.120	0.141	0.079	0.130	0.065	0.147

The reflection coefficient data of Tables IV.1 and IV.2 are averages over those wavelengths of the solar spectrum of importance to the specific semiconductor when used as a solar cell. Later in the chapter we will determine these wavelengths. It is clear from these tables that the

---

[*] To apply silicon dioxide to silicon simply heat clean silicon surfaces in an oxidizing ambient (sometimes pure oxygen, but more frequently oxygen with water vapor added, as this produces an oxide of a given thickness in less time). The temperatures involved are normally between 1,000 and 1,200°C.

use of an antireflection coating can, significantly, improve the transmission, T', of light into a semiconductor; T' being given by:

$$T' = (1 - f)\Omega, \qquad\qquad (IV.4)$$

where Ω is the intensity of the incident light.

A final comment on the reflection of light from semiconductor
surfaces needs to be made. The discussion in this section, so far, has been
for normal incidence. If the light is incident at some angle, Ψ', to the
surface (see Figure IV.2), then only a fraction, $\cos\{\Psi'\}$, of the light is
incident normally. The portion of the light which is transmitted into the
semiconductor, and which is therefore available for the generation of hole-
electron pairs, T', will now be reduced to:

$$T' = (1 - f)\cos\{\Psi'\}\Omega, \qquad\qquad (IV.5)$$

where f is the normal incidence reflection coefficient.

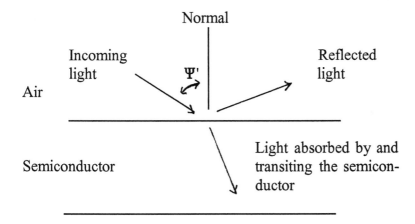

Figure IV.2. Light incident on a semiconductor at other than normal
incidence.

Light Interaction

The solar spectrum was presented in Figure II.1 for both air-mass-
zero (AMO) and air-mass-one (AM1) conditions [8-13]. The difference

in spectral irradiance under the two conditions is due to the filtering effect of a clear, dry atmosphere. The power density is reduced from 135.3 mW/cm^2 (AM0) to 107 mW/cm^2 (AM1) by this filtering. In Chapter II we also discussed the enhancement in energy reception experienced at a given location when tracking systems, rather than fixed solar energy collectors, are utilized. All of these effects must be considered in the design of practical solar cell systems. Not only do they affect the total energy received, but they influence the spectral and temporal distribution of the power density. We will avoid the chaotic situation caused by fluctuations in atmospheric filtering by restricting our examination of the interaction of light and semiconductors to AMO and AM1 spectral conditions only. Hence, the values for performance of solar cells and solar cell systems arrived at in this, and following chapters, will be maximum values or upper bounds[*].

Once the photons have entered the semiconductor, many of the individual photons strike electrons within the semiconductor and surrender their energy to the electrons. If the electron that receives this energy is in the valence or conduction band of the semiconductor and remains in the same band after receiving the energy, the electron will rapidly lose the excess energy gained from the photon. It does this by collisions with the lattice atoms. The net effect of photon absorption in this instance is an increase in the temperature of the semiconductor. If, however, the electron suffering the collision is in the valence band (and the vast majority of electrons are in the valence band) and receives enough energy to leave that band, then two "things" can happen. (1) If the electron is repositioned to somewhere in the forbidden gap, then the electron will "fall" back into the valence band; often emitting a photon similar to the initial photon. This photon will often be reabsorbed and then a third photon will be emitted by another electron as it too falls back into the valence band. Eventually a photon will emerge from the semiconductor. (2) Should the

---

[*] AMO refers, of course, to the maximum solar input outside the earth's atmosphere and AM1 is the maximum value for solar energy at the earth's surface. There are other air-mass conditions, existing because of water vapor and other atmospheric constituents and all of these other air-mass conditions have greater filtering effects on sunlight reducing the solar insolation from air-mass-one.

electron receive enough energy to reach the conduction band, the electron will remain in the conduction band and there will be a hole in the valence band (representing the electron that has moved to the conduction band). The electron-hole pair thus created carries a significant amount of stored energy. Note that, in general, the electron will have more than the minimum amount of energy required to enter the conduction band, and, therefore, will be positioned at some elevated energy level. As described in Figure IV.3, the excess energy of such an electron is the difference between its total energy and the energy of the conduction band edge. This energy is rapidly lost by collisions with the atoms of the semiconductor resulting in a net increase in temperature for the semiconductor. Once the excess energy is lost, the electron still has an effective energy greater than before its collision with the photon and is now situated at the bottom of the conduction band. The hole half of the electron-hole pair will, in general, also have some excess energy (see Figure IV.3). It will lose this excess energy through collisions with the semiconductor atoms, coming to "rest" at the valence band edge. The electron-hole pair now possesses an energy equal to the energy band gap energy, or E_g.

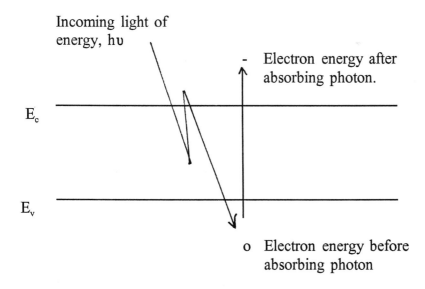

Figure IV.3. The simplified energy-band diagram of a semiconductor, indicating a typical interband transition as a result of photon-electron interaction. The difference in energy levels for the electron is $h\upsilon$. The band gap, E_g, is the minimum photon energy for absorption.

As discussed in Chapter III, the electron-hole pairs will eventually recombine, converting their energy (E_g) to heat. However, this process, which involves interband recombination, is slow compared to the intraband loss of energy described above. There is time to separate the charge carriers (the electrons and holes of the electron-hole pairs). For our present purposes, we will assume that the hole-electron pairs can be and are separated. The excess energy, E_g, of the hole-electron pairs can, therefore, be transferred to some external load.

There are many potential interactions of light with the interior of a semiconductor, other than the intra- and interband interactions previously discussed in this section. Many are used to probe the electronic structure of semiconductors, and others play major roles in solid state lasers and LEDs (light emitting devices). These phenomena play, at best, a secondary role in the operation of the solar cell and will not be discussed further here. For readers interested in these phenomena, an excellent introductory survey can be found in Pankove [14].

We have established that, in order to convert the energy inherent in a photon to useful electrical energy in an electron-hole pair, it is necessary that the photon have an energy greater than or equal to that of the energy gap of the semiconductor involved. Figure II.1 presented the solar spectral irradiance in terms of the wavelength of the solar photons and in terms of the energy of the photons, E_{ph}, where:

$$E_{ph} = hc/\lambda. \tag{IV.6}$$

Here, h is Planck's constant, c is the speed of light and λ is the wavelength of the photon.

From a knowledge of the energy in sunlight at a given wavelength it is possible to determine the number of photons possessing that or shorter wavelengths. In particular, it is possible to determine the number of photons present with an energy greater than some stated value. This has been done in Figure IV.43 for both AMO and AM1 solar conditions.

It is clear, from Figure IV.4, that the smaller the energy gap of the semiconductor, the greater the number of photons with an energy adequate to excite electron-hole pairs. In converting the number of incident photons to the number of generated electron-hole pairs, the simplest assumption to use is:

One electron-hole pair per absorbed photon. (IV.7)

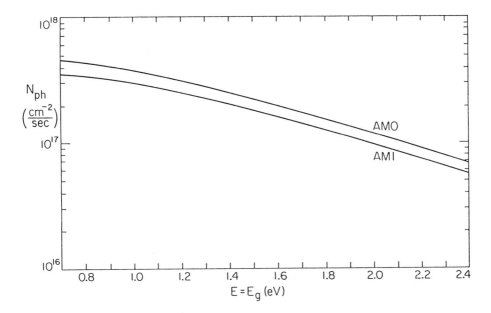

Figure IV.4. The number of photons in the solar spectrum per square centimeter per second with an energy greater than the given value, for both AMO and AM1 conditions.

Equation IV.7, while an excellent approximation, can be subject to a number of second-order corrections in unusual situations. Under the conditions of very high photon flux, such as experienced in lasers (or, perhaps, in extremely high optical concentration solar cell systems), two (or more) low energy photons can simultaneously collide with a single electron and jointly supply the required energy to "jump" the energy gap. This phenomenon has been observed in GaAs, CdSe and CdTe [15] as well as in CdS and ZnS lasers [16, 17]. This occurrence is of little importance for most solar cell energy systems and will not be considered further in our computations. Another secondary effect arises from those photons whose energies are considerably in excess of the semiconductor energy gap. Provided sufficient energy is available, it is possible that two or more electrons can be excited across the energy gap by collision with a single, very energetic, photon. This process must compete with the excitation of a single, highly energetic, electron, and is unlikely for solar cells exposed to light from our sun. Study of the quantum mechanical requirements of this situation leads to the finding that a photon must have in excess of three times the energy value of the band gap if it is to produce two hole-electron pairs. The combination of the low probability of this interaction, with the exceedingly small flux of photons of the required high energy, leads us to exclude this phenomenon in our first order design approach to photovoltaic cells.

Assuming, then, that we obtain one electron-hole pair for each incoming photon with an energy equal to or greater than the semiconductor energy gap, we can calculate the maximum generated photocurrent for a solar cell by assuming that all charge pairs are utilized. The photocurrent density, J_{ph}, is:

$$J_{ph} = qN_{ph}, \tag{IV.8}$$

where q is the absolute value of the charge on an electron and N is the number of photons with requisite energy.

Furthermore, each hole-electron pair produced in the solar cell carries a net effective energy of an amount, E_g. This energy can, theoretically, be extracted from the solar cell and transferred to some external load. The maximum rate of transfer of this energy, the maximum solar cell delivered power, P_{max}, is given by:

$$P_{max}/area = E_g N_{ph}. \tag{IV.9}$$

In Figure IV.5 the maximum potential photocurrent density, J_{ph}, and the maximum potential solar power per unit area, $P_{max}/area$, are plotted as functions of the energy gap of a semiconductor. Note that, whereas the current density is a steadily declining function of the energy gap, there is a distinct peak in the maximum obtainable photopower density. This is a direct result of the potentially greater energy which is "extractable" per electron-hole pair at higher band gap energies. Another point to be noted concerning the information in Figure IV.5, is that the maximum theoretical efficiency of a solar cell is less than 50% for the range of band gap energies considered[*].

---

[*] Subsequent work in this volume will amply demonstrate that, for the Earth's sun and for its location in the solar system, the range of energy gaps considered in Figure IV.5 includes all practical cases. It should be noted that a hotter star, such as Sirius, will emit more high energy photons and wider energy gap semiconductors can make better use of its high energy photons.

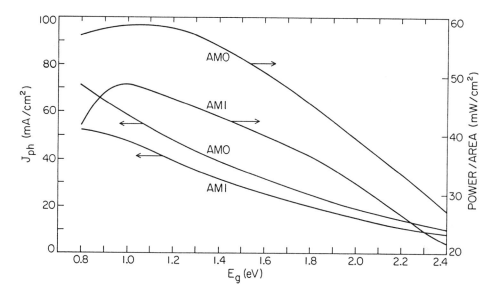

Figure IV.5. The maximum potential photocurrent density, J_{ph}, and photopower density, P_{max}/area, obtainable from a solar cell as a function of the semiconductor energy gap and the solar spectral conditions.

Preliminary Material Selection

From Figures II.1 and IV.5 we can make a preliminary selection of those semiconductors that have potential use as the photon-absorber-converter in a solar cell. Table IV.3 lists some semiconducting materials that are potential candidates, together with their energy gap and refractive index. Table IV.3 is not meant to be all inclusive, merely representative. It must be remembered that combinations of semiconducting materials are often also semiconductors. For example, the compound semiconductor, $GaAs_xP_{1-x}$, is constructed from the compound semiconductors gallium arsenide and gallium phosphide and exhibits an energy gap between the 1.39 eV of GaAs and the 2.24 eV of GaP. Such a semiconductor might well make an excellent solar cell. However, it should also be kept in mind that such complex semiconductors are difficult, and expensive to fabricate, and the technology involved in making solar cells from them is both complex and difficult. There are, in addition, a number of polycrystalline solar cell materials which have been used with success for a

Table IV.3

Some potential semiconductors for use in solar cells. The energy gap values (at 300°K), refractive index and some commentary about each listed semiconductor are included

Semiconductor	Energy Gap (eV)	Refractive Index	Remarks
Ge	0.67	4.00	Well developed technology, but the energy gap is small
Si	1.107	3.42	Close to the maximum potential power in Figure IV.4, a plentiful material with a well developed technology
InP	1.35	3.10	Low lifetime and expensive
GaAs	1.42	3.30	Backbone of the LED-solid state laser field, well developed technology, expensive
CdTe	1.50	2.67	Though made from scarce materials shows promise
AlSb	1.58	3.18	Indirect energy gap
CdSe	1.70	2.75	Long lifetime, expensive and "primitive" technology
GaP	2.26	3.32	Long lifetime, in use in lasers and LEDs; energy gap is probably too wide

considerable period of time [18-20]. Efficiencies of a few percent (approaching 10%) have been reported for these materials. The basic motivation for the use of solar cells made from polycrystalline materials lies in their low relative cost. Such thin film solar cells are very inexpensive to fabricate; the process being allied to painting at times. Solar cells made from amorphous materials are also potential economically prudent candidates. Such materials are less expensive to fabricate than single crystal materials since no effort is required to form single crystals of semiconductor. Amorphous silicon solar cells containing hydrogen have exhibited efficiencies in the mid-teens and now account for about one watt of solar cell produced electrical solar power in four [21-23]. As will be shown later when we discuss economic viability and general system constraints, the use of optical concentration tends to reduce the cost of the solar cells as a major factor in the overall solar energy system cost. This text will concentrate on single crystal solar cells. Moreover, the theory of operation for semiconductor diodes put forward in the last chapter and utilized in this and subsequent chapters is basically adapted for single crystal solar cells. Some attention will, however, be spent on amorphous solar cells and solar cells constructed from polycrystalline materials in a subsequent chapter. The latter tend to exhibit instabilities at high temperatures, have significantly lower operating efficiencies than single crystal solar cells and their operating characteristics tend to be dominated by imperfectly understood crystal interface effects.

Consider Table IV.3 in light of Figure IV.5. The use of germanium in a solar cell does is not practical owing to the small band gap and resultant poor power output per unit area. Silicon appears to have a band gap which is close to the optimum for power conversion of sunlight. As we continue with our study of solar cells we will be considering a number of other factors which bear on solar cell performance; factors such as internal resistance losses and lifetime effects. These factors produce a shift in the location of the optimum energy gap for power conversion from its location in Figure IV.5 to a higher value. However, they do not rule out silicon solar cells. It does appear, however, that solar cells made from single crystal semiconductors with energy gaps in excess of two electron volts perform poorly when illuminated by the solar spectrum. We will be discussing alternative structures for solar cells in which some of these wide energy gap semiconductors play a role. For example, very large gap materials such as gallium phosphide and aluminum phosphide have been used as semitransparent layers overlaying a standard solar cell made from silicon and protecting it from ambient conditions.

Absorption

In constructing Figure IV.5 the assumption was made that all of the photons entering the semiconductor with an energy greater that the energy gap were absorbed and converted to hole-electron pairs or to heat. As pointed out in the introductory section of this chapter, this is not necessarily so. If the semicondcutive material is thin enough, there may well be a number of photons which pass entirely through the semiconductor. This section discusses the absorption of light in a semiconductor by virtue of band-to-band excitation, as a function of the thickness of the semiconductor, the properties of the semiconductor and the wavelength of the light.

The electric field, of a plane wave of frequency, v, traveling in one direction, x, is given by:

$$\mathscr{E} = \mathscr{E}_o \exp\{i2\pi vt\}\exp\{-i2\pi n/c\}\exp\{-2\pi vkx/c\}, \tag{IV.10}$$

where $\mathscr{E}_o$ is the amplitude of the electric field at x = 0 (the surface of the semiconductor), x is the displacement into the semiconductor, c is the speed of light, v is the photon frequency, n is the index of refraction of the semiconductor, k is the extinction coefficient and t is the time.

The optical power remaining after the photons propagate a distance, x, through a material of conductivity, σ, is given by:

$$\frac{P(x)}{P(o)} = \frac{\sigma\mathscr{E}^2(x)}{\sigma\mathscr{E}^2(o)} = \exp\{-4\pi vkx/c\}. \tag{IV.11}$$

The absorption coefficient, α, is defined by:

$$\frac{P(x)}{P(o)} = \exp\{-2\alpha x\}. \tag{IV.12}$$

This leads to the following relationship between the absorption coefficient, α, and the extinction coefficient, k:

$$\alpha = 4\pi vk/c. \tag{IV.13}$$

After allowance for internal reflections, the fraction of the light, T" which is transmitted entirely through a semiconductor "slab" can be written as:

$$T'' = \frac{(1 - f)^2 \exp\{-\alpha x\}}{1 - f^2 \exp\{-2\alpha x\}}, \qquad (IV.14)$$

where f is the fraction of light reflected from the illuminated surface. When 2α is large, as it will be in the case of a highly absorbing semiconductor, then Equation IV.14 is reduced to:

$$T'' \simeq (1 - f)^2 \exp\{-\alpha x\}. \qquad (IV.15)$$

Let us briefly examine the absorption coefficient. From Equation IV.12 we can also define α as a function of the light intensity, $\Omega(h\nu)$:

$$\alpha(h\nu) = \{1/\Omega(h\nu)\}\{\partial\Omega(h\nu)/\partial x\}. \qquad (IV.16)$$

In absorbing a photon, exciting an electron across the forbidden gap of a semiconductor, and thus creating a hole-electron pair, both the total energy and the momentum of the system (the particles involved may include electrons, holes, phonons, and atoms as well as photons) must be conserved. Additionally, any applicable quantum mechanical "selection rules" for the energy shifts must be satisfied. Including all these factors, we can write for the absorption coefficient:

$$\alpha(h\nu) = A\Sigma\{P_{if}n_i n_f\}, \qquad (IV.17)$$

where n_i is the density of electrons in the initial location (within the valence band) and n_f is the density of available final states (empty states in the conduction band), P_{if} is the probability of the required energy and momentum transfers taking place and A is a numerical constant.

In a direct gap semiconductor the conduction band minimum of energy possesses the same momentum as does the valence band maximum Therefore, no transfer of momentum is required when providing the valence band electrons with sufficient energy to transit the energy gap to the conduction band. Under these conditions it has been shown [24] that the absorption coefficient is:

$$\alpha(h\upsilon) = q^2[2m^*_{ce}m^*_{ve}/(m^*_{ce} + m^*_{ve})/nch^2m^*_{ce}]^{3/2}[h\upsilon - E_g]^{1/2}, \qquad \text{(IV.18)}$$

where m^*_{ce} and m^*_{ve} refer to the effective masses of electrons in the conduction band and hole in the valence band, respectively, and h is Planck's constant. Note that this expression is only valid for photon energies, $h\upsilon$, in excess of the energy gap, E_g. For photon energies less than the energy gap, the absorption coefficient is effectively zero. Additionally, the derivation of this expression assumes a quadratic dependence on energy for the densities of electrons in the valence band ($n_{i'} \propto (E_v - E)^2$) and for the empty states in the conduction band ($n_f \propto (E - E_c)^2$). Because of this assumption, the expression in Equation IV.18 is only valid at photon energies relatively close to the absorption edge*.

In some semiconductors, quantum mechanical selection rules forbid direct interband excitations at k = 0 (see Chapter III), but do allow them for k ≠ 0, with a probability increasing as k^2. In these materials the absorption coefficient, in the neighborhood of the absorption edge, is of the form [24]:

$$\alpha(h\upsilon) = [4q^2/3][m^*_{ce}m^*_{ve}/(m^*_{ce} + m^*_{ve})]^{5/2}/[nch^32m^*_{ce}m^*_{ve}\upsilon][h\upsilon - E_g]^{3/2}. \qquad \text{(IV. 19)}$$

Note the differing dependencies of the absorption coefficient upon the photon energy as the selection rules change.

Furthermore, many semiconductors are not direct gap materials. The most widely used semiconductor, silicon, is, in fact, an indirect gap semiconductor. In indirect gap materials, the electron in the valence band maximum must acquire not only the energy necessary to reach the conduction band minimum, but must also acquire a significant amount of momentum. The average amount of required momentum change is the momentum corresponding to the conduction band minimum, since the momentum of the valence band maximum is defined as being zero. The momentum of the incoming photon, P_{ph}, is given in terms of its wavelength, λ_{ph}:

---

* The absorption edge is the wavelength at which the semiconductor begins to rapidly absorb light. The shift in absorption is quite abrupt in direct gap semiconductors, but requires a significant change in wavelength for indirect gap semiconductors.

$$P_{ph} = h/\lambda_{ph}. \qquad \text{(IV.20)}$$

This value of momentum is far less than the momentum difference between electrons in the valence band maximum and electrons sited in the conduction band minimum. The interband transition requires a conservation of momentum. Therefore, a third particle is required to make the interband transition possible. The electron receives essentially all the energy it requires from the photon, but must also either emit or absorb a phonon* to allow for the necessary conservation of momentum.

At low temperatures there are few phonons [25], and the interband excitation process involving the emission of phonons is dominant for all optical flux levels. Since an emitted phonon requires some energy (on the order of kT), this requires that the incident photons must have an energy in excess of E_g before photon absorption can commence. As the temperature of the semiconductor increases, the number of phonons increases and the energy required of the incident photons decreases. We can write for the absorption coefficient of an indirect semiconductor near the absorption edge:

$$\alpha(h\upsilon) = \alpha_e(h\upsilon) + \alpha_a(h\upsilon), \qquad \text{(IV.21)}$$

where α_e is that portion of the absorption coefficient due to phonon emission processes and α_a is that portion of the absorption ceifficient resulting from phonon absorption. The absorption coefficient due to phonon emission is:

$$\alpha_e(h\upsilon) = B[h\upsilon - E_g + E_p]^2/[1 - \exp\{-E_p/kT\}], \qquad \text{(IV.22)}$$

where E_p is the energy of the phonon and B is a numerical constant. For photon absorption with phonon absorption involved, the absorption coefficient is:

---

* A phonon is generally considered to be the particle representation of a lattice vibration. Such lattice vibrations are generally the result of the energy owing to the temperature of the crystal lattice. Unlike electrons which obey Fermi-Dirac statistics, phonons obey Bose-Einstein statistics and any number of phonons may have a given energy.

$$\alpha_a(h\upsilon) \propto (h\upsilon - E_g - E_p)^2/[\exp\{E_p/kT\} - 1]. \qquad (IV.24)$$

Since the number of phonons available for absorption is finite at any temperature, but the number of phonons which can be emitted is not limited, at some energy and photon flux (dependent on the ambient temperature) the phonon emission process will dominate.

The discussion of the absorption coefficient which we have carried out to date is for the absorption coefficient at photon energies close to the energy gap; i.e., near to the absorption edge. These values are of considerable use in determining the energy gap of semiconductors. In practice, for solar cells, we need to also consider photon energies considerably removed from the minimum for absorption ($h\upsilon \sim E_g$). For example, silicon has an energy gap close to 1.1 eV, while the solar spectrum contains photons that possess energies in excess of four eV. To determine the absorption coefficient as a function of the photon energy over a complete range of solar photon energies we must have recourse to the literature.

As examples of the absorption coefficients of semiconductors, consider those semiconductors listed in Table IV.3. In particular, consider those materials that the ensuing discussion indicated as being of particular interest: e.g., Si, InP, GaAs, CdTe, AlSb and CdSe. The absorption coefficient for all of these materials features a complex dependence of photon wavelength. As an example, consider the absorption coefficients for silicon and gallium arsenide as presented in Figure IV.6.

Using the absorption coefficients for silicon and gallium arsenide from Figure IV.6, coupled with absorption data in the literature on cadmium telluride [26-29], aluminum antimonide [30,31], indium phosphide [32], and cadmium selenide [33, 34] we can compute the optical energy absorbable in these semiconductors as a function of the wavelength of light. In Table IV.4, the solar spectral power density (in mW/cm^2) under AMO and AM1 conditions is presented along with the absorbed power density potentially obtainable under these conditions when the six sample semiconductors listed above are used in solar cells; listed as a function of the wavelength of light. The maximum efficiency for conversion from light to electrical energy is also given for each semiconductor under the two spectral conditions. Note that it is assumed that, in each case, the semiconductor is sufficiently thick to absorb all of the photons possessing an energy greater than the energy gap and that all photons incident upon the semiconductor surface do enter that surface.

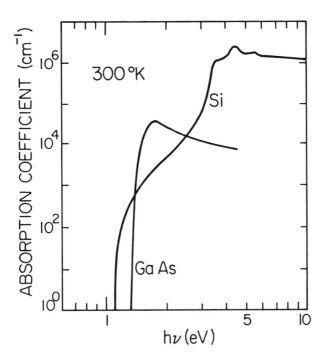

Figure IV.6. The absorption coefficients for gallium arsenide and silicon as a function of photon energy [35-39].

Now consider the situation which obtains when, still assuming that the reflection coefficient at the semiconductor surface is zero, the semiconductor is of finite thickness. In this situation some of the photons will pass entirely through the semiconductor. From Equation IV.12 and the absorption coefficients for our six example semiconductors, it is possible to derive the percentage of solar energy available for conversion to electrical energy as a function of location within the semiconductor. This data is presented in Figures IV.7 and IV.8. In these figures the power available for conversion to electrical energy is provided as a function of the thickness of the semiconductor, for the six example semiconductors. Note that the direct energy gap semiconductors require a smaller thickness of material to absorb the available solar energy. This is a result of the less complicated nature of the photon-driven interband transition for electrons in these materials.

Study of Figures IV.7 and IV.8 leads to a number of important design considerations for solar cells. The required solar cell thickness for 100% absorption for materials such as GaAs, InP, CdTe, AlSb and

Table IV.4

AMO (AO) and AM1 (A1) spectral power density and potential solar cell provided power levels (mW/cm^2) for selected semiconductors as a function of the wavelength (μm)

λ (μm)	Sun AO/A1	Si AO/A1	InP AO/A1	GaAs AO/A1	CdTe AO/A1	AlSb AO/A1	CdSe AO/A1
∞	31.77	00.00	00.00	00.00	00.00	00.00	00.00
1.15	25.04	00.00	00.00	00.00	00.00	00.00	00.00
1.14	09.51	06.20	00.00	00.00	00.00	00.00	00.00
1.00	08.41	05.43	00.00	00.00	00.00	00.00	00.00
0.99	08.29	06.98	04.97	00.00	00.00	00.00	00.00
0.90	06.02	05.34	03.66	00.00	00.00	00.00	00.00
0.89	09.93	07.41	08.76	08.24	06.98	00.00	00.00
0.80	08.35	05.95	06.93	06.85	04.82	00.00	00.00
0.79	12.37	08.29	09.63	10.36	10.46	07.35	03.90
0.70	08.05	05.21	06.10	06.58	06.82	04.38	02.72
0.69	15.15	08.66	10.20	10.97	11.37	12.75	13.38
0.60	13.25	07.49	08.79	09.49	09.82	11.02	11.58
0.59	17.70	08.64	10.12	10.88	11.16	12.62	13.25
0.50	14.30	07.21	08.40	09.06	09.31	10.53	11.00
0.49	18.77	07.53	08.80	09.49	09.85	11.02	11.58
0.40	15.10	05.88	06.87	07.40	07.67	08.61	09.05
0.39	10.17	03.24	03.78	04.08	04.24	04.75	04.99
0.30	07.91	02.00	02.34	02.52	02.61	02.94	03.07
0.29	01.63	00.42	00.49	00.53	00.55	00.62	00.65
0.20	00.56	00.13	00.15	00.16	00.17	00.19	00.20

Table IV.4 (continued)

AMO (AO) and AM1 (A1) spectral power density and potential solar cell provided power levels (mW/cm^2) for selected semiconductors as a function of the wavelength (μm)

λ (μm)	Sun AO/A1	Si AO/A1	InP AO/A1	GaAs AO/A1	CdTe AO/A1	AlSb AO/A1	CdSe AO/A1
0.19	00.01	00.00	00.00	00.00	00.00	00.00	00.00
0.00	00.01	00.00	00.00	00.00	00.00	00.00	00.00
Tot.	135.3	57.37	56.75	54.55	54.65	49.11	47.75
	107.0	44.64	43.24	42.06	41.32	37.67	37.62
Percent (%) of Possible Power							
	100	42.40	41.82	40.32	40.39	36.30	35.29
	100	41.72	40.41	39.31	38.61	35.21	35.16

CdSe is less than 25 to 50 μm, while for silicon the required thickness for complete photon absorption is on the order of 1000 μm. The cost of a solar cell is dependent on the amount of semiconductor used, and it is impractical, both for reasons of economy and for processing technology, to fabricate solar cells of a thickness much less than 100 μm or much greater than 150 μm. Furthermore, solar cells constructed from the compound semiconductors under consideration must be thicker (to avoid breakage during processing) than the minimum value considered earlier. In the case of silicon, a photovoltaic cell which is 150 μm thick can convert, at most, 38.6% of the energy in AMO sunlight or 37.6% of the energy in AM1 sunlight. These values represent a loss of some 10% from the potential power levels listed in Table IV.4.

Another factor that must be taken into consideration in the construction of solar cells arises from a combination of the absorption rate of photons with distance into a semiconductor and the processing technologies used in the fabrication of solar cells. The processing techniques utilized in the construction of solar cells reduce, in general, the lifetime exhibited by the hole-electron pairs near the illuminated surface of the solar cells. Once a hole-electron pair recombines, the energy that

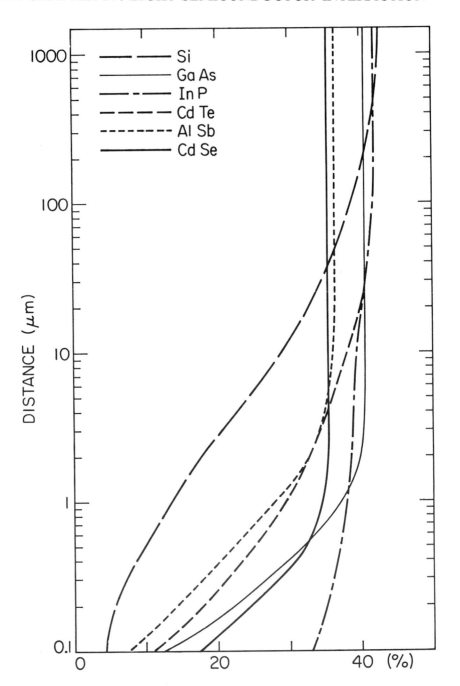

Figure IV.7. The percentage of solar energy which can be converted to electrical energy as a function of the thickness of specific semiconductors--under AMO condtions.

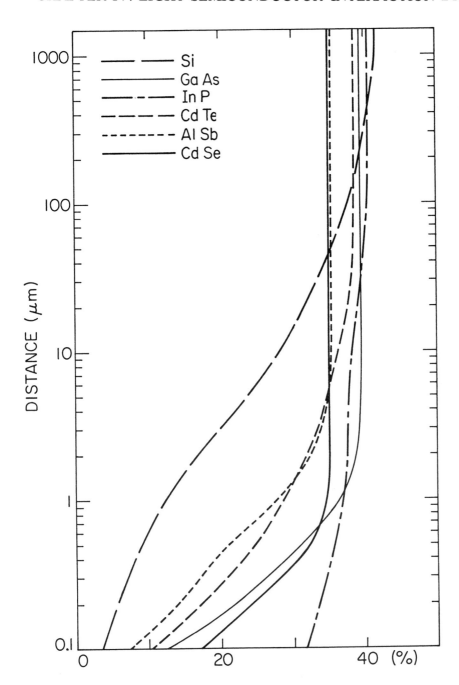

Figure IV.8. The percentage of solar energy available for conversion to electrical energy as a function of solar cell thickness with the semiconductor as a parameter--under AM1 conditions.

the hole-electron pair carried is converted into heat within the solar cell and is no longer potentially available as energy for the external load[*]. Should the lifetime of the electron-hole pairs be sufficiently short, the electron-hole pairs generated by absorption of photons will recombine before they can be separated and, to all effects and purposes, the region of low lifetime will be optically dead. In a material such as gallium arsenide, if the first micron of the solar cell below the illuminated surface is "optically dead" as a result of low lifetime, then over 90% of the potentially available power, as given in Table IV.4, is lost, turned into heat and is unavailable electrically.

In subsequent chapters we will return to the lifetime in the front surface regions (the illuminated regions) of solar cells, and the relations between this lifetime and overall solar cell efficiency. For the present, Figures IV.9 and IV.10 will suffice. In these figures the potential optical power (energy) convertible to electrical power (energy) is presented for the six semiconductor examples, as a percentage of the total power density in sunlight. This is done subject to certain conditions: (1) the thickness of the solar cells considered is assumed to be 150 μm, and (2) the photons are assumed to enter the solar cell via a layer of low lifetime material. This layer is known as the "dead layer" and is assumed to have a bulk lifetime so low and a surface recombination velocity so high as to preclude the effective generation[#] of any holes and electrons. Any light absorbed in this region results, then, in heat rather than in useful electrical current.

An increase in temperature in a solar cell is typically accompanied by a decrease in performance (see the following chapters for details). The heat generated in a "dead layer" will increase the temperature of a solar cell. This is another reason for optimizing the conversion of solar energy to electrical energy. Study of Figures IV.9 and IV.10 indicates that the "dead layer" in silicon could be as much as 0.5 μm and still permit high operating efficiencies. Three of our example semiconductors, CdTe,

---

[*] See Chapter III and following chapters for discussions of the techniques employed in the separation of electron-hole pairs prior to their recombination.
[#] Effective generation assumes that the photons are absorbed and the resulting electron-hole pairs are separated into their component parts (holes and electrons) prior to recombination occurring.

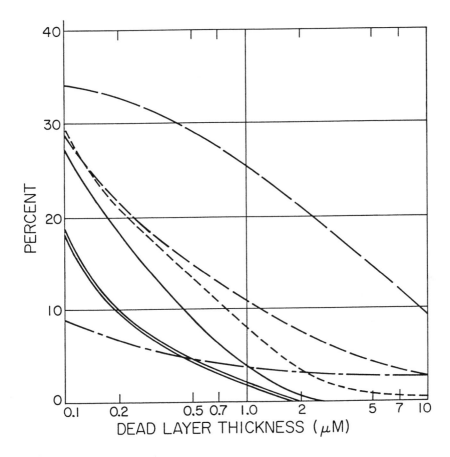

Figure IV.9. The percentage of solar power convertible by the six sample semiconductors as a function of "dead layer" thickness. The semiconductors are 150 μm thick and the input is AMO (135.3 mW/cm^2) with no reflection.

Matr.	Ideal %	Symbol	Matr.	Ideal %	Symbol
Si.	38.6	--- --- ---	CdTe	40.3	--- --- ---
InP	41.8	--- - ---	AlSb	36.3	- - - - - - - -
GaAs	40.3	------------	CdSe	35.2	=======

GaAs and AlSb, when used in solar cells, might be permitted a "dead layer" as wide as 0.15 μm, but the remaining semiconductors, CdSe and InP are at a severe disadvantage for "dead layers" as thin as 0.05 μm.

It is this phenomenon that has led to the consideration of heterojunction solar cells. In a heterojunction solar cell it is theoretically possible to reduce the "dead layer" thickness to zero, while maintaining

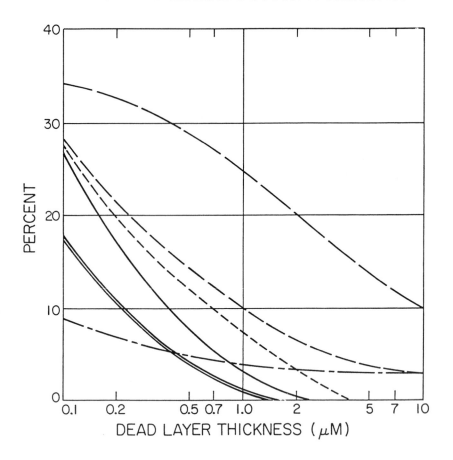

Figure IV.10. The percentage of solar power convertible by the six sample semiconductors as a function of "dead layer" thickness. The semiconductors are 150 μm thick and the input is AM1 (107 mW/cm^2) with no reflection.

Matr.	Ideal %	Symbol	Matr.	Ideal %	Symbol
Si.	38.6	--- --- ---	CdTe	40.3	--- --- ---
InP	41.8	--- - ---	AlSb	36.3	- - - - - - - - -
GaAs	40.3	------------	CdSe	35.2	════════

the electrical characteristics necessary to separate the hole-electron pairs and to supply the load current. There are other techniques for achieving the reduction of this "dead layer" and they will be discussed later. However, Figures IV.9 and IV.10 clearly indicate the need for "dead layer" thickness minimization.

Reflection and Absorption

We have considered the reflection of light from a semiconductor surface, both in the presence of an antireflection coating and without one. We have also studied the absorption of photons by semiconductors, as a function of the frequency of the photons, the nature of the semiconductor, and the thickness of the semiconductor. In this section we combine these two phenomena and make a number of predictions concerning the six semiconductors being used as examples.

In terms of reflection there are three conditions of potential interest: (1) we can obtain values of available power for uncoated semiconductors (clearly, the worst case), (2) we can obtain available power values for semiconductors with optimal surface treatments which would exhibit no energy loss by photon reflection (clearly the best case[*]), or (3) we can obtain available power values for some surface treatment lying between the above cases, a situation which is practical. The indices of reflection provided in Tables IV.1 and IV.2 are averaged values; averaged over the optical wavelengths of interest for each semiconductor. With care, a practical manufacturing process can reduce the reflection coefficient values of Table IV.2 by in excess of two-thirds. Using these reduced values, with the data from Figures IV.9 and IV.10 we can obtain potential output power densities for solar cells made from the six potential example semiconductors. To be practical, let us assume solar cells of 150 μm thickness. This reduces the power available from silicon solar cells by approximately 10%, but has no material effect on the other semiconductors being considered. Let us consider solar cells with "dead" layers of 0, 0.1, 0.2 and 0.5 μm, leaving until a later time an analysis of why the thickness and reflection values used here are practical.

Consider the ideal antireflection coating. There is no initial surface reflection and, therefore, the input power density to the solar cell is 135.5 mW/cm^2 under AMO conditions and 107 mW/cm^2 for AM1 conditions. In Table IV.5 we present the power density potentially available from solar cells made from the six sample materials and having an ideal, non-reflecting, antireflection coating on the illuminated surface.

---

[*] Such a surface would require more than a single layer of antireflection coating. Such coatings and associated treaments will be discussed in detail later.

Table IV.5

The electrical power density (mW/cm^2) potentially available as a function of the spectral type of illumination and "dead layer" thickness for a 150 μm thick solar cell constructed from selected semiconductors. An ideal antireflection coating is assumed

"Dead Layer" Thickness (μm)	Si	InP	GaAs		CdTe	AlSb	CdSe
				AMO			
0.0	52.6	56.7	54.54		54.65	49.1	47.74
0.1	46.1	11.5	37.3		38.7	38.5	23.4
0.2	44.1	09.2	24.1		29.2	28.8	13.5
0.5	38.7	06.5	11.3		19.8	18.0	05.7
				AM1			
0.0	40.2	43.2	42.0		41.25	37.71	37.66
0.1	36.4	09.0	28.1		30.1	29.52	18.88
0.2	34.6	07.2	18.22		22.68	21.6	11.02
0.5	30.3	05.0	08.83		14.97	13.32	04.22

From the known energy band gaps of these materials, as provided in Table IV.3, we can estimate the photocurrent density using Equation IV.8 and Figure IV.4. These, the maximum potential photocurrents, are presented in Table IV.6.

As discussed earlier in connection with Figures IV.9 and IV.10, the data of Table IV.5 provide a strong reason for the minimization of the "dead layer" thickness. However, the photocurrent information in Table IV.6 is the initial indication of a phenomenon that has a significant effect on the performance of solar cells. This effect concerns the magnitude of the photocurrent. Since the photocurrent must flow through the internal (parasitic) resistance of the solar cell, as well as through the external load, there are internal energy losses that are dependent on the photocurrent. These losses are considered in the following two chapters. For now, it is sufficient to note that a device with a smaller photocurrent has a potential for smaller internal losses and, as a result, larger energy gap materials are favored for solar cells.

Table IV.6

The photon generated electric current density (mA/cm^2) potentially available as a function of the spectral type of illumination and "dead layer" thickness for a 150 μm thick solar cell made from selected semiconductors. An ideal antireflection coating is assumed

"Dead Layer" Thickness (μm)	Si	InP	GaAs		CdTe	AlSb	CdSe
				AMO			
0.0	47.0	43.9	39.2		37.7	30.1	27.4
0.1	41.7	08.9	26.8		26.9	23.6	13.4
0.2	39.9	07.1	17.3		20.3	17.7	07.7
0.5	35.0	05.0	08.1		13.8	11.0	03.2
				AM1			
0.0	38.0	33.5	30.2		28.6	23.1	21.6
0.1	32.8	07.0	20.3		20.9	18.1	10.9
0.2	31.3	05.5	13.1		15.8	13.3	06.2
0.5	27.4	03.8	06.3		10.4	08.2	02.4

Rather than consider the worst case (no antireflection coating) in the detail of Tables IV.5 and IV.6, let us consider this situation only for the no "dead layer" case. Under this "ideal" condition, the electrical power and photocurrent potentially available are provided by Table IV.7.

Clearly, the potential performance of photovoltaic cells without an antireflection coating is so poor, as compared to the ideal situation of Tables IV.5 and IV.6, that no additional consideration of this configuration is warranted. In practice, of course, we cannot achieve the ideal results of Tables IV.5 and IV.6. Assuming a reasonably practical surface coating for solar cells, the potentially obtainable power and photocurrent densities will be approximately those predicted in Figures IV.11 and IV.12.

Study of Figure IV.11 reconfirms our earlier finding that, for solar cells with no "dead layer", the performance of all six example semiconductors when used in solar cells is roughly equivalent, but that with increasing "dead layer" thickness it would seem that silicon is to be preferred. The effects of the photocurrent on photovoltaic cell performance will be treated in later chapters. Note that silicon does have a fairly high photocurrent; a property which is not necessarily an asset owing to resistance losses within solar cells.

Table IV.7

The potential delivered power density (mW/cm^2) and photocurrent density (mA/cm^2) available for solar cells with a thickness of 150 μm, no antireflection coating and no "dead layer", for selected semiconductors under AMO and AM1 spectral conditions

	Si	InP	GaAs		CdTe	AlSb	CdSe
				AMO			
Power Density	36.5	41.8	38.9		42.8	35.7	39.0
Current Density	32.9	32.4	28.0		29.7	21.9	22.4
				AM1			
Power Density	28.1	31.9	30.0		32.7	27.4	30.7
Current Density	26.6	24.7	21.6		22.7	16.8	17.6

Study of Figure IV.12, for AM1 conditions, yields results somewhat similar to those of Figure IV.11, for AMO conditions. Note that the power and current densities under AM1 conditions are consistently smaller than those under AMO conditions (Figures IV.11 and IV.12, and Tables IV.5 through IV.7). When we consider orbiting solar power stations, the smaller power to be expected from solar cells under AM1 conditions is balanced by the larger energies expended in orbiting solar cells in order that they can be exposed to AMO conditions. Thus, when solar energy is considered, orbiting versus ground based solar systems are not too different in performance.

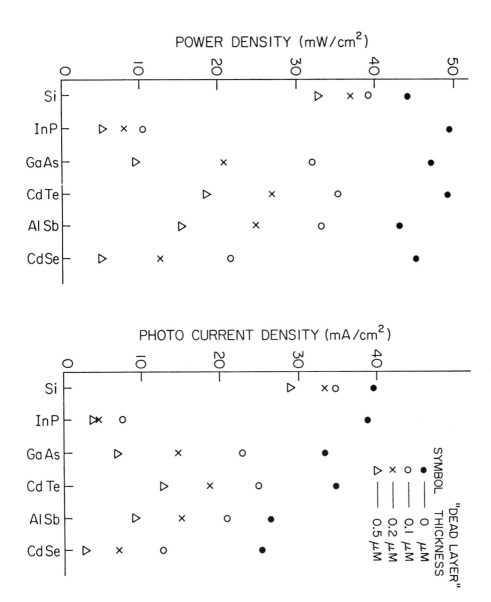

Figure IV.11. Potential solar power density from a photovoltaic cell and potential photocurrent density for cells that are 150 μm in thickness, under AMO illumination conditions, with an anitreflection (n ̃ 1.5) on the illuminated surface and with "dead layer" thickness as a parameter.

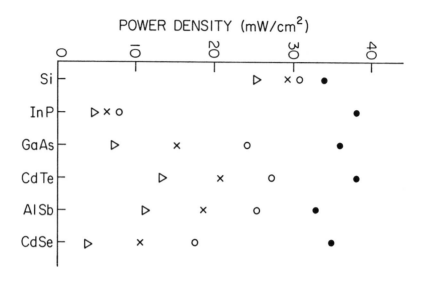

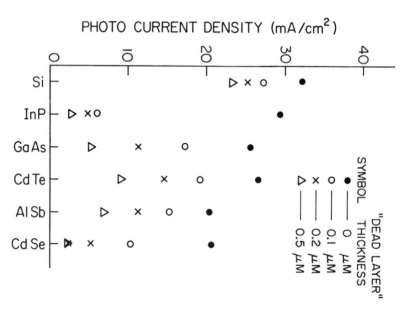

Figure IV.12. Potential solar power density from a photovoltaic cell and potential photocurrent density for cells that are 150 μm in thickness, under AM1 illumination conditions, with an antireflection coating (n ~ 1.5) on the illuminated surface and with "dead layer" thickness as a parameter.

References

1 J. I. Pankove, Optical Processes in Semiconductors, Prentice-Hall, Englewood Cliffs, NJ, 1971, Chap. 4.

2 G. Brackley, K. Lawson and D. W. Satchess, "Development of Integral Covers for Silicon Solar Cells, National Technical Information Service, U. S. Department of Commerce, Springfield, VA, report #N75-20567/4ST.

3 J. A. Scott-Monock, P, M. Stella, and J. E. Avery, "Development of Processing Procedures for Advanced Silicon Solar Cells", National Technical Information Service, U. S. Department of Commerce, Springfield, VA, report #N-75-17085/3ST.

4 G. Haake, "Exploratory Development of Transparent Conductor Material", National Technical Information service, U. S. Department of Commerce, Springfield, VA, report #AD-A008783/3ST.

5 H. W. Rauch, Sr. and D. R. Ulrich, "Integral Glass Covers for Silicon Solar Cells, National Technical Information Service, U. S. Department of Commerce, Springfield, VA, report #AD/AOO2846/4ST.

6 M. M. Koltun, "Selective Surfaces and Coatings in Solar Radiation Engineering", Natinal Technical Inforamtion Service, report # N73-15598. This is an English language translation of an article in Geliotech, Akad Nauk UZB, SSR, Vol. 5, 1971, p. 70.

7 Many useful articles can be found in the National Technical Inforamtion Service report #PS-75/692, "Optical Coatings for Solar Cells and Solar Collectors--A Bibliography, U. S. Department of Commerce, Springfield, VA.

8 E. G. Gibson, The Quiet Sun, NASA, 1973, p. 14.

9 E. Petit, in Astrophysics, J. A. Hynek, editor, McGraw-Hill, New York, 1951, Chap. 6.

10 M. P. Thekaekara, in Solar Energy, Vol. 14, 1973, p. 109.

11 International Symposium on Solar Radiation Simulation, Institute of Environmental Science, 1965.

12 S. I. Rasool, Physics of the Solar System, NASA, 1972.

13 F. S. Johnson, in Journal of Meteorology, Vol. 11, 1954, p. 431.

14 J. I. Pankove, Optical Processes in Semiconductors, Prentice-Hall, Englewood Cliffs, NJ, 1971, Chaps. 3, 6, 7, 9 and 10.

15 N. G. Basov et. al., in Journal of Physical Society of Japan, Supp. 21, 1966, p. 277.

16 R. Braunstern and N. Ockman, in Physical Review, Vol. 134, 1964, p. A499.

17 S. Wang and C. C. Chang, in Applied Physics Letters, Vol. 12, 1968, p. A499.

18 K. W. Böer, Proceedings of International Workshop on CdS Solar Cells and other Abrupt Heterojunctions, Newark NJ, 1975, NSF/RANN-AER-75-15858.

19 K. W. Böer, Proceedings of the 1977 Annual Meeting of the American Section of the International Solar Energy Society.

20 J. duBow, in Electronics, 11 Nov. 1976, p. 86.

21 An early report in the Wall Street Journal (6 July 1977, p. 8) reports efficiencies of approximately five percent.

22 A. H. Pawlikiewicz, in IEEE Transactions on Electron Devices, Vol. 37, 1990, p. 1758.

23 Y. Arai, et. al., in IEEE Electron Device Letters, Vol. 12, 1991, p. 460.

24 J. Bardeen, F. J. Blatt and L. H. Hall, in Proceedings of the Atlantic City Photoconductivity Conference, 1954, Wiley and Chapman and Hall, 1956, p. 146.

25 S. Wang, Solid-State Electronics, McGraw-Hill, 1966, p. 46.

26 S. Marinuzzi, M. Perrot and J. Fourmy, in Journal de Physics, Vol. 25, 1964, p. 203.

27 L. R. Ladd, in Infrared Physics, Vol. 6, 1966, p. 145.

28 C. Konak, in Physics Status Solidi, Vol. 3, 1963, p. 1274.

29 K. Mitchell, A. L. Fahrebrach and R. H. Bube, in Journal of Applied Physics, Vol. 48, 1977, p. 829.

30 J. P. David, et. at., in Revue de Physics Applications, Vol. 1, 1956, p. 172.

31 A. J. Noreika, M. H. Farnlabe and S. A. Zeitmann, NASA report CR-86396, 1969.

32 R. R. Howson, in Journal De Physics, Vol. 25, 1964, p. 212.

33 K. Rogge and A. A. Smotarkov, in Physics Status Solidi, Vol. 4, 1971, p. K65.

34 M. Cardona and G. Herbeke, in Physical Review, Vol. 137, 1965, p. 1467.

35 H. R. Philipp and E. A. Taft, in Physics Review, Vol. 113, 1959, p. 1002.

36 H. R. Phillipp and E. A. Taft, in Physics Review Letters, Vol. 8, 1963, p. 13.

37 D. E. Hill, in Physics Review, Vol. 133, 1964, p. A866.
38 V. S. Bagaer and L. I. Paduchikh, in Soviet Phyics, Solid Sate, Vol. 11, 1970, p. 2676.
39 D. Bois and P. Pinard, in Physics Status Solidi, Vol. 7, 1971, p. 85.

CHAPTER V: BASIC THEORETICAL PERFORMANCE

Introduction

In earlier chapters we have seen that photons entering a semiconductor can, if of sufficient energy, produce, by collision with valence band electrons, hole-electron pairs in the semiconductor. These hole-electron pairs will, if undisturbed, recombine in a time dictated by the bulk lifetime and surface recombination velocity of the semiconductor. Under these conditions the energy represented by the absorbed photons is converted into heat within the semiconductor. However, if a local electric field is present within the semiconductor it can be used to separate the constituent parts (holes and electrons) of the hole-electron pairs. Once separated, the collected charge carriers (holes and electrons) produce a space charge that results in a voltage across the semiconductor. This voltage is known as the photovoltage. If the separated charge carriers are allowed to flow through an external load before eventually recombining, they constitute a photocurrent. The product of the photovoltage and the photocurrent represents a net flow of energy (when integrated over time) from the solar cell (semiconductor) to the external load. This energy originated in the sun and was converted from photon-energy to electrical energy when the photons were absorbed within the solar cell and the resultant hole-electron pairs were separated by the internal electric field.

This chapter will consider various methods of generating the internal electric field required for separation of hole-electron pairs, and then derive the equations necessary to predict the magnitude of the electrical power generated and the efficiency with which the solar cells convert solar to electrical energy. The solar cell output will then be examined as a function of the energy gap. In keeping with the general approach of this volume, the semiconductors used as examples in determining the performance of solar cells will, generally, be considered to be single crystal. However, polycrystalline solar cell construction will be considered where appropriate and a later chapter will be devoted to the performance of amorphous and polycrystalline based solar cells.

Local Electric Fields

Electric fields can be generated within a semiconductor by: (1) altering the composition or impurity concentration of the semiconductor (a heterotransition), (2) by varying the type of impurity doping within a semiconductor (a pn junction), or (3) by varying both (a heterojunction). Another widely used means of generating an internal electric field is the Schottky junction. In a Schottky junction one half of a pn junction is replaced by a metal or metal-semiconductor alloy. This produces a potential barrier at the metal-to-semiconductor interface. This type of barrier is thus a subclass of the heterojunction.

Other possible techniques which can be employed in the generation of local electric fields are: (1) the introduction of mechanical strains, (2) the introduction of impurity atoms which produce strains, (3) the Dember effect (this is a result of the differential diffusivities of the mobile charge carriers), (4) the photomagnetoelectric effect[*], and (5) a phenomenon known as the anomalous photovoltaic effect. The introduction of strains or other impurity variations (vacancies, interstials, etc.) in a semiconductor produces small voltage changes, frequently exhibiting self-cancellation due to interactions between adjacent defects. Additionally, such imperfections act as recombination sites, decreasing the lifetime, and, therefore, reducing the photocurrent and overall solar cell performance. The Dember and photomagnetoelectric effects are small [2-5]. The anomalous photovoltaic effect frequently produces voltages much greater than the energy gap [6-8]. The voltages displayed by this phenomenon are often a function of the illuminating spectrum and of the techniques employed in the preparation of the semiconductor. No complete explanation exists at present for the anomalous photovoltage. Current theories rely on the summation of the photovoltages produced by a number of small cells that may, or may not, be polycrystallites [7]. In summation, none of the techniques surveyed in this paragraph appears promising when used in solar cells, and we shall not consider them further.

---

[*] This effect is a close relative of the well-known Hall effect [1].

Let us return to the most useful techniques for generating internal electric fields. (1)The simplest technique has long been the homojunction, or pn junction, diode, as considered in Chapter III. Only a single semiconducting material is required for this type of solar cell. This reduces the required technology to a minimum, but does demand that the semiconductor under consideration be one that can be "doped" with both donors and acceptors. From our studies in Chapter III, it is clear that this condition is not always met. In many II-VI semiconducting compounds the equilibrium stoichiometry of the crystal precludes either a net hole or a net electron concentration. For example, CdSe is available only as an n-type semiconducting crystal. It is possible that the technology known as ion implantation will provide a means to circumvent this difficulty, but this has not yet been demonstrated. (2) Potentially, the least expensive technique for constructing solar cells is that heterojunction method specifically known as a Schottky barrier. The semiconductor involved is but a single type (n or p) involving a minimum of processing. The metal film required can be evaporated or sputtered and any subsequent processing (to form a semiconductor-metal alloy, for example) is done at relatively low temperatures (below 800°C). (3) A third method, involves the use of two semiconductors (one n-type and one p-type) to form a heterojunction. This has the major disadvantage that the technologies for two semiconductors are required. A potential advantage for this kind of junction derives from the optical properties of the two semiconductors. The semiconductors through which the light initially passes can be possessed of a sufficiently wide energy gap to enable the semiconductor to be effectively transparent to sunlight. The other layer, the substrate, can be constructed of a semiconductor which efficiently absorbs sun light. This advantage can be significant as evidenced in the $GaAs/Al_xGa_{1-x}As$ photovoltaic cell [9]. (4) The heterotransition method for electric field generation does not employ a pn junction, but, normally, depends on the differences in crystal structure and, hence, energy band configuration of two semiconducting materials of the same type [10, 11]. The generated electric fields are smaller than the heterojunction or homojunction methods and the technical difficulties in fabrication are equivalent. This technique will not receive any further discussion in this work. (5) Solar cells made from polycrystalline materials such as Cu_2S/CdS promise to be relatively inexpensive to fabricate [12]. For this reason alone, thin-film polycrystalline solar cells deserve attention. However, the efficiencies observed for solar cells made from thin-film polycrystalline semiconductors   are low (5-10%) and they often exhibit performance degradation when exposed to ambient conditions [13]. Space does not permit discussion of this type of solar cell here, but there is ample technical literature [14-17].

We will, in this work, consider solar cells of the heterojunction, Schottky barrier and pn junction types. The electrical characteristics of heterojunctions, Schottky barriers and pn junction diodes are of the same general form, with magnitudes depending on the material properties of the semiconductors employed. We start our analysis with a study of pn junctions.

PN Junction Electrical Characteristics

Consider the pn junction of Figure IV.1. The generation of hole-electron pairs by photons interacting with valence band electrons near the space charge region leads to a separation of the mobile charge carriers

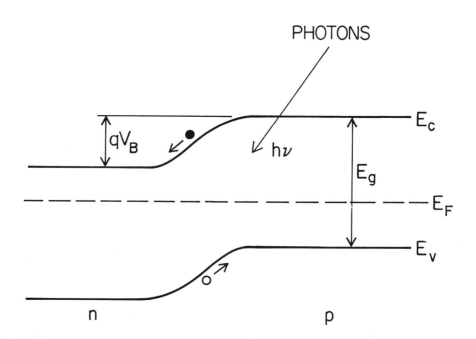

Figure IV.1. A pn junction under illumination. E_g is the energy gap, E_v is the upper edge of the valence band, E_c is the lower edge of the conduction band, E_f is the Fermi level and V_B is the built-in voltage.

with the electrons flowing towards the n-side (the n-side is the side of the junction with minimum energy for electrons) and holes flowing towards the p-side (the side exhibiting minimum energy for the holes)*.

In Figure V.2, some of the hole-electron pairs, generated in Figure V.1 by the absorption of photons, have been divided and the individual charge carriers have been separated and are now on the opposite sides of the pn junction. As shown in Figure V.2 the presence of excess holes on the p-side of the junction and excess electrons on the n-side leads to a reduction of the built-in energy barrier (qV_B in Figure V.2), by an amount, qV_p, where V_p is known as the photovoltage.

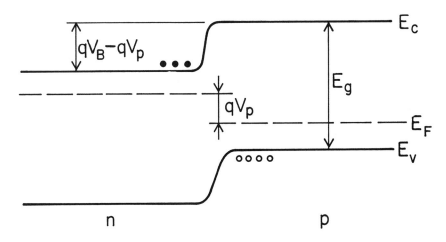

Figure V.2. The origin of the photovoltage, V_p, in a pn junction diode. The vertical scale is energy with electron energy increasing upward and hole energy increasing downward.

The current versus voltage characteristic of a pn junction, in the absence of light was qualitatively described in Chapter III and is quantitatively represented as [18]:

---

* The space charge region of a diode is a region close to the physical boundary where the p- and n-type regions meet. It is characterized by an electric field, which is responsible for the bending of the conduction and valence band edges in Figure V.1.

$$J = J_s[\exp\{qV/kT\} - 1] \,, \qquad\qquad (V.1)$$

where J is the current density, J_s is the saturation current density of the junction, V is the applied voltage, q is the absolute value of the electronic charge, k is Boltzmann's constant and T is the absolute temperature. J_s is dominated by the diffusion of minority carriers and is given by:

$$J_s = \frac{qD_{pn}p_{no}}{L_{pn}} + \frac{qD_{np}n_{po}}{L_{np}} \,, \qquad\qquad (V.2)$$

where p_{no} is the thermal equilibrium density of holes on the n-type side of the junction, n_{po} is the thermal equilibrium density of electrons on the p-type side, D_{pn} and L_{pn} are the diffusion constant and diffusion length of holes on the n-type side, and D_{np} and L_{np} are the diffusion constant and diffusion length of electrons on the p-type side. From the Einstein relationship between diffusion constant and mobility:

$$D_{pn} = \{kT/q\}\mu_{pn} \,,$$

and $\qquad\qquad\qquad\qquad\qquad\qquad\qquad\qquad\qquad\qquad (V.3)$

$$D_{np} = \{kT/q\}\mu_{np} \,,$$

where μ_{pn} is the hole mobility on the n-type side and μ_{np} is the electron mobility on the p-type side of the junction. The thermal equilibrium densities of the minority carriers are given by:

$$p_{no} = n_i^2/N_D \,,$$

and $\qquad\qquad\qquad\qquad\qquad\qquad\qquad\qquad\qquad\qquad (V.4)$

$$n_{po} = n_i^2/N_A \,,$$

where N_A and N_D are the acceptor and donor impurity concentrations on the p-side and n-type side of the junction and n_i is the intrinsic carrier concentration. Finally, the minority carrier diffusion lengths are given by:

$$L_{pn} = \sqrt{(\tau_{pn}D_{pn})} \,,$$

and   (V.5)

$$L_{np} = \sqrt{\{\tau_{np} D_{np}\}} \ ,$$

where τ_{pn} and τ_{np} are the hole lifetime on the n-type side and the electron lifetime on the p-type side, the minority carrier lifetimes.

As is clear from the preceding, Equation V.1 is heavily dominated by minority carrier properties. Equation V.1 generally accurately describes the current versus voltage characteristics of pn junctions where the semiconductor has a reasonably narrow energy gap (see below for details) and the current level is small. Even when the current becomes large, Equation V.1 is valid provided the voltage, V, is taken to be the voltage actually imposed across the junction, V_D, where:

$$V_D = V - Ir_D \ ,$$ (V.6)

where r_D is the sum of the lead and internal (semiconductor) diode series resistances and I is the current through the diode.

The most serious occasions for departure from Equation V.1 arise from the presence of a significant thermal generation-recombination current in the saturation current, J_S, of pn junctions. The presence of an electric field in the vicinity of a pn junction sweeps away any mobile charge carriers (holes and electrons) that are present. From Equation III.48 this implies that the generation-recombination mechanism now produces hole-electron pairs ($np < n_i^2$). These pairs, in turn, are separated and swept across the junction by the electric field, to be replaced by still more hole-electron pairs, etc., etc.. The generation rate for carriers in the space charge region of a pn junction has been shown to depend inversely on the minority carrier lifetimes (τ_{pn} and τ_{np}), on the intrinsic carrier concentration, n_i and on the energy levels of the trap sites involved in recombination [19, 20]. The current density, J_G, due to this source is given, approximately, by [21]:

$$J_G = qW_c\{\partial n'/\partial t\},$$ (V.7}

where W_c is the width of the space charge (or depletion) region surrounding the junction and n' is the carrier concentration in the space charge region.

Following the approach of Moll [21] and assuming the quasi-Fermi levels to be constant through the depletion region and that the electric field is constant:

$$J_G = q \int_0^{W_c} \{\partial n'/\partial t\}\{\partial x/\psi_i\}\partial \psi_i \; , \tag{V.8}$$

where ψ_i is the intrinsic Fermi level.

It has been shown [22] that if the reverse current is dominated by generation-recombination current, then, for small forward voltages, the forward current is also dominated by generation-recombination current, and:

$$J_G = J_{GS}\sinh\{qV/2kT\} = J_{GS}\sinh\{\lambda V/2\} \; , \tag{V.9}$$

where J_{GS} is the saturation current density for generation-recombination current and depends on temperature, the impurity concentrations and the nature of the traps present in the semiconductor. For a single deep level recombination center [21]:

$$J_{GS} \stackrel{\sim}{\sim} \{qn_i/\tau\}\{kT/q\mathscr{E}_{max}\} \stackrel{\sim}{\sim} (n_i kTW_c/\tau V_B) \; , \tag{V.10}$$

where τ is the minority carrier lifetime in the space charge region, $\mathscr{E}_{max}$ is the maximum electric field strength in the space charge region, and V_B is the built-in voltage for the pn junction (see Equation III.67).

Because the generation-recombination current (Equation V.9) depends on the intrinsic carrier concentration, n_i, whereas the diffusion current (Equation V.1) is dependent on n_i^2, the reverse current in a wide energy gap[*] semiconductor is dominated by generation-recombination current. However, at some forward voltage, the diffusion current will again dominate the current versus voltage characteristic since the exponential in Equation V.1 depends on λV and that of Equation V.9 depends on $\lambda V/2$. For a p+n step junction, the forward voltage, V', at

---

[*] A wide energy gap in this context is one with a forbidden energy gap in excess of 0.8 eV. This, clearly, includes all practical solar system solar cell semiconductors.

which the minority carrier injection current (Equations V.1 and V.2) equals the generation-recombination current (Equations V.9 and V.10) is given by the solution to[*]:

$$\frac{qD_{pn}n_i^2}{N_D\sqrt{\{D_{pn}\tau_{pn}\}}}(e^{\lambda V'} - 1) = \frac{qn_iW_c}{\tau\lambda V_B}(e^{\lambda V'/2} - e^{-\lambda V'/2}) .$$ (V.11)

The use of a p+n step junction implies that the electron diffusion current (the second term in Equation V.2) is effectively zero. In the case of such a step junction we may also take, where ϵ is the dielectric constant:

$$W_c = \sqrt{\{2\epsilon/qN_D\}}\sqrt{\{V_B - V_D\}} ,$$ (V.12)

and, from Chapter III, the built-in voltage is:

$$V_B = (1/\lambda\}\ln\{N_AN_D/n_i^2\}.$$ (V.13)

At room temperature, λ is approximately 38. With this value for λ, the results of the following chapters, and Equation V.11, the photovoltage of an operating solar cell will generally be in excess of V' and we will be able to use Equation V.1 for the description of the current versus voltage characteristic of the pn junctions being used as solar cells.

Heterojunction Electrical Characteristics

The heterojunction has an energy band diagram under thermal equilibrium generally similar to that of Figure V.3. The "notch" in the lower conduction band edge and the discontinuity in the upper valence band edge arise as a result of the dissimilarities in the crystal structure of

---

[*] A p+n step junction exhibits a much higher impurity concentration on the p-type side than on the n-type side and the transition from p-type to n-type is abrupt (much much less distance is required to go from p-type to n-type than a charge carrier diffusion length).

the two semiconductors. The height of the conduction band energy notch discontinuity, ΔE_c, is given by:

$$\Delta E_c = \chi_2 - \chi_1 , \qquad\qquad (V.14)$$

where χ_1 and χ_2 are the electron affinities of the two semiconductors. The valence band discontinuity is given by:

$$\Delta E_v = E_{g1} - E_{g2} - \Delta E_c.$$

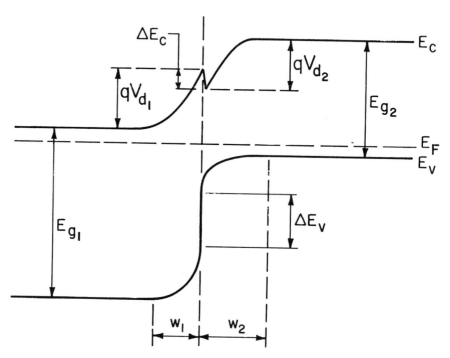

Figure V.3. The energy band diagram of a np heterojunction diode in thermal equilibrium. E_{g1} and E_{g2} are the band gap energies of the two semiconductors and E_F is the Fermi level.

The advantages of a heterojunction solar cell arise, not from any increase in the theoretical maximum power level obtainable, but from the use of an optically transparent "front" layer (the side of the solar cell through which the sunlight first passes). This implies that all of the photons absorbed generate hole-electron pairs in the substrate region. This region can then be prepared very carefully in order to maximize the minority carrier lifetime and, hence, the separation efficiency. As indicated in the preceding chapter, the "front" layer in a pn junction solar cell is generally of low lifetime and the carriers generated therein by photon absorption are normally ineffective as sources of photocurrent [23]. With a very wide gap material ($E_g > 2.0$ eV) as the "front" layer there is little or no absorption in this region. the "front" layer can now be doped heavily, reducing the series resistance of the solar cell. As we will see, subsequently, these changes enhance the performance of a solar cell by a significant amount.

The current versus voltage characteristic of a heterojunction may contain components due to three principal current mechanisms: (1) diffusion current due to the injection of minority carriers from each side of the junction, similar to a pn junction, (2) generation-recombination current arising within the space charge region, and (3) current arising from tunneling. In determining the diffusion current we must include the effects of the actual crystal-to-crystal interface [24]. With the "front" layer heavily doped to reduce series resistance and its much lower carrier generation rate, it is possible to use a step junction approximation for the diffusion current. The current, for a np step heterojunction[*], is, therefore, given by:

$$J = J_{SH}[\exp\{\lambda V_D\} - 1] , \tag{V.16}$$

where the heterojunction saturation current, J_{SH}, is given by:

$$J_{SH} = \{qD_{np}n_i^2/L_{np}N_A\}X_T . \tag{V.17}$$

This expression differs from that for an n+p step junction as given in Figure V.2 only in the factor, X_T. This factor is empirical, accounting for the imperfect nature of carrier transmission from the crystal structure of one semiconductor to the crystal structure of the other semiconductor.

The current terms arising from the generation-recombination mechanism in heterojunctions are similar to those of Equations V.9 and

---

[*] The n region is the wide gap energy semiconductor and the p-type region is the narrow energy gap semiconductor.

V.10. However, the exact magnitude is difficult to determine owing to uncertainties in the lattice matching at the metallurgical interface between the two semiconductors. Following the same line of reasoning used in pn junctions, we can state that for the operational conditions anticipated for heterojunction solar cells the generation-recombination current should be small compared to the diffusion current. In those instances where it is not, the structural mismatch will, generally, be too large to make the heterojunction device a feasible solar cell.

Tunneling is the dominant current mechanism for the vast majoritiy of semiconductor combinations in heterojunctions, the exceptions being those semiconductors with a very small lattice mismatch. The tunneling current density is of the form [25, 26]:

$$J_T = K_I N_t \left[(m^* q/N)^{1/2} V_D \exp\{4 \hbar/3\} \right], \tag{V.18}$$

where K_I is a constant which depends upon the effective mass of the charge carriers, the dielectric constant of the semiconductor and the built-in potential. N_t depends upon the density of available states within the forbidden gap, at, or near, the metallurgical junction (including interface states [27]). N is the impurity density, $\hbar$ is Planck's constant divided by 2π and m^* is the effective mass of the majority charge carrier of the wide gap semiconductor.

It is difficult to predict the magnitude of the tunneling current owing to the variability in and instability of the interface states between the two semiconducting materials. These effects arise, in part, as a result of our present inability to provide a reliable processing technology for heterojunctions. In general, the current-voltage characteristics of most heterojunctions appear to be consistent with tunneling current as the principal mechanism, yielding large dark currents and solar cells with poor efficiency. However, there are notable exceptions. Among these are heterojunctions made from ZnSe and Ge, GaAs and Ge and, most significantly, $Ga_{1-x}Al_xAs$ and GaAs [28-31][*]. Heterojunctions made from these combinations of semiconductors have electrical characteristics

---

[*] Useful heterojunctions have been fabricated with many other material combinations [32, 33], including those using polycrystalline substrates or polycrystalline deposits [32-36].

consistent with diffusion current (Equation V.16), and can, in principle, make "useful" solar cells.

Electrical Characteristics of Schottky Junctions

In Schottky, or metal-semiconductor junctions, when used in solar cells, the space charge electric field used to separate the hole-electron pairs is generated by the work function difference between the metal and the semiconductor. Schottky barrier current versus voltage characteristics are primarily due to majority carrier transport [37]. Figure V.4 depicts a typical Schottky barrier diode.

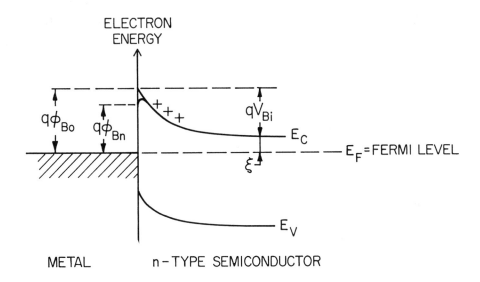

Figure V.4. The energy diagram for a Schottky barrier diode. ϕ_{Bo} is the barrier energy, $q\phi_{Bo}$ is the barrier voltage, $q\phi_{Bn}$ is the effective barrier voltage including image forcing lowering, and ξ is the energy difference between the bottom of the conduction band, E_c, and the Fermi level.

The emission of electrons from the semiconductor (for the metal-on-n-type-semiconductor shown in Figure V.4) into the metal constitutes forward current. (For a metal-on-p-type-semiconductor the forward current would consist of holes.) This forward current is a combination of thermionic emission and diffusion [38-40]. We can write for the current density versus voltage, for a metal-on-n-type-semiconductor as a function of the barrier energy and temperature:

$$J = A^*T^2 \exp\{-\lambda\phi_{Bn}\}[\exp\{\lambda V_D\} - 1] = J_{ss}[\exp\{\lambda V_D\} - 1] ,$$ (V.19)

where A*, the Richardson constant, is given by:

$$A^* = 120\{m^*/m\} ,$$ (V.20)

where m is the free electron mass.

A subclass of the Schottky barrier type of solar cell is the MOS or metal-oxide-semiconductor solar cell. A typical example, utilizing p-type silicon and aluminum as the metal, is shown in cross section in Figure V.5. The thin (20 to 40 Å) oxide layer is processed carefully to insure that the metal and semiconductor are nowhere in contact. This processing involves an elaborate annealing sequence to prevent micro-cracking of the oxide.

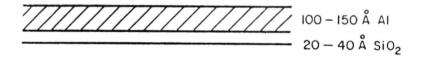

100 – 150 Å Al

20 – 40 Å SiO$_2$

p – TYPE Si

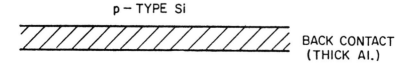

BACK CONTACT
(THICK Al.)

Figure V.5. The cross section of an aluminum-oxide-on-p-type silicon MOS solar cell, after Charlson and Lien [41].

The most useful effect of the oxide interfacial layer is the effective increase in the barrier voltage. For example, the barrier voltage for aluminum on silicon is normally on the order of 0.4 volts [42], but with an interfacial oxide a value of 0.85 volts is observed [41]. If we assume that the current mode through the interface layer is tunneling [41] and use the tunneling theory of Green, et. al. [43], and the thermionic emission-diffusion theory of Crowell and Sze [38] for the Schottky barrier proper, then, the expression for the current, in MOS solar cell similar to that of Figures V.5 and its energy counterpart of Figure V.6, is:

$$J = A^{**}T^2 \exp\{-\lambda\phi_{Bo}\}[\exp\{\lambda V_D\} - 1] . \tag{V.21}$$

In Equation V.21, A^{**} is the effective Richardson constant, equal to 32 amperes/cm^2/°K^2 for aluminum on p-type silicon. The ideality factor , n, is experimentally found to be 1.42 for the al-oxide-p-type silicon MOS solar cell [41]. The difference between A^{**} and the true Richardson constant, A^*, which is equal to 71 amperes/cm^2/°K^2 for p-type silicon, and the existence of the non-unity ideality factor in Equation V.21 are illustrative of the differences between the simple Schottky barrier and MOS cells. These differences arise because the current flows, not only over the barrier, but through the oxide interface layer in a MOS solar cell.

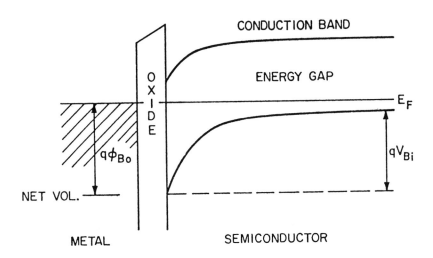

Figure V.6. The energy band diagram for an aluminum-oxide-p-type silicon MOS solar cell.

The entire subject of metal-oxide-semiconductor solar cells has received attention only recently. Operating solar cell efficiencies in excess of 24% have been predicted for these devices [44]. For the reader who desires to further pursue this subject, an excellent starting point is Shewchun, et. al. [45], which treats the theory of performance for MOS solar cells in terms of tunneling under nonequilibrium carrier concentrations. This work arrives at a maximum operating efficiency of 21% for a MOS solar cell. Salter and Thomas [46] have also treated this subject, arriving at an expression similar, in voltage sensitivity, to Equation V.21. Their data, taken with a complex nickel-copper-gold metal layer, indicate an ideality factor of 1.05.

The basic element affecting the behavior of Schottky and MOS Schottky junctions is the interface between the semiconductor and the metal or between the oxide and the semiconductor. A complex literature exists covering these interfaces; too voluminous for our space limitations. The reader is referred to [47-51] as an introduction to the topic. Here, we summarize by saying that the MOS type of device is not well understood at present and depends to such a significant extent on surface interface effects (including crystal orientation for the semiconductor [45]) that are not completely clear, that we will simplify its treatment and consider the MOS device as a variety of Schottky barrier, with current versus voltage characteristics following Equation V.21.

Open Circuit Voltage and Short Circuit Current

When a solar cell is illuminated, the incoming photons generate hole-electron pairs by collision with valence band electrons. The hole-electron pairs are separated into individual holes and electrons and these individual charge carriers are collected, becoming a photocurrent density, J_{ph}. The process of collecting these holes and electrons also gives rise to a photovoltage that is so biased as to generate a forward diode current density, J_D (see Figure V.2 for details on the charge carrier collection). The photocurrent is in what is normally considered to be the reverse direction so that the current densities, J_D and J_{ph}, are opposing. The expression for diode current density is provided by Equation V.1 for pn junctions, by Equation V.16 for heterojunctions, by Equation V.19 for Schottky barrier devices and by Equation V.21 for the MOS type of solar cell. In general, the current versus voltage characteristic of any of these solar cell types under illumination resembles that of the dashed line of Figure III.15. The steady state equivalent circuit of a solar cell under illumination, coupled to an external load, R_L, is provided in Figure V.7.

In Chapter IV we discussed the generation of electron-hole pairs by the absorption of photons. Let us term this generation rate G_L. Then, the absolute value of the photocurrent density may be written as:

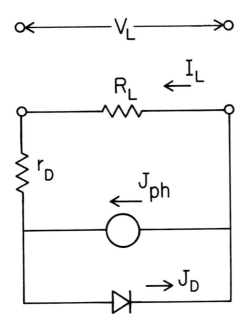

Figure V.7. The electrical equivalent circuit for an illuminated solar cell and an external load, R_L. The series resistance of the solar cell, including lead resistance, is r_D.

$$J_{ph} = qX_L G_L. \tag{V.22}$$

This photocurrent depends strongly on the factor, X_L, that is, on the collection distance[*] for photon-generated charge carriers. In a pn junction or in a heterojunction the collection distance will depend on the minority carrier lifetimes in both the n-type and p-type sides of the solar cell, upon the surface recombination velocity, the carrier mobility and upon the impurity concentration in the solar cell. In Schottky barrier and MOS solar cells, the semiconductor side of the solar cell is the controlling factor and the metal or oxide portions of the solar cell are important only in-so-far as they affect the interface recombination of holes and

---

[*] By collection distance we mean the distance from the metallurgical junction over which we can effectively separate and collect the individual components of the generated hole-electron pairs.

electrons. Later in this, and subsequent chapters, we will consider the nature of the collection distance, X_L, in detail. For the present, let us note that the maximum possible photocurrent occurs when the entire volume of the solar cell lies within the collection distance.

Considering our expressions for diode forward current density, the general expression for the net current density in a pn or heterojunction, solar cell under illumination is given by:

$$J = J_S[\exp\{\lambda V_D\} - 1] - J_{ph} , \tag{V.23}$$

where for a pn junction solar cell, J_S is provided by Equation V.2 and V_D is the voltage appearing across the junction. It is assumed that sufficient illumination is present to guarantee operation in the diffusion (often known as the minority carrier injection mode) mode as opposed to the generation-recombination current mode. Equation V.2 is frequently modified owing to the proximity of the ohmic contacts at the solar cell surface. This modified expression is considered in Appendix C. For heterojunction solar cells, the characteristics of the narrower band gap side of the junction dominate. For situations where the factor, X_T, is an ideal unity and the n-type side has the wider energy gap, from Equation V.17, the saturation current density is:

$$J_S = J_{SH} = qD_{np}n_i^2/(L_{np}N_A), \tag{V.24a}$$

while for situations where the crystal interface factor, X_T, is unity and the p-type side of the junction has the wider energy gap:

$$J_S = J_{SH} = qD_{pn}n_i^2/(L_{pn}N_D) . \tag{V.24b}$$

The general expression for the current versus voltage characteristic of a Schottky or MOS solar cell under illumination is:

$$J = J_{SS}[\exp\{\lambda V_D/n\} - 1] - J_{ph} , \tag{V.25}$$

where for Schottky barrier solar cells, n is unity and for MOS solar cells, n = two [52]. See Equations V.19 and V.21 to determine J_{SS} for these solar cell types.

The short circuit current density, J_{SC}, for an ideal solar cell is dependent primarily on the intensity of the illumination and the material parameters that control the collection distance. This, frequently quoted performance indicator for solar cells, is given, for pn and heterojunction solar cells, by:

$$J_{SC} = -J_{ph}/(1 + \lambda A_D r_D J_S/n) \ ,$$

or, for Schottky and MOS solar cells,   (V.26)

$$J_{SC} = -J_{ph}/(1 + \lambda A_D r_D J_{SS}/n) \ .$$

Here, A_D is the area of the solar cell, J_S is the saturation current density for p+n, n+p or heterojunctions (see Equation V.7 or V.24), J_{SS} is the saturation current of Schottky and MOS type barriers (see Equation V.19 or V.21 depending on the solar cell type). The ideality factor, n, in this situation of no effective bias across the diode, is approximately two for the pn junction* and heterojunction classes of solar cells, is approximately unity for a Schottky barrier type of solar cell and lies between one and two for MOS solar cells.

Note that the short circuit current appears on a current versus voltage diagram (such as Figure III.15) at the boundary between the third and fourth quadrants. Note, also, that with no voltage, the short circuit current condition is one in which the diode (solar cell) delivers no power to the external load.

The open circuit voltage indicates a situation in which no current is fed to a load and all of the separated holes and electrons remain in the solar cell until they recombine. This produces a significant voltage across the junction, but no power is delivered to the load. The open circuit voltage for a solar cell is given by:

$$V_{OC} = \{n/\lambda\}\{\ln[\{J_{ph} + J_S\}/J_S]\} \ ,$$ (V.27)

where J_S is the generalized saturation current density appropriate to the type of junction (barrier) in use in the solar cell. In Equation V.27, n, the

---

* For a shorted pn junction the generation-recombination current dominates (the voltage across the junction is far less than the v' of Equation V.11) for both pn and heterojunction devices.

ideality factor, lies between one and two for MOS solar cells and is essentially unity for all of the other solar cell junction types under consideration. Since the open circuit voltage is larger than v' in Equation V.11, the saturation current density, J_s is given by, or is derivable from, Equations V.2, V.19, V.21 or V.24, depending upon the type of barrier utilized in the solar cell.

Consider Equation V.27. As written, this expression assumes that the photocurrent, J_{ph}, is much larger than the saturation current, J_S. In this case, the minority carrier diffusion current dominates the voltage versus current characteristic for both pn and heterojunction devices. Equation V.27 also imposes no upper bound on the value of the open circuit voltage. In practice, once the open circuit voltage has risen to the internal barrier voltage height, V_B, the internal device barrier to current flow disappears and the diode current increases dramatically [53-55]. These internal barrier voltages are given by:

(a) for a pn junction:

$$V_B = \{1/\lambda\}\ln\{N_A N_D/n_i^2\} , \tag{V.28a}$$

(b) for a heterojunction:

$$V_B = \Delta E_g - (\chi_2 - \chi_2) , \tag{V.28b}$$

and, (c) , for a Schottky barrier or MOS solar cell:

$$V_B \leq [\phi_{Bn} - \{1/\lambda\}\ln\{N_{eff}/n\}] < \{\chi_2 - \chi_1\} . \tag{V.28c}$$

Here ΔE_g is the difference in band gap energies between the two semiconductors involved in a heterojunction and N_{eff} is the effective density of states in the semiconductor in a Schottky or MOS barrier solar cell (either in the conduction band for an n-type semiconductor, or in the valence band, for a p-type semiconductor. χ is the electron affinity for a semiconductor (the difference between the free electron level and the lower conduction band edge).

In all types of solar cell considered, an absolute upper limit on the internal potential barrier is the band gap energy of the semiconductor (the heterojunction case is somewhat more complex, but the effective limit is approximately the energy gap of the semiconductor with the smaller band gap). In practice, doping considerations, to be discussed later, and surface recombination phenomena [56] reduce the obtainable open circuit photvoltage below this maximum value.

Optimum Power Conditions

There are a number of papers in the literature that treat the power performance of solar cells [57-65] including charge carrier collection effects, solar cell series resistance and saturation current among other performance variables. The following approach is simple and provides a circuit-and-physics oriented approach to solar cell performance and modeling. Starting with the equivalent circuit of Figure V.7, the power, P, delivered to the load resistor, R_L, is:

$$P = -V_L R_L = -[V_D + r_D I_L]I_L , \qquad (V.29)$$

where V_L is the load voltage, V_D is the voltage across the solar cell junction, r_D is the series resistance of the solar cell, and I_L is the load current, which is given by:

$$I_L = A_D[J_S\{\exp(\lambda V_D) - 1\} - J_{ph}] , \qquad (V.30)$$

where A_D is the solar cell junction area and the saturation current, J_S, is taken from Equation V.2 for pn junctions; Equation V.24 for heterojunctions; Equation V.19 for Schottky barrier solar cells; and Equation V.21 for mos junction type solar cells.

Maximizing the power delivered to the load resistor leads to the following expression, which must be satisfied by V_D', the diode junction voltage when the solar cell is delivering maximum power, and the factor, K' :

$$[\{\lambda V_D'/n\} + 1 + K']\exp\{\lambda V_D'/n\} = [\{J_{ph} + J_S\}/J_S][1 + K'] . \qquad (V.31)$$

In Equation V.31, the factor K' is known as the loss factor and is given by:

$$K' = 2(r_D A_D)J_S(\lambda/n)\exp\{\lambda V_D'/n\} . \qquad (V.32)$$

Note that the loss factor depends upon the saturation current density and the series resistance-area product of the solar cell, as well as the voltage across the junction and the junction temperature.

Combination of expressions V.29, V.30, V.31 and V.32 yields the maximum power density which can be transferred to the load resistor from the solar cell:

$$\frac{P_{max}}{A_D} = \frac{\{J_{ph} + J_S\}V_D'\{ \lambda V_D'/n\}}{[\lambda V_D'/n + 1 + K']} [1 - \{K'/2\}/\{1 + K'\}] . \tag{V.33}$$

Study of Equation V.33 reveals that, as the loss factor, K', approaches infinity, the maximum power transferred to the load approaches zero, and as K' approaches zero, the delivered power density approaches a maximum.

Before examining the implications inherent in Equations V.31 through V.33, let us determine the optimum load resistance, R_L'. Utilizing Equations V.29 through V.32, we find that the load resistance for maximum power transfer is:

$$R_L' = \{2r_D/K'\}\{1 + 3K'/2\} , \tag{V.34}$$

or, in terms of diode voltage, photocurrent density and saturation current density:

$$R_L' = \frac{\lambda V_D'/n + 1 + K'}{A_D(J_{ph} + J_S)\lambda/n} \frac{(1 + 3K'/2)}{(1 + K')} , \tag{V.35}$$

The efficiency of the solar cell is the ratio of the power transferred to the load to the incoming solar energy. From Chapter IV, the incoming solar energy density under AMO and AM1 conditions is:

$$I_{ns} = 0.1353 \text{ watts/cm}^2 \text{ (AMO)} ,$$

and

$$\tag{V.36}$$

$$I_{ns} = 0.107 \text{ watts/cm}^2 \text{ (AM1)} .$$

The maximum efficiency, η_s, for a solar cell is given by:

$$\eta_s = [\{P_{max}/A_D\}/\ I_{ns}] \times 100\% \ . \tag{V.37}$$

In Chapter IV we considered the potentially available photocurrent as a function of the energy gap of the semiconductor. In Figure IV.4 both the photocurrent and the maximum potential power density were presented. They were presented as a function of the energy gap and with AMO and AM1 illumination conditions as a parameter. This figure assumes that: (1) all of the light falling on the semiconductor is absorbed (i.e., no reflection from the front surface and no through-transmission of photons; (2) all of the electron-hole pairs generated by the absorbed photons are subsequently delivered to the external circuit (i.e., there is no surface or volume recombination; and (3) there are no internal losses (i.e., the solar cell series resistance is zero ohms). Using these values of photocurrent (the maximum possible values for photocurrent) in Equations V.31, V. 32, V.33 and V.37, we can obtain the maximum potential output power density and solar cell efficiency, provided values of the diode series resistance, r_D, and of the saturation current density, J_S, are known.

In the following chapter we will obtain values for the solar cell series resistance as a function of the design philosophy for the cell, the semiconductor chosen and the construction techniques employed in fabricatingl. Additionally, in the following chapter we will consider the saturation current using of Equation V.2 for pn junctions, V.24 for heterojunctions, V.19 for Schottky barriers or Equation V.21 for mos type solar cells. From these values, output power and efficiency for solar cells for specific semiconductors with a particular design, can be determined.

First, however, it is desirable to pursue a more general approach to the performance of solar cells. Let us select values of K', the loss factor, and J_S, the saturation current density. These values can then be used to compute the effective diode junction voltage, the potential maximum output power density and the efficiency of the solar cells. In specifying values of the loss factor, it is clear that we will require values that are relatively small, if the resultant output power density is to be close to the desired maximum. From the discussion of specific examples in the following chapters, let us select values of K' equal to zero (the ideal, no series resistance for the solar cell situation), 0.25, 1.0 (medium size values that are typical of those encountered in many practical situations) and 10 (a relatively large value).

Selection of the desired values of saturation current density, J_S, is somewhat more difficult and, clearly, is dependent on the type of barrier employed. When considering saturation current density for a pn junction, it is important to recall, from Chapter III, that the intrinsic carrier concentration in a semiconductor is given by:

$$n_i^2 = N_c N_v \exp\{-E_g/kT\} \ , \qquad (V.38)$$

where N_c and N_v are the conduction and valence band densities of states and E_g is the semiconductor energy gap. Observing Equations V.2, V.17, V.19 and V.21, the general form of saturation current density for pn junctions, heterojunctions, Schottky barriers and MOS barriers is:

$$J_S = \{Const.\}\exp\{-\lambda \underline{V}\} \ , \qquad (V.39)$$

where the constant and $\underline{V}$ are dependent upon the type of barrier.

In general, the pn junction has the smallest saturation current density of the junction types under consideration. Recall that it has been shown that the maximum value of $\underline{V}$ that is "reasonably" possible, is the energy gap voltage. Therefore, in making estimates for the minimum values of saturation current density to be encountered, we will use the saturation current density of a solar cell employing a pn step junction. From Equations V.2 and V.38 we will use:

$$J_S(min) = \{qD/N_S\}\{N_c N_v/L\}\exp\{-E_g/kT) \ , \qquad (V.40)$$

where N_S is the impurity concentration of the lightly doped side of the pn step junction and D and L are the diffusion constant and diffusion length for the minority carriers in this lightly doped side.

In practice, phenomena such as surface recombination, bulk lifetime, and the generation-recombination mechanism will tend to increase the saturation current density at least an order of magnitude above that predicted by Equation V.40.

Let us now estimate the practical minimum values of saturation current density as predicted by Equation V.40. Referring to Appendix G, Chapter III and Chapter IV; let us assume an impurity concentration level, N_S, in the lightly doped side of $10^{16}/cm^3$; conduction and valence band densities of states of approximately $10^{19}/cm^3$; a diffusion constant of approximately 10 cm^2/second; and a diffusion length of 10^{-2} cm. The resultant estimates for minimum saturation current density at 300°K are presented in Table V.1.

Table V.1

An estimation of the minimum theoretical saturation current density experienced for semiconductor step junction solar cells as a function of the energy gap, at a temperature of 300°K[*]

E_g(eV)	J_S(A/cm^2)	E_g(eV)	J_S(A/cm^2)	E_g(eV)	J_S(A/cm^2)
0.8	1.1×10^{-08}	1.4	9.3×10^{-19}	2.0	8.0×10^{-29}
0.9	2.3×10^{-10}	1.5	2.0×10^{-20}	2.1	1.7×10^{-30}
1.0	4.8×10^{-12}	1.6	4.1×10^{-22}	2.2	3.5×10^{-32}
1.1	1.0×10^{-13}	1.7	8.6×10^{-24}	2.3	7.5×10^{-34}
1.2	2.1×10^{-15}	1.8	1.8×10^{-25}	2.4	1.6×10^{-35}
1.3	4.4×10^{-17}	1.9	3.8×10^{-27}		

Following the discussion in Chapter IV on the potential useful range of semiconductor energy gaps in solar cells, we need to consider, at most, a range of band gap energies between 0.9 and 2.0 eV. In turn, this implies that a range of saturation current densities from 10^{-28} to 10^{-8} A/cm^2 is appropriate. Note that the saturation current density has been increased by approximately an order of magnitude in order to account for such effects as generation-recombination current and Schottky tunneling.

Consider Figures V.8 through V.15. In these figures the estimated maximum potential power density as a function of the band gap energy, with saturation current density as a parameter, is presented. Each figure either assumes an input under AMO conditions (135.3 mW/cm^2) or under AM1 conditions (107 mW/cm^2). The loss factor upon which a particular

---

[*] In many examples in the following chapters we will use other values for N_S (typically, 10^{15}/cm^3) based on considerations to be brought out in later chapters, considerations having to do with the collection distance. In this chapter we shall use the value of 10^{16}/cm^3 for N_S in order to minimize the saturation current density and maximize the potential power performance.

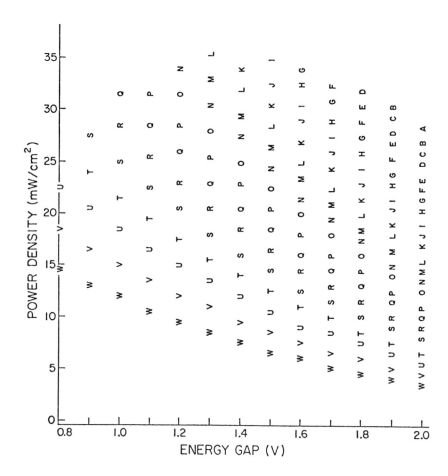

Figure V.8. The potential output power density (mW/cm²) under AMO conditions as a function of the energy gap (eV) for solar cells with a loss factor, K', equal to zero.

** Key **

(J_S, the saturation current density, A/cm²)

A => $J_S = 1 \times 10^{-28}$ B => $J_S = 1 \times 10^{-27}$ C => $J_S = 1 \times 10^{-26}$

D => $J_S = 1 \times 10^{-25}$ E => $J_S = 1 \times 10^{-24}$ F => $J_S = 1 \times 10^{-23}$

G => $J_S = 1 \times 10^{-22}$ H => $J_S = 1 \times 10^{-21}$ I => $J_S = 1 \times 10^{-20}$

J => $J_S = 1 \times 10^{-19}$ K => $J_S = 1 \times 10^{-18}$ L => $J_S = 1 \times 10^{-17}$

M => $J_S = 1 \times 10^{-16}$ N => $J_S = 1 \times 10^{-15}$ O => $J_S = 1 \times 10^{-14}$

P => $J_S = 1 \times 10^{-13}$ Q => $J_S = 1 \times 10^{-12}$ R => $J_S = 1 \times 10^{-11}$

S => $J_S = 1 \times 10^{-10}$ T => $J_S = 1 \times 10^{-09}$ U => $J_S = 1 \times 10^{-08}$

V => $J_S = 1 \times 10^{-07}$ W => $J_S = 1 \times 10^{-06}$

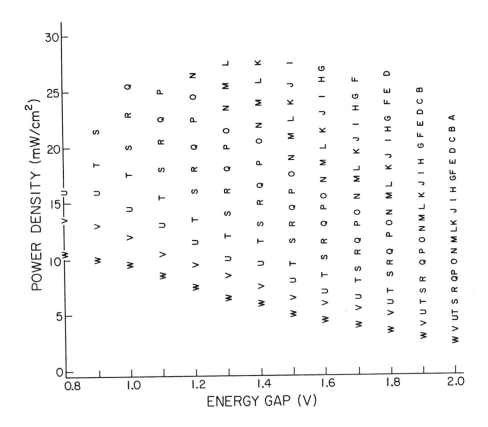

Figure V.9. The potential output power density (mW/cm²) under AM1 conditions as a function of the energy gap (eV) for solar cells with a loss factor, K', equal to zero.

** Key **
(J_S, the saturation current density, A/cm²)

A => $J_S = 1 \times 10^{-28}$ B => $J_S = 1 \times 10^{-27}$ C => $J_S = 1 \times 10^{-26}$

D => $J_S = 1 \times 10^{-25}$ E => $J_S = 1 \times 10^{-24}$ F => $J_S = 1 \times 10^{-23}$

G => $J_S = 1 \times 10^{-22}$ H => $J_S = 1 \times 10^{-21}$ I => $J_S = 1 \times 10^{-20}$

J => $J_S = 1 \times 10^{-19}$ K => $J_S = 1 \times 10^{-18}$ L => $J_S = 1 \times 10^{-17}$

M => $J_S = 1 \times 10^{-16}$ N => $J_S = 1 \times 10^{-15}$ O => $J_S = 1 \times 10^{-14}$

P => $J_S = 1 \times 10^{-13}$ Q => $J_S = 1 \times 10^{-12}$ R => $J_S = 1 \times 10^{-11}$

S => $J_S = 1 \times 10^{-10}$ T => $J_S = 1 \times 10^{-09}$ U => $J_S = 1 \times 10^{-08}$

V => $J_S = 1 \times 10^{-07}$ W => $J_S = 1 \times 10^{-06}$

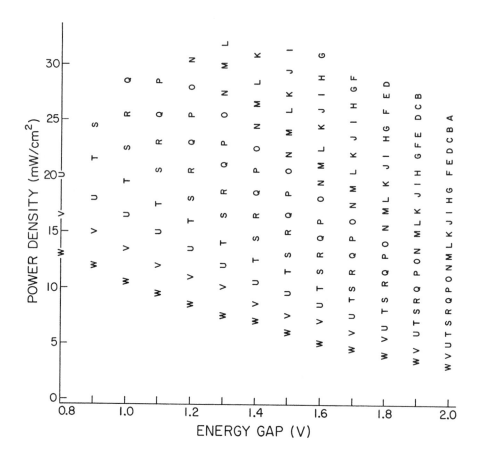

Figure V.10. The potential output power density (mW/cm²) under AMO conditions as a function of the energy gap (eV) for solar cells with a loss factor, K', equal to 0.25.

** Key **
(J_S, the saturation current density, A/cm²)

A => $J_S = 1 \times 10^{-28}$ B => $J_S = 1 \times 10^{-27}$ C => $J_S = 1 \times 10^{-26}$
D => $J_S = 1 \times 10^{-25}$ E => $J_S = 1 \times 10^{-24}$ F => $J_S = 1 \times 10^{-23}$
G => $J_S = 1 \times 10^{-22}$ H => $J_S = 1 \times 10^{-21}$ I => $J_S = 1 \times 10^{-20}$
J => $J_S = 1 \times 10^{-19}$ K => $J_S = 1 \times 10^{-18}$ L => $J_S = 1 \times 10^{-17}$
M => $J_S = 1 \times 10^{-16}$ N => $J_S = 1 \times 10^{-15}$ O => $J_S = 1 \times 10^{-14}$
P => $J_S = 1 \times 10^{-13}$ Q => $J_S = 1 \times 10^{-12}$ R => $J_S = 1 \times 10^{-11}$
S => $J_S = 1 \times 10^{-10}$ T => $J_S = 1 \times 10^{-09}$ U => $J_S = 1 \times 10^{-08}$
V => $J_S = 1 \times 10^{-07}$ W => $J_S = 1 \times 10^{-06}$

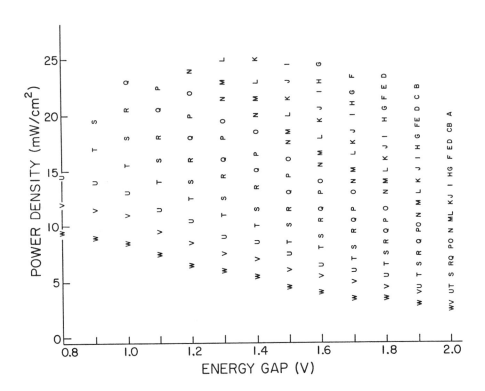

Figure V.11. The potential output power density (mW/cm^2) under AM1 conditions as a function of the energy gap (eV) for solar cells with a loss factor, K', equal to 0.25.

** Key **
(J$_S$, the saturation current density, A/cm^2)

A => J$_S$ = 1 x 10^{-28} B => J$_S$ = 1 x 10^{-27} C => J$_S$ = 1 x 10^{-26}
D => J$_S$ = 1 x 10^{-25} E => J$_S$ = 1 x 10^{-24} F => J$_S$ = 1 x 10^{-23}
G => J$_S$ = 1 x 10^{-22} H => J$_S$ = 1 x 10^{-21} I => J$_S$ = 1 x 10^{-20}
J => J$_S$ = 1 x 10^{-19} K => J$_S$ = 1 x 10^{-18} L => J$_S$ = 1 x 10^{-17}
M => J$_S$ = 1 x 10^{-16} N => J$_S$ = 1 x 10^{-15} O => J$_S$ = 1 x 10^{-14}
P => J$_S$ = 1 x 10^{-13} Q => J$_S$ = 1 x 10^{-12} R => J$_S$ = 1 x 10^{-11}
S => J$_S$ = 1 x 10^{-10} T => J$_S$ = 1 x 10^{-09} U => J$_S$ = 1 x 10^{-08}
V => J$_S$ = 1 x 10^{-07} W => J$_S$ = 1 x 10^{-06}

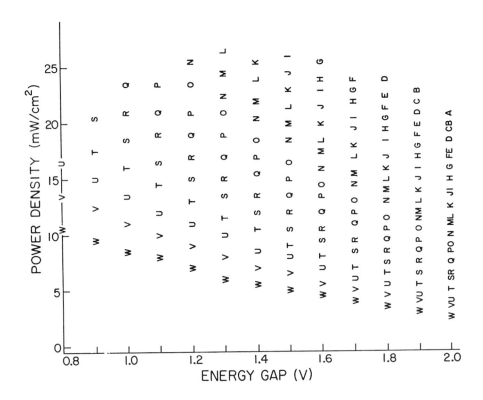

Figure V.12. The potential output power density (mW/cm²) under AMO conditions as a function of the energy gap (eV) for solar cells with a loss factor, K', equal to unity.

** Key **
(J_S, the saturation current density, A/cm²)

A => J_S = 1 x 10^{-28} B => J_S = 1 x 10^{-27} C => J_S = 1 x 10^{-26}
D => J_S = 1 x 10^{-25} E => J_S = 1 x 10^{-24} F => J_S = 1 x 10^{-23}
G => J_S = 1 x 10^{-22} H => J_S = 1 x 10^{-21} I => J_S = 1 x 10^{-20}
J => J_S = 1 x 10^{-19} K => J_S = 1 x 10^{-18} L => J_S = 1 x 10^{-17}
M => J_S = 1 x 10^{-16} N => J_S = 1 x 10^{-15} O => J_S = 1 x 10^{-14}
P => J_S = 1 x 10^{-13} Q => J_S = 1 x 10^{-12} R => J_S = 1 x 10^{-11}
S => J_S = 1 x 10^{-10} T => J_S = 1 x 10^{-09} U => J_S = 1 x 10^{-08}
V => J_S = 1 x 10^{-07} W => J_S = 1 x 10^{-06}

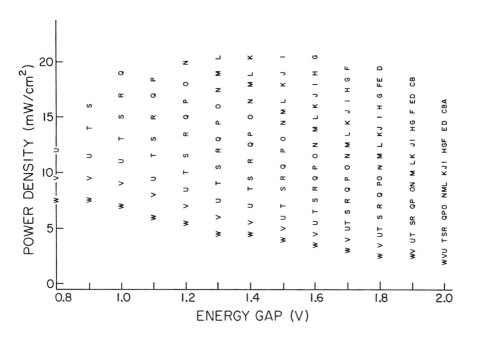

Figure V.13. The potential output power density (mW/cm^2) under AM1 conditions as a function of the energy gap (eV) for solar cells with a loss factor, K', equal to unity.

** Key **
(J$_S$, the saturation current density, A/cm^2)

A => J$_S$ = 1 x 10^{-28} B => J$_S$ = 1 x 10^{-27} C => J$_S$ = 1 x 10^{-26}
D => J$_S$ = 1 x 10^{-25} E => J$_S$ = 1 x 10^{-24} F => J$_S$ = 1 x 10^{-23}
G => J$_S$ = 1 x 10^{-22} H => J$_S$ = 1 x 10^{-21} I => J$_S$ = 1 x 10^{-20}
J => J$_S$ = 1 x 10^{-19} K => J$_S$ = 1 x 10^{-18} L => J$_S$ = 1 x 10^{-17}
M => J$_S$ = 1 x 10^{-16} N => J$_S$ = 1 x 10^{-15} O => J$_S$ = 1 x 10^{-14}
P => J$_S$ = 1 x 10^{-13} Q => J$_S$ = 1 x 10^{-12} R => J$_S$ = 1 x 10^{-11}
S => J$_S$ = 1 x 10^{-10} T => J$_S$ = 1 x 10^{-09} U => J$_S$ = 1 x 10^{-08}
V => J$_S$ = 1 x 10^{-07} W => J$_S$ = 1 x 10^{-06}

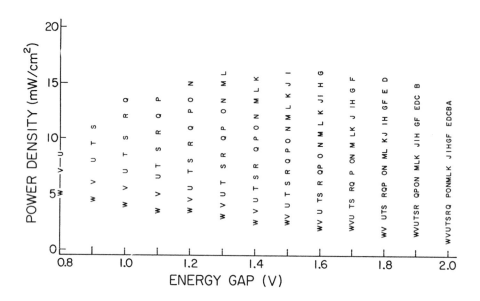

Figure V.14. The potential output power density (mW/cm²) under AM0 conditions as a function of the energy gap (eV) for solar cells with a loss factor, K', equal to ten.

** Key **
(J_S, the saturation current density, A/cm²)

A => $J_S = 1 \times 10^{-28}$ B => $J_S = 1 \times 10^{-27}$ C => $J_S = 1 \times 10^{-26}$
D => $J_S = 1 \times 10^{-25}$ E => $J_S = 1 \times 10^{-24}$ F => $J_S = 1 \times 10^{-23}$
G => $J_S = 1 \times 10^{-22}$ H => $J_S = 1 \times 10^{-21}$ I => $J_S = 1 \times 10^{-20}$
J => $J_S = 1 \times 10^{-19}$ K => $J_S = 1 \times 10^{-18}$ L => $J_S = 1 \times 10^{-17}$
M => $J_S = 1 \times 10^{-16}$ N => $J_S = 1 \times 10^{-15}$ O => $J_S = 1 \times 10^{-14}$
P => $J_S = 1 \times 10^{-13}$ Q => $J_S = 1 \times 10^{-12}$ R => $J_S = 1 \times 10^{-11}$
S => $J_S = 1 \times 10^{-10}$ T => $J_S = 1 \times 10^{-09}$ U => $J_S = 1 \times 10^{-08}$
V => $J_S = 1 \times 10^{-07}$ W => $J_S = 1 \times 10^{-06}$

solar cell study is based, is indicated in each figure. Recall that in Figure IV.4, the maximum power density output for a solar cell occurred for band gap energies of approximately one to one point one electron volts. In these Figures, V.8 through V.15, we continue to assume that all of the absorbed photons contribute to the photocurrent, but the output now

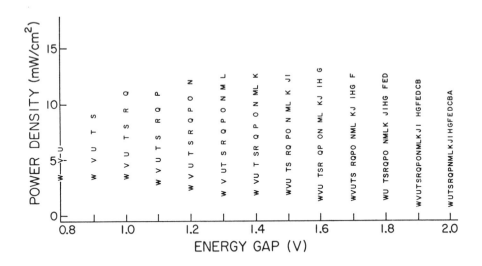

Figure V.15. The potential output power density (mW/cm²) under AM1 conditions as a function of the energy gap (eV) for solar cells with a loss factor, K', equal to ten.

** Key **
(J_S, the saturation current density, A/cm²)

A => $J_S = 1 \times 10^{-28}$	B => $J_S = 1 \times 10^{-27}$	C => $J_S = 1 \times 10^{-26}$
D => $J_S = 1 \times 10^{-25}$	E => $J_S = 1 \times 10^{-24}$	F => $J_S = 1 \times 10^{-23}$
G => $J_S = 1 \times 10^{-22}$	H => $J_S = 1 \times 10^{-21}$	I => $J_S = 1 \times 10^{-20}$
J => $J_S = 1 \times 10^{-19}$	K => $J_S = 1 \times 10^{-18}$	L => $J_S = 1 \times 10^{-17}$
M => $J_S = 1 \times 10^{-16}$	N => $J_S = 1 \times 10^{-15}$	O => $J_S = 1 \times 10^{-14}$
P => $J_S = 1 \times 10^{-13}$	Q => $J_S = 1 \times 10^{-12}$	R => $J_S = 1 \times 10^{-11}$
S => $J_S = 1 \times 10^{-10}$	T => $J_S = 1 \times 10^{-09}$	U => $J_S = 1 \times 10^{-08}$
V => $J_S = 1 \times 10^{-07}$	W => $J_S = 1 \times 10^{-06}$	

also depends on internal losses, due to the series resistance and to the effects of the counter current, resulting from the optically generated photovoltage. As a result of these considerations, the maximum power densities are found to occur for semiconductors with energy gaps between 1.3 and 1.6 eV. Additionally, the inclusion of saturation current, forward biasing effects and internal resistance losses have reduced the available power densities. For example, under AMO conditions with K' equal to 0.25 we see that for an 0.8 eV gap material the available power density, at a saturation current density of 10^{-08} amperes/cm^2, is 20 mW/cm^2; whereas in Figure IV.4 we found a value of 56.7 mW/cm^2. Similarly, a 1.4 eV energy gap material under AMO conditions yields a potential 34.2 mW/cm^2 if we include diode photobiasing and internal losses; whereas Figure IV.4 predicts 55 mW/cm^2.

Clearly, the loss factor plays a major role in the potential performance of these solar cells. The estimated maximum power density performance of a semiconductor with an energy gap of 1.1 eV under AM1 conditions is 25 mW/cm^2 when K' is zero, but when K' reaches 10, the maximum obtainable theoretical power density is only 12 mW/cm^2; a value less than half of the K' equal zero value. The operational value of the saturation current density is even more critical. Consider, again, a semiconductor with an energy gap of 1.1 eV (essentially that of silicon). For a saturation current density of 0.1 picoamperes/cm^2, the potential output power density is approximately 25 mW/cm^2 (corresponding to an efficiency of 23%), but for a saturation current density of one microamperes/cm^2, the delivered power density is only approximately eight mW/cm^2 (corresponding to an efficiency of only 7.5%).

In the following chapter we will investigate the relationships between manufacturing techniques, loss factor, and saturation current; making use of the specific semiconductors we have selected as examples. That K' and J_S are related is clear from the examination of Equations V.31 and V.32. We will see in Chapter VI that the value of the diode resistance, r_D, depends strongly on the design and construction of a solar cell, and that the overall performance is, necessarily, a strong function of the photocurrent. It is clear that we cannot attain, in a physically practical solar cell, a loss factor of zero. Equally clear is the fact that the smaller the value of the loss factor, the higher the performance.

While valuable in assessing the power density performance of solar cells, Figures V.8 through V.15 are not the complete picture. They assume that all of the electron-hole pairs generated by absorption of photons are collected and, therefore, contribute to the photocurrent. From our previous discussions in Chapters III and IV is clear that the bulk and surface lifetimes contribute significantly to a reduction in the photocurrent. Without considering specific examples, it is impossible to judge the total effect of recombination on the photocurrent [66]. In keeping with the approach of this chapter, we can, however, estimate the performance of solar cells wherein the photocurrent is a percentage of that value theoretically obtainable, as suggested in Figure IV.4.

The following Tables, V.2 and V.3, were prepared to show the effect on operating diode junction voltage for maximum voltage, V_D', and maximum delivered power density, P/A_D, of a variation in photocurrent. In these tables the photocurrent is considered to be: first, 100% of the ideal maximum photocurrent as provided in Figure IV.4; second, 90% of the ideal maximum; third, 80% of the ideal maximum; and fourth, a poor 50% of the ideal value. In keeping with our earlier discussion, a range of energy gaps, ranging from 1.0 eV to 1.8 eV in value, is selected. The saturation currents used for each energy gap are also given in these tables, using values three times higher than the minimum values of Table V.1. This adjustment makes the results provided in Tables V.2 and V.3 considerably more realistic than Figures V.8 through V.15 and in line with the practical impurity concentrations selected in the analyses of the following chapters. A non-zero value of loss factor is selected in order to make the situation more realistic. The value for K' is 0.25, which, while being non-zero, is of a reasonable size (see the analyses of Chapter VI). We will encounter much larger values of K' in subsequent chapters.

Study of Tables V.2 and V.3 reconfirms the earlier estimate that the maximum performance is to be expected for solar cells constructed from semiconductors with energy gaps lying between 1.3 and 1.6 eV. Furthermore, these tables provide graphic evidence that it is important to minimize the recombination losses (i.e., to maximize the collected electron-hole pairs). The relatively small values of junction voltage in Tables V.2 and V.3 are indicative of one characteristic of solar cells. The solar cell is typically a high current-low voltage source of direct current electricity. Therefore, if solar cells are to be used as power sources in electronic systems, it is necessary to wire them in series, if one wishes to operate with any reasonable magnitude of voltage. For example, if we utilize a semiconductor with an energy gap of 1.1 eV and operate under AM1 illumination conditions with 90% of the electron-hole pairs contributing to the photocurrent, then a series combination of in excess of 20 solar cells would be required to charge a 12 volt battery.

Table V.2

The potential delivered power density, P/A_D, (mW/cm^2), junction voltage, V_D', (V) and maximum operating efficiency, η_S, (%) for solar cells as a function of energy gap, E_g, (eV), saturation current density, J_S, (A/cm^2), and photocurrent density, J_{ph}, (mA/cm^2). Computed for K' = 0.25 and under AMO conditions

E_g	J_S	100% J_{ph}	V_D' (100% J_{ph})	P/A_D (100% J_{ph})	η_S (100% J_{ph})	V_D' (90% J_{ph})	P/A_D (90% J_{ph})
1.0	1.40×10^{-11}	58.98	0.5002	24.94	18.43	0.4976	22.32
1.1	3.00×10^{-13}	53.59	0.5937	27.16	20.07	0.5910	24.35
1.2	6.30×10^{-15}	48.51	0.6875	28.66	21.18	0.6849	25.70
1.3	1.32×10^{-16}	43.67	0.7816	29.50	21.80	0.7790	26.45
1.4	2.79×10^{-18}	39.30	0.8758	29.87	22.08	0.8732	26.80
1.5	6.00×10^{-20}	35.34	0.9699	29.85	22.06	0.9672	26.79
1.6	1.23×10^{-21}	31.17	1.0649	28.99	21.43	1.0622	26.02
1.7	2.58×10^{-23}	28.16	1.1601	28.60	21.14	1.1574	25.68
1.8	5.40×10^{-25}	25.14	1.2552	27.69	20.46	1.2525	24.87

E_g	η_S (90% J_{ph})	V_D' (80% J_{ph})	P/A_D (80% J_{ph})	η_S (80% J_{ph})	V_D' (50% J_{ph})	P/A_D (50% J_{ph})	η_S (50% J_{ph})
1.0	16.50	0.4947	19.72	14.57	0.4831	12.02	08.88
1.1	18.00	0.5882	21.51	15.90	0.5765	13.17	09.73
1.2	18.99	0.6819	22.24	16.44	0.6702	13.96	10.32
1.3	19.55	0.7760	23.43	17.31	0.7642	14.41	10.65
1.4	19.81	0.8702	23.74	17.55	0.8584	14.63	10.81
1.5	19.80	0.9643	23.74	17.55	0.9524	14.65	10.83
1.6	19.23	1.0593	23.07	17.05	1.0474	14.25	10.53
1.7	18.98	1.1545	22.77	16.83	1.1426	14.08	10.41
1.8	18.38	1.2496	22.09	16.33	1.2377	13.65	10.09

Table V.3

The potential delivered power density, P/A_D, (mW/cm^2), junction voltage, V_D', (V) and maximum operating efficeincy, η_S, (%) for solar cells as a function of energy gap, E_g, (eV), saturation current density, J_S, (A/cm^2), and photocurrent density, J_{ph}, (mA/cm^2). Computed for K' = 0.25 and under AM1 conditions

E_g	J_S	100% J_{ph}	V_D' (100% J_{ph})	P/A_D (100% J_{ph})	η_S (100% J_{ph})	V_D' (90% J_{ph})	P/A_D (90% J_{ph})
1.0	1.40×10^{-11}	48.03	0.4952	20.09	18.78	0.4926	17.98
1.1	3.00×10^{-13}	42.96	0.5882	21.56	20.15	0.5856	19.31
1.2	6.30×10^{-15}	38.27	0.6816	22.41	20.95	0.6789	20.09
1.3	1.32×10^{-16}	34.14	0.7754	22.87	21.37	0.7728	20.51
1.4	2.79×10^{-18}	31.95	0.8706	24.14	22.56	0.8680	21.66
1.5	6.00×10^{-20}	27.87	0.9638	23.39	21.85	0.9612	20.99
1.6	1.23×10^{-21}	24.93	1.0592	23.06	21.55	1.0565	20.70
1.7	2.58×10^{-23}	22.64	1.1546	22.89	21.39	1.1519	20.55
1.8	5.40×10^{-25}	20.85	1.2505	22.87	21.38	1.2478	20.35

E_g	η_S (90% J_{ph})	V_D' (80% J_{ph})	P/A_D (80% J_{ph})	η_S (80% J_{ph})	V_D' (50% J_{ph})	P/A_D (50% J_{ph})	η_S (50% J_{ph})
1.0	16.81	0.4896	15.88	14.84	0.4781	09.68	09.05
1.1	18.05	0.5827	17.08	15.96	0.5710	10.45	09.76
1.2	18.77	0.6760	17.77	16.61	0.6643	10.91	10.20
1.3	19.77	0.7698	18.16	16.97	0.7581	11.17	10.44
1.4	20.24	0.8650	19.18	17.93	0.8532	11.82	11.05
1.5	19.62	0.9583	18.61	17.39	0.9464	11.48	10.73
1.6	19.35	1.0535	18.34	17.14	1.0416	11.33	10.59
1.7	19.20	1.1489	18.21	17.02	1.1370	11.27	10.53
1.8	19.10	1.2448	18.11	16.97	1.2329	11.22	10.50

References

1 S, M. Sze, <u>Physics of Semiconductor Devices</u>, Wiley-Interscience, New York, 1981, p. 35 and 38.
2 H. Dember, in Physiche Zeitung, Vol. 32, 1931, p. 554 and 856.
3 H. Dember, in Physiche Zeitung, Vol. 33, 1931, p. 209.
4 W. Van Roosbroeck, in Physics Review, Vol. 101, 1956, p. 1731.
5 J. I. Pankove, <u>Optical Processes in Semiconductors,</u> Prentice-Hall, Englewood Cliffs, NJ, 1971, p. 320-322.
6 Reference [5], p. 323-329.
7 E. J. Adirovich, et. al., in Soviet Physics, Solid State, Vol. 7, 1966, p. 2946.
8 B. Goldstein and L. Pensak, in Journal of Applied Physics, Vol. 30, 1959, p. 155.
9 L. W. Jones and R. C. Moon, IEEE Photovoltaic Specialists Conference, Phoenix, AZ, 1975.
10 Y. Marfaing and J. Chevallier, in IEEE Transactions on Electron Devices, Vol. 18, 1971, p. 465.
11 R. M. Raymond and R. E. Hayes, In Journal of Applied Physics, Vol. 48, 1957, p. 1359.
12 Spray-on techniques for the fabrication of thin film solar cells have a long history of investigation. See, for example, AD 609204 (NTIS), a report by R. R. Chamberlin and J. S. Skarnos in U. S. Air Force contract AF 33(615)-1578, 1964.
13 See, for example, a report by J. Bernard, et. al., "Partial Results on CdS and Thin Film Solar Cells: Spatial Reliability, DERTS, 51 Rue Caranon, 31 Toulouse, France, NT 102/19/70.
14 J. Monassen, G. Hodes and D. Cohen, JECS Meeting, 1977, # 332.
15 The Institute of Energy Conversion at the University of Delaware, directed by Dr. K. W. Boer, has conducted research into polycrystalline solar cells for a number of years. Publications by staff members of the IEC appear regularly in the Proceedings of the IEEE Photovoltaics Specialists' Conference and the Proceedings of the American and International Solar Energy Societies.
16 H. J. Hovel, <u>Semiconductors and Semimetals, Volume II</u>, Academic Press, San Francisco, CA., 1975, p. 195-198.
17 P. R. Zampieri, et. al., "Obtention of Polycrystalline Solar Cells from Metallurgical Grade Silicon", in the Proceedings of the 8th. Alternative Energy Sources Conference in Miami, FL., 1987, p. 639.

18 W. Shockley, Electrons and Holes in Semiconductors, Van Nostrand, New York, 1950, Chap. 12.
19 R. Hall, in Physics Review, Vol. 83, 1951, p. 228.
20 W. Shockley and T. Read, in Physics Review, Vol. 87, 1952, p. 835.
21 J. L. Moll, Physics of Semiconductors, McGraw-Hill, New York, 1964, p. 119.
22 C. T. Sah, R. N. Noyce and W. Shockley, in Proceedings of IRE, Vol. 45, Sept. 1957.
23 H. J. Hovel and J. M. Woodall, Conference Record of the 10th. IEEE Photovoltaic Specialists' Conference, Nov. 1973, p. 25.
24 A. G. Milnes and D. L. Feucht, Heterojunctions and Metal-Semiconductor Junctions, Academic Press, New York, 1972.
25 Reference [16], p. 134.
26 G. H. Parker, "Tunneling in Schottky Barriers", Thesis, California Institute of Technology, 1969.
27 H. J. Hovel, IBM Research Report RC 2786, 1970.
28 H. J. Hovel and A. G. Milnes, in IEEE Transactions on Electron Devices, Vol. 16, 1969, p. 766.
29 D. K. Jadus and D. L. Feucht, in IEEE Transactions on Electron Devices, Vol. 16, 1969, p. 102.
30 W. P. Dumke, J. M. Woodall and V. L. Rideout, in Solid State Electronics, Vol. 15, 1972, p. 1339.
31 Electronics, 29 May 1975, p. 41.
32 D. A. Cusano, in Solid-State Electronics, Vol. 6, 1963, p. 217.
33 F. Buch, A. L. Fahrebruch and R. H. Bube, in Journal of Applied Physics, Vol. 48, 1977, p. 1596.
34 T. S. Te Velde, in Solid State Electronics, Vol. 16, 1973, p. 1305.
35 M. K. Maikherjee, et. al., in Journal of Applied Physics, Vol. 48, 1977, p. 1538.
36 M. Bettini, et. al., in Journal of Applied Physics, Vol. 48, 1977, p. 1603.
37 A. M. Cowley and S. M. Sze, in Journal of Applied Physics, Vol. 36, 1965, p. 3212.
38 C. R. Crowell and S. M. Sze, in Solid State Electronics, Vol. 9, 1966, p. 1035.
39 C. R. Crowell, in Solid State Electronics, Vol. 8, 1965, p. 395.
40 D. C. Scharfetter, in Solid State Electronics, Vol. 8, 1965, p. 299.
41 E. J. Charlson and J. C. Lein, in Journal of Applied Physics, Vol. 46, 1975, p. 3982.

42 A. Y. C. Yu and C. A. Mead, in Solid State Electronics, Vol. 13, 1970, p. 97.

43 M. A. Green, F. D. King and J. Shewchun, in Solid State Electronics, Vol. 17, 1974, p. 551.

44 D. L. Pulfrey and R. F. McQuat, in Applied Physics Letters, Vol. 24, 1974, p. 167.

45 J. Shewchun, R. Singh and M. A. Green, in Journal of Applied Physics, Vol. 48, 1977, p. 765.

46 G. C. Salter and R. E. Thomas, in Solid State Electronics, Vol. 20, 1977, p. 95.

47 R. K. Jain, et. al., in Journal of Applied Physics, Vol. 48, 1977, p. 1543.

48 J. G. McCaldin, T. C. McGill and C. A. Mead, in Physics Review Letters, Vol. 36, 1976, p. 67.

49 J. E. Roe, in Journal of Vacuum Science and Technology, Vol. 13, 1976, p. 798.

50 G. Louie, J. R. Cheligorsky and M. L. Cohen, in Journal of Vacuum Science and Technology, Vol. 13, 1976, p. 790.

51 C. Lanza and H. J. Hovel, in IEEE Transactions on Electron Devices, Vol. 24, 1977, p. 392.

52 The value of n used in Schottky barriers is not precisely unity, but is a function of the image force lowering (see R. C. Neville and C. A. Mead, in Journal of Applied Physics, Vol. 41, 1970, p. 3795). By apportioning voltage drops between the oxide and semiconductor portions of a mos diode an equivalent derivation for n can be obtained, F. Ordung, et. al., in Journal of Applied Physics, Vol. 50, 1980, p. 450.

53 S. R. Dhariwal, L. S. Kothari and S. C. Jain, in IEEE Transactions on Electron Devices, Vol. 23, 1976, p. 504.

54 J. E. Parrott, in IEEE Transactions on Electron Devices, Vol. 21, 1974, p. 89.

55 J. G. Fossan and F. A. Lindholm, in IEEE Transactions on Electron Devices, Vol. 24, 1977, p. 325.

56 P. K. Dubey and V. V. Paranjape, in Journal of Applied Physics, Vol. 48, 1977, p. 324.

57 M. Wolf and H. Rauschenback, Advanced Energy Conversion 3, 1963, p. 455.

58 M. Wolf, in Energy Conversion, III, 1971, p. 63.

59 J. L. Loferski, in Journal of Applied Physics, Vol. 27, 1956, p. 277.

60   E. S. Rittner, in Journal of Energy, Vol. 1, 1977, p. 9.
61   F. A. Lindholm and C. T. Sah, in IEEE Transactions on Electron Devices, Vol. 24, 1977, p. 299.
62   J. R. Hauser and P. M. Dunbar, in IEEE Transactions on Electron Devices, Vol. 24, 1977, p. 305.
63   R. C. Neville, in Proceedings of the Annual Meeting of the American section of the International Solar Energy Society, Denver, CO.,1978.
64   B. Hoeneisen and R. C. Neville, 3rd Miami International Conference on Alternative Energy Sources, Miamı, FL., 1980.
65   T. U. Townsend, S. A. Klein and W. A. Beckman, in Proceedings of the 1989 Annual Conference of the American Solar Energy Society, Denver, CO., 1989, p. 206.
66   In this work I have not been very detailed in considering charge carrier collection mechanisms, owing to lack of space. This topic can be quite complex, as is obvious from the discussion in this work. The reader is referred to: H. C. Card, In Journal of Applied Physics, Vol. 47, 1976, p. 4964; A DeVos and H. J. Pauwels, in IEEE Transactions on Electron Devices, Vol. 24, 1977, p. 393; and P. F. Ordung, A Adibi, D. Heald, R. Neville, and J. Skalnik, Photovoltaic Conference, Luxembourg, 1977.

CHAPTER VI: SOLAR CELL CONFIGURATION AND PERFORMANCE

Introduction

The nature of the solar spectrum was described in Chapter II. In Chapter IV we discussed the maximum energy that was potentially available for conversion to useful electrical energy within a solar cell--as a function of the energy gap of the semiconductor employed. In Chapter V we discussed the effects, on solar cell performance, of solar cell internal series resistance and internal junction biasing. This was done in a general way by assigning values of saturation current, J_S, and diode loss factor, K', to solar cells as a function of the semiconductor energy gap. The later portions of Chapters IV and V also considered the effects of bulk and surface recombination on the potentially available photocurrent and output power from solar cells.

The characteristics of pn and heterojunction solar cells depend primarily on the properties of the semiconductor(s) employed. The most important semiconductor property in these cases is, clearly, the width of the energy gap. When Schottky junction or mos solar cells are employed, the barrier height, which is a function of the semiconductor and metal employed, is of prime importance. For all types of solar cells the charge carrier mobility and lifetime play major roles in overall performance. Both the collection/separation of electron-hole pairs (and, hence, the photocurrent) and the internal series resistance depend on the design philosophy of the solar cell and upon the techniques used in its construction.

In this chapter we will consider various solar cell optical orientations and construction techniques. With specific semiconductors as examples, we will determine solar cell series resistance, saturation current, optimum junction voltage, maximum power density delivered to the external load, and the value of the optimized external load. This will be done assuming room temperature (300°C) operation. The net result will be a set of design relationships, indicating the potential output for solar cells of various types, as well as providing the reader with an appreciation of the various design trade-offs.

In keeping with the discussions of Chapters IV and V, let us select semiconductors with energy gaps lying between 1.0 and 1.8 electron volts. From these earlier discussions, we shall consider only single crystal semiconductors[*]. As earlier, let us select a number of semiconductors as specific examples, indicated in Table VI.1.

Table VI.1

Example semiconductors

Semiconductor	Symbol	Energy Gap at 300°C (eV)
Silicon	Si	1.106
Indium Phosphide	InP	1.35
Gallium Arsenide	GaAs	1.42
Cadmium Telluride	CdTe	1.50
Aluminum Antimonide	AlSb	1.58
Cadmium Selenide	CdSe	1.70

The preceding table of semiconductors does not exhaust the list of possibilities. It does, however, span the range of energy band gap widths which is considered to be most promising and provides a distribution of single element and compound semiconductors with a broad distribution of available technology and potential performance and problems. Clearly any one of these materials is also a potential semiconductor in a heterojunction or Schottky barrier solar cell. For example, GaAs has been

---

[*] Besides the additional complexity involved in the analysis of polycrystalline based solar cells, single crystal based solar cells exhibit a higher operating efficiency. We will, in a later chapter, consider amorphous and polycrystalline semiconductor based solar cells.

used for some time as the substrate semiconductor in a GaAs-AlGaAs heterojunction solar cell [1-3] and the other materials have been used as well [4-8].

Let us now proceed to a discussion of design philosophy and construction technology as applied to solar cells made from these example materials.

Optical Orientation

Light can enter a solar cell from a number of directions. Consider a pn junction or heterojunction solar cell. In Figure VI.1 we display three general configurations of solar cell-light interaction. In the top illustration of Figure VI.1 the solar cell is securely mounted on a heat sink. This heat sink also acts as an electrical contact to the larger of the two regions of the solar cell, the substrate. The light enters the solar cell via the other region, the "front layer". Metallic contacts to the upper region mask portions of the solar cell and thereby reduce the effective optically absorbing area. The middle illustration of Figure VI.1 presents a second configuration, known as the vertical configuration. Here, the light enters the solar cell unhampered by the presence of metallic contacts. The electrical contacts are at the non-illuminated ends of the solar cell and the junction is both parallel to the incoming light and vertical, in orientation, to the heat sink. Note that care must be taken to insure that the heat sink does not short out the pn junction. This limits the effectiveness of this configuration of solar cell.

There are two types of inverted configuration solar cell, displayed in the bottom illustration of Figure VI.1. In one case, there are electrical contacts on the illuminated surface, a situation which restricts the photon input. In the other inverted configuration the electrical contacts are on the down side and do not interfere with the passage of photons. In both instances the inverted configuration solar cell has a heat sink which is on the non-illuminated side and is positioned very close to the junction of the solar cell. This positioning assists in maintaining junctions at a low temperature and, as we will see, enhances the operating efficiency. Note that, if the electrical contacts to both solar cell regions (substrate and "front layer") are on the lower or heat sink side, it is necessary to configure the electrical contacts and heat sink in order to avoid shorting the junction.

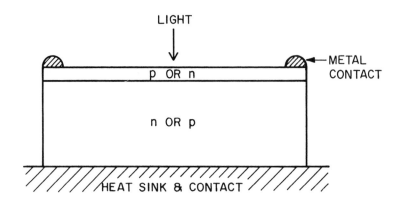

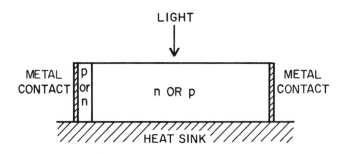

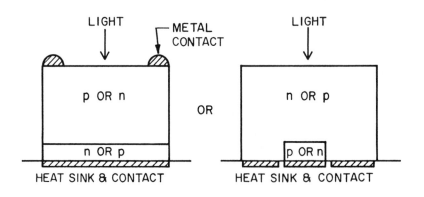

Figure VI.1. PN and heterojunction solar cell-light configurations.

Figure VI.1 is drawn for a pn or heterojunction cell with two semiconductor regions per solar cell. If we replace the thinner of the two regions by either a thin metal or thin metal-oxide layer, the same optical configuration classifications can be seen to apply to Schottky and mos cells. For standard configurations, it is clear that care is required to form metal layers thin enough to pass light and thick enough to carry current.

Before any numerical computations are made, consider some of the advantages and disadvantages of each configuration of solar cell. The standard configuration (top illustration of Figure VI.1) solar cell is simple to construct and has good contact from solar cell to heat sink. This last item is important for all semiconductor solar cells and, most particularly so, in terms of those systems where optical concentration is employed to increase the intensity of the photon flux. Recall, from Chapter V, that the saturation current density for the four types of junctions (pn junction, heterojunction, Schottky junction and mos junction--Equations V.2, V.17, V.19 and V.21). In every case, for every junction type, the saturation current density rises sharply with increasing temperature. From Equations V.31 through V.33 an increase in saturation current density results in a decrease in the solar cell delivered power density. Since practical solar cell efficiencies are rarely in excess of 25% this means that some 75%, or more, of the solar energy ($\sim$80 mW/cm^2 under AM1 conditions) is converted into heat within the semiconductor. This heat must be removed from the solar cell to avoid further increase in the operating temperature.

The standard configuration solar cell has a number of drawbacks as well. Once the electron-hole pairs are generated, they must be split into their component electrons and holes and the individual charge carriers must be collected by the junction. In the upper, or "front layer", the charge carriers must flow laterally to the electrical contacts. This upper region is a very thin (optically transparent) conductor (or semiconductor). Because this upper region is of low lifetime[*], it acts, in part, as a "dead layer" (see Chapter IV). To minimize the loss of photon generated charge carriers in this "dead layer" we must make it thin. This thinness creates high internal resistance with attendant decreased power performance. This drawback is particularly severe in Schottky barrier and mos solar cells.

---

[*] This low lifetime is the result of the construction techniques employed. We will consider these techniques later in this work.

PN junction solar cells also show ill effects from this phenomenon, but heterojunction solar cells avoid the problem by constructing the "front layer" of the solar cell of a semiconductor which is optically transparent. This advantage is, in turn, counterbalanced by excess recombination at the heterojunction proper and by a more complicated fabrication technology. Another major drawback in standard configuration solar cells is the presence, on the illuminated surface, of the thick metal electrical contacts (the thickness is required to minimize the series resistance). This results in a reduction of the produced photocurrent.

The vertical configuration solar cell can be very well heat sunk, has no metal on the optical surface to reflect light and reduce optical input and photocurrent, and possesses a major circuit advantage. In Chapter V the low voltage output of solar cells was discussed in connection with the need for "stacking" or series connecting solar cells in order to obtain practical operating voltages (12 to 24 volts). The vertical configuration solar cell is particularly suited to stacking. The principal disadvantages of vertical solar cells are: (1) the need to passivate (protect) the large illuminated surface (which, in this configuration contains one "end" of the junction) from ambient conditions[*] and (2) the large number of electron-hole pairs generated at some considerable distance from the junction. When remote carrier generation occurs, the likelihood of charge carrier separation by the junction is reduced, decreasing the photocurrent. Both of these problems arise from the methods required for construction of a vertical solar cell.

Referring to Figure VI.2, note that a vertical solar cell is often fabricated as one would construct a standard configuration solar cell--by taking a large slice of semiconductor[#] and placing a junction parallel to one side. A standard configuration solar cell would next be mounted on a heat sink and the "front layer" electrical contacts applied. In a vertical solar cell, the large slice is now diced into a large number of small cells (see Figure VI.2c). The small solar cells are rotated through 90° and

---

[*] The incoming photons are absorbed close to the illuminated surface and the passivation is required to minimize surface recombination and saturation current, and so to maximize the photocurrent.

[#] Presently, the slices of semiconductor are circular with diameters of some 15 to 20 centimeters. Other shapes and sizes will be discussed in a later chapter.

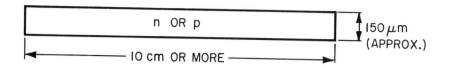

(a) A large slice, rectangular or round, of semiconductor

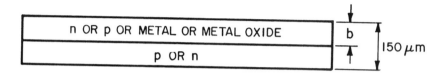

(b) Slice of semiconductor after junction fabrication

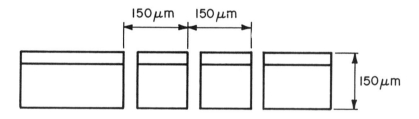

(c) Separation of slice into individual cells, two of which are detailed

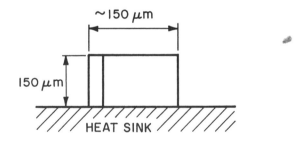

(d) Attachment of vertical solar cell to heat sink

Figure VI.2. A brief outline of the fabrication process for a vertical configuration solar cell. In Figure VI.2b dimension b is between one and 10 μm for a heterojunction solar cell, less than one for a pn junction and on the order of 500 Å for Schottky or mos solar cells.

mounted upon a heat sink--very carefully, to avoid shorting the junction. The passivation and antireflection coating must now be applied to the illuminated surface. The size of a completed vertical configuration solar cell is dictated, in part, by the discussion in Chapter IV. There it was noted that a solar cell need only be some 150 μm thick, in the direction parallel to the incoming photons, to absorb in excess of 95% of the incoming optical energy. Any increase in thickness beyond this value increases construction cost with little or no corresponding increase in electrical power produced. Any decrease in thickness makes for substantially increased handling difficulty during fabrication, as a result of the fragile nature of semiconductors. This increases the cost of fabrication while diminishing the output power. The length of a vertical solar cell, in a direction at right angles to the photon flux, is limited by the mobility and lifetime of the minority charge carriers. Direct energy gap semiconductors have low minority carrier lifetimes and, as a result, hole-electron pairs generated far from the junction have a low probability of separation and collection.

The inverted solar cell configurations in the lower illustrations of Figure VI.1 have the advantage that no "dead layer" exists near the illuminated surface. The two types of inverted configurations presented represent two approaches to solving circuit, optical and thermal difficulties. In one, the contacts to the bulk (or substrate) region are on the illuminated surfaces. This reduces the optically absorbing area and increases the difficulty encountered in electrically "stacking" the solar cells to achieve higher system voltage, but does improve the heat sinking of the solar cell. The other approach places both sets of contacts on the down side, reducing "stacking" and photon collection problems, but making assembly more expensive and increasing the difficulty of adequately heat-sinking the solar cell.

In this chapter we will examine both standard and inverted configurations in detail, making use of our six example semiconductors. The results may be extended quite simply to vertical solar cells, provided that the semiconductor has sufficient lifetime and the charge carriers possess adequate mobility to enable them to reach the electrical contacts. Study of the lifetime and mobility data in Appendix B indicates that silicon is, by far, the most promising material for inverted and vertical solar cells. For an second examination of inverted solar cells, the reader is referred to Lammert and Schwartz [9].

Device Design - Minority Carrier Collection

When fabricating pn junction or heterojunction solar cells of any configuration, the bulk or substrate region of the semiconductor device is lightly or moderately doped with impurities while the second region (the top or "front layer" or illuminated region in a standard configuration cell, the bottom or non-illuminated region in an inverted solar cell and a thin, side-illuminated region in a vertical solar cell) is heavily doped. This heavy concentration of impurities is required to reduce the series resistance of the solar cell and to permit ohmic contact to be made to this region. Note that the large impurity concentration reduces the carrier lifetime and creates a "dead layer". This "dead layer" is not a problem in inverted or vertical configuration solar cells, but is of major concern for standard configuration solar cells. Therefore, for reasons of cost (the thicker a region of a solar cell, the greater the expense of fabrication) and, in standard solar cells, to minimize the "dead layer" width, it is desirable to keep this second region to minimum thickness.

From Chapter III we have the following expressions for the extent of the junction electric field into the p- and n-regions of a pn step junction solar cell.

$$x_n = \sqrt{\{2\epsilon V_D/qN_D\}} \ \sqrt{\{1/(1 + N_D/N_A)\}} \ ,$$

and $\hspace{12cm}$ (VI.1)

$$x_p = (N_D/N_A)x_n \ ,$$

where x_n is the space charge region expansion distance into the n-region from the junction (taken as $x = 0$), x_p is the expansion into the p-region, ϵ is the permittivity of the semiconductor, N_D is the impurity concentration in the n-region, and N_A is the impurity concentration in the p-region. The quantity, V_D, is the net voltage across the junction and is given by:

$$V_D = V_B - V_p \ ,$$ $\hspace{10cm}$ (VI.2)

where V_p is the photovoltage and V_B is the built-in voltage across the step junction and is determined from:

$$V_B = (kT/q)\ln\{N_A N_D / n_i^2\} \; , \tag{VI.3}$$

where k is Boltzmann's constant, T is the absolute temperature of the junction and n_i is the intrinsic carrier concentration[*].

Consider a pn step junction with $N_L \gg N_S^{\#}$. In this situation, the value of X_L', the space charge (or electric field) width in the "front layer" will be zero (see Equation VI.1). In a heterojunction solar cell this is also effectively the case since the difference in energy gap widths for the two materials forces the electric field to expand only into the semiconductor with the lower energy gap. In metal/semiconductor or metal-oxide/semiconductor junctions the electric field also expands principally into the semiconductor substrate layer. The value of X_S, the space charge width in the substrate, is, at most, X_S', where:

$$X_S' = \sqrt{\{2\epsilon E_g / q n_i\}} \; . \tag{VI.4}$$

Here, X_S' has been calculated assuming a built-in voltage equal to the band gap width of the semiconductor and a carrier concentration equal to the intrinsic value, n_i.

Table VI.2 presents a value of X_S' for the six sample semiconductors.

Table VI.2

Maximum extent of the junction electric field (the depletion or space charge layer), X_S'

Semiconductor	Si	InP	GaAs	CdTe	AlSb	CdSe
X_S' (cm)	0.031	0.891	2.08	3.45	19.66	53.1

---

[*] From Chapter V we expect a positive value for the photovoltage when a solar cell is converting solar to electrical energy.

[#] Here N_L refers to the "front layer" (which is the illuminated layer in standard configuration solar cells) and N_S refers to the substrate region (the illuminated region in an inverted solar cell). One of these two regions will be n-type and the other will be p-type.

Note that, under the assumptions made, this electric field (the depletion region) extends into the semiconductor in mos and Schottky junctions, into the smaller energy gap semiconductor in heterojunctions, and into the lightly doped substrate region in pn step junctions. Recall, also, that we are considering solar cells having a practical total thickness of 150 μm[*]. The electric field width as given in Table VI.2 is sufficient to completely fill the substrate region for all standard, vertical and inverted configuration solar cells. However, Table VI.2 is based on three assumptions; none of which can be fully realized in a realistic solar cell. The first assumption is that the substrate is intrinsic. In practice, the technology of the early 1990s cannot meet this requirement. A reasonable technology limited minimum value for N_S is on the order of $10^{14}/cm^3$. Moreover, in order to lower the saturation leakage current of the solar cell diode, N_S generally needs to be on the order of $10^{15}/cm^3$. The second assumption is that the junction voltage is only the built-in voltage of the step junction and is equal to the energy gap of the pn junction (or the energy gap of the substrate semiconductor in a heterojunction, mos or Schottky barrier solar cell. In practice, the junction voltage is always less than the energy gap width (see Equation VI.3). A third factor that must be included in any discussion of the substrate depletion layer width is that the solar cell becomes forward biased, owing to the separation/collection of the hole-electron pairs. As a result, the voltage, V_D, in Equation VI.2 is reduced, rapidly, towards some fraction of a volt[#].

To estimate the junction voltage that actually exists in a Schottky barrier solar cell, consider the maximum built-in potential, ϕ_{Bo}, for a Schottky barrier solar cell. Given this value, we can substitute for V_B in Equation VI.2, and so determine the depletion layer thickness, for built-in voltage alone (i.e., for short circuit conditions), X_{SS}:

$$X_{SS} = \sqrt{\{2\epsilon\phi_{Bo}/qN_S\}} \ . \tag{VI.5}$$

---

[*] This thickness is almost entirely due to the semiconductor in mos and Schottky junctions, the smaller energy gap semiconductor in heterojunctions and the lightly doped semiconductor substrate in pn step junctions.

[#] Examples, using the six sample semiconductors, to be worked later in this chapter and in subsequent chapters, will demonstrate the exact values to be expected.

Table VI.3 presents values of Schottky barrier voltages for the six example semiconductors under consideration and selected metals.

Table VI.3

Metal-semiconductor barrier energies, ϕ_{Bo}, in eV for the six example semiconductors (10-15)

Semiconductor	Si	InP	GaAs	CdTe	AlSb	CdSe
Metal		n-type semiconductor				
Pt	0.90	0.60	0.84	0.76	0.60	0.37
Au	0.80	0.52	0.90	0.71	0.51	0.49
Ag	0.78	0.54	0.88	0.81	0.52	0.43
Al	0.72	0.51	0.80	0.76	------	0.36
Pd	0.81	0.55	0.85	0.74	0.55	0.42
		p-type semiconductor				
Pt	-----	0.74	0.48	0.75	0.58	-----
Au	0.34	0.76	0.42	0.73	0.55	-----
Al	0.58	-----	0.67	0.54	-----	-----
Ti	0.61	0.74	0.53	-----	0.53	-----
Cu	0.46	-----	-----	-----	0.44	-----

Observe that the barrier values listed in Table VI.3 depend upon the metal, the semiconductor, and upon the semiconductor type. The values provided are all less than one volt and are less than the energy gap.

The mos junction is less well understood than the Schottky junction. As of 1993, this solar cell type has been constructed primarily on silicon because of the ease in fabricating the required thin oxide layer (see Chapter V) with this semiconductor. For this type of barrier, values of barrier energy of 0.85 (aluminum-silicon dioxide-on p-type silicon) and 0.67 (chromium-silica-on p-type silicon) have been reported [16]. Data from other sources [17-20] for mos barriers on both silicon and gallium arsenide substrates indicate similar values. Note that reduced leakage currents, resulting from the oxide layer, make these devices promising; even if, as yet, insufficiently understood.

The purpose of this chapter is to estimate the efficiencies of solar cells of "practical" construction. To this end, let us consider Schottky and mos junctions under a single heading (Schottky) and select the "best" barrier energies from Table VI.3 and the literature. Then, the maximum barrier energies to be encountered, in practice, for Schottky junctions may be taken as those in Table VI.4.

Table VI.4

Practical maximum Schottky junction barrier energies (eV) and the specific metal employed for the six example semiconductors

Semiconductors	Si	InP	GaAs	CdTe	AlSb	CdSe
			n-type semiconductor			
Energy	0.90	0.60	0.90	0.81	0.60	0.49
Metal	Pt	Pt	Au	Ag	Pt	Au
			p-type semiconductor			
Energy	0.95	0.76	0.67	0.75	0.58	*
Metal	Hf	Au	Al	Pt	Pt	*

To compute pn junction built-in potentials we utilize Equation VI.3. As stated earlier, the minimum potential value for substrate impurity concentration, N_S, is an impurity concentration of $10^{14}/cm^3$. The value for the "front layer" concentration, N_L, depends, in part, on whether this region is introduced by diffusion or ion implantation. An effective value for N_L of $5 \times 10^{19}/cm^3$ is commonly encountered. Combining these values, with those for n_i^2 at 300°K from Chapter III, we have for the built-in voltage, the values of Table VI.5.

It is difficult to predict the effective barrier potential of a heterojunction. A rough estimate may be effected by observing the open circuit voltage of a heterojunction solar cell. From Sreedhar [21] and Sahi

---

* In Chapter III we discussed the fact that p-type CdSe has not been practically fabricated to date. Thus, neither metal-semiconductor (Schottky) junctions on p-type CdSe nor CdSe pn junctions are feasible. It is possible to construct heterojunction devices using n-type CdSe as one side of the junction. The values given in Table VI.6 are estimates for this case.

Table VI.5

Estimated practical maximum built-in voltages for pn junctions constructed from the example semiconductors (in volts)

Semiconductor	Si	InP	GaAs	CdTe	AlSb	CdSe
V_B	0.76	1.08	1.18	1.23	1.41	*

and Milnes [22] some values of open circuit heterojunction solar cell voltages are: (1) n-type GaP on p-type Si, 0.67 V; (2) n-type GaP on p-type GaAs, 0.82 V; (3) p-type GaP on n-type GaAs, 1.05 V; and (4) n-type ZnSe on p-type GaAs, 0.925 V. Note that these values are on the order of those of Table VI.5 for pn junctions. Calculations of substrate depletion layer width using these barrier voltages lead to results similar in magnitude to those using of the results of Table VI.4 in Equation VI.5 for Schottky and Table VI.5 in Equation VI.3 for pn junctions.

For a substrate impurity concentration of $10^{14}/cm^3$ we can obtain an estimate of the substrate depletion layer width in a solar cell under short circuit conditions (the photovoltage equal to zero). These depletion widths, for the example semiconductors are given in Table VI.6.

Table VI.6

The "practical" maximum depletion layer width (in μm) in the semiconductor substrates for the six example semiconductors, as a function of the various junction types and at a temperature of 300°K

Semiconductor	Si	InP	GaAs	CdTe	AlSb	CdSe
Metal-semiconductor barrier on n-type semiconductor substrate						
	3.457	3.049	3.547	3.126	2.703	2.398
Metal-semiconductor barrier on p-typ semiconductor substrate						
	3.552	3.432	2.983	2.999	2.657	*
Step pn junction or heterojunction						
	3.192	4.093	4.036	3.860	4.129	4.21

* P-type CdSe is not available so there is no Schottky barrier on p-type CdSe, but there can be a heterojunction in an n-type semiconductor.

Observe that the depletion layer widths of Table VI.6 are not only very much smaller than the electric field widths of Table VI.2, but they are also much less than the optical absorbing thickness of the semiconductor (~150μm). If electron-hole pair separation/collection depended solely on the depletion layer width, the performance of standard and inverted configuration solar cells would be largely negated. Fortunately, there are other phenomena which can assist in the production of photocurrent. These phenomena are used to bring the optically generated carriers within range of the electric field in the depletion layer of a solar cell junction. First, consider the diffusion length in a semiconductor and to what extent it effectively extends the collecting range of the depletion layer.

Once generated by photon absorption in the bulk regions (areas of no electric field) of the solar cell, hole-electron pairs move randomly through the semiconductor. Should there be a junction in the semiconductor crystal, there will be, of course, an electric field in the vicinity of the junction. This field serves to collect electron-hole pairs and to separate them, thus producing a concentration gradient in electron-hole pairs. Now consider the p-type region of a solar cell. Electrons in this region, close to the depletion region often, randomly, move into the electric field. When this occurs, the electrons are accelerated across the junction to the n-type side. A similar process occurs, of course, to the holes randomly moving on the n-type side as they are accelerated towards the p-type side. The effect of this minority carrier removal is to create an electron concentration gradient between the bulk region on the p-type side and the edge of the depletion region. Thus, an electron within a diffusion length of the junction on the p-type side will be collected (the same goes for holes within a diffusion length of the junction on the n-type side). The diffusion length, L, is given by:

$$L = \sqrt{\{D\tau\}} , \qquad (VI.6)$$

where, from Chapter III:

$$D = \{kT/q\}\mu . \qquad (VI.7)$$

The lifetimes, τ, and mobilities, μ, for the semiconductors being used as examples in this work were discussed in Chapter III. Recall that these material properties are functions of temperature and impurity concentration. In this chapter we are considering solar cell operation at room temperature (27°C). The preceding discussion of depletion layer width used a substrate impurity concentration of $10^{14}/cm^3$ and a high impurity concentration "front layer" of 5 x $10^{19}/cm^3$. A few additional words concerning the "practicality" of these concentrations are in order. The "front layer" concentration varies with distance into the semiconductor. If the "front layer" is the result of a diffusion process, the impurity concentration at the surface is much higher than that at the junction. Typically, $N_L(x)$ will follow an error function curve [23, 24] with a surface concentration well in excess of either the conduction band or the valence band density of states (see Appendix B and [25]). If the "front layer" is the result of ion implantation, the impurity density reaches a peak at some distance into the semiconductor; the distance determined by the semiconductor, its crystal orientation, the impurity species and the energy of the implant [26]. Utilizing modern technologies, such as molecular beam epitaxy [27, 28], it is possible to keep the "front layer" concentration at approximately the density of states level, which is roughly 5 x $10^{19}/cm^3$. This impurity concentration is high enough to adversely affect the lifetime of the "front layer", but it also is high enough to support a thin "front layer" without excessive resistance.

The substrate impurity concentration must be small in order to enhance the diffusion length and the depletion layer width, but needs to be sufficiently high to reduce the bulk series resistance of the solar cell. This bulk series resistance, r_D, is given by:

$$r_D = \{l/A_D\}\{1/q\mu_{Sm}N_S\} \ , \tag{VI.8}$$

where l is the length of the substrate (generally taken to be 150 μm in this work); A_D is the junction area of the solar cell, which we assume is equal to the cross-sectional area of the substrate; μ_{Sm} is the substrate majority carrier mobility; and N_S is the impurity concentration in the substrate. In Chapter V, in connection with saturation current, we used a substrate impurity concentration of $10^{16}/cm^3$. This produced a low value of saturation current density. Earlier in this chapter we utilized a substrate impurity concentration of $10^{14}/cm^3$ because this value yields a wider depletion layer width, at the cost of increased saturation current density. In practice, a carrier concentration of approximately $10^{15}/cm^3$ provides a satisfactory balance between series resistance, diffusion length, saturation current and processing technology.

Using a value of N_S equal to $10^{15}/cm^3$ and a value of N_L equal to $5 \times 10^{19}/cm^3$, in conjunction with the mobilities and lifetime values of Appendix B, the literature and Chapter III, we have the data provided in Table VI.7. This will be used as input for the computation of the minority carrier diffusion lengths in the substrates of solar cells made from our example semiconductors.

Table VI.7

Estimated values of impurity concentration, minority carrier mobility and lifetime, as functions of semiconductor for a temperature of 300°K and the six example semiconductors [22, 29, 30]

Semiconductor	Si	InP	GaAs	CdTe	AlSb	CdSe
"Front Layer"						
Concentration (cm^{-3})				-5×10^{19}		
"front layer" mobility $(cm^2/volt\text{-}second)$						
p-type layer	135	450	1000	700	140	-----
n-type layer	80	150	100	50	180	450
"front" layer lifetime (seconds)						
p-type layer	10^{-7}	10^{-10}	10^{-10}	10^{-9}	10^{-10}	-----
n-type layer	10^{-7}	10^{-10}	10^{-10}	10^{-9}	10^{-10}	10^{-10}
Substrate						
Concentration (cm^{-3})				-1×10^{15}		
substrate mobility $(cm^2/volt\text{-}second)$						
p-type layer	1500	3500	6500	950	200	-----
n-type layer	500	600	350	90	400	600
substrate lifetime (seconds)						
p-type layer	8×10^{-5}	6×10^{-8}	6×10^{-8}	2×10^{-6}	1×10^{-7}	------
n-type layer	8×10^{-5}	3×10^{-8}	3×10^{-8}	1×10^{-7}	9×10^{-8}	1.5×10^{-9}

In a "realistic" solar cell, both the minority carrier mobilities and lifetimes may well be less than the values provided in Table VI.7, particularly if the processing involved in fabrication of the solar cell is substandard. However, the mobilities and lifetimes furnished in Table VI.7 are achievable and lead to the diffusion lengths of Table VI.8.

Table VI.8

Estimated minority carrier diffusion lengths for n- and p-type regions of solar cells employing the example semiconductors, at 300°K

Semiconductor	Si	InP	GaAs	CdTe	AlSb	CdSe
	"front layer" diffusion length (µm)					
p-type layer	5.91	0.341	0.509	1.35	0.191	-----
n-type layer	4.55	0.197	0.161	0.36	0.216	0.341
	substrate diffusion length (µm)					
p-type layer	577	145	198	436	44.7	-----
n-type layer	322	42.4	32.4	30	60	9.49

From the discussion relating to "dead layer" thickness in Chapter V, the heavily doped "front layer" region in standard or vertical configuration step junction and heterojunction solar cells should be small, with a maximum thickness under a micron. Since this region has a low lifetime (see Table VI.7), and the surface recombination velocity of such heavily doped regions is high, it is unlikely that a large percentage of carriers will be collected and separated in this region. The diffusion lengths for the "front layer" provided in Table VI.8 are, therefore, adequate. The substrate, however, is another matter. For any configuration of solar cell, the electron-hole pairs will be generated by photon absorption within some distance of the illuminated surface. From Figures IV.7 and IV.8, this distance is given in Table VI.9.

Table VI.9

The approximate depth beneath the illuminated surface at which electron-hole pair optical generation ceases (µm)

Semiconductor	Si	InP	GaAs	CdTe	AlSb	CdSe
Depth	1000	0.3	2	20	8	1

From practical considerations we have set the maximum solar cell thickness to a value of 150 µm. This results in a loss in the potentially convertible solar energy of approximately 5% for silicon based solar cells. Note that, for the other example semiconductors, the absorption occurs so rapidly that this limited thickness of solar cell has no effect. Comparing the values of optical absorption depth in Table VI.9 with the diffusion lengths presented in Table VI.8, we see that, for 150 µm thick solar cells the diffusion lengths for all six example semiconducting materials are adequate to collect all of the optically generated charge carriers for standard configuration solar cells*.

Consider inverted and vertical configuration solar cells. From our discussions in connection with Figures VI.1 and VI.2 and Tables VI.9 it is possible, in these configurations, for optical hole-electron pair generation to take place at a distance approximating 150 µm from the junction. Considering the minority carrier diffusion lengths of Table VI.8 we note that, if the substrate is p-type, then silicon, indium phosphide, gallium arsenide and cadmium telluride have the potential of collecting almost all of the hole-electron pairs. Not all, because, even in the case of silicon with its 557 µm minority carrier diffusion length, the diffusion length in any of these example semiconductors is never more than four times the 150 µm limit. For indium phosphide and gallium arsenide, the diffusion length is approximately equal to the maximum generation distance of 150 µm. In the case of n-type substrates only silicon has a sufficiently large minority carrier diffusion length, long enough to assure the collection of most of the hole-electron pairs.

There is an additional source of charge carrier loss. The illuminated surface of vertical junction configuration solar cells is formed by sectioning the original wafer (see the discussion in association with Figure VI.2). This procedure enhances the surface recombination velocity and reduces the photocurrent for these devices. Note that this problem is not so severe with inverted configuration solar cells. For this configuration of device, the fabrication process is tailored to minimize surface recombination velocity. In standard configuration solar cells the surface recombination contributes to the "dead layer" and, hence, has already been taken into consideration. Finally, note that, at the substrate contact, the surface recombination is assumed to be essentially infinity (see the discus-

---

* This, clearly, does not include those carriers generated by optical absorption within the "dead layer".

sion in Chapter III). This produces a minority concentration gradient in the vicinity of the substrate contact which funnels charge carriers in the wrong direction. Study of Figure VI.1 will demonstrate to the reader that this problem is unimportant for standard configuration solar cells and those inverted configuration solar cells with their substrate contacts on the non-illuminated surface. It is, however, of importance to vertical junction solar cells, resulting in a "dead layer" near the substrate contact and reducing overall performance for the solar cell.

There is a solution to all of these problems; a solution which has the additional advantage of reducing the substrate series resistance. Consider the energy-versus-distance diagram for the solar cell displayed in Figure VI.3.

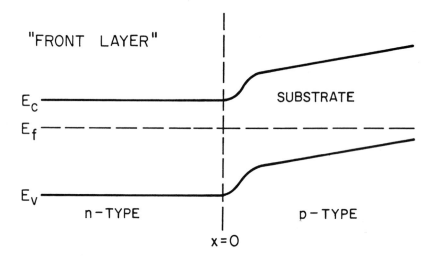

Figure VI.3. A solar cell with a variable impurity concentration in the substrate. E_c is the lower edge of the conduction band, E_F is the Fermi level and E_v is the upper edge of the valence band.

In Figure VI.3 the heavily doped "front layer" is assumed to have a constant impurity concentration of approximately $5 \times 10^{19}/cm^3$. At $x = 0$ (the junction) the substrate is relatively lightly doped (an impurity concentration of approximately $10^{19}/cm^3$), but the impurity concentration (acceptor in the example of Figure VI.3) of the substrate is increased as the distance from the junction is increased. The result is a built-in electric field which impels minority carriers towards the junction. This electric field is given by:

$$\mathcal{E}(x) = \frac{kT}{qN_S(x)} \frac{\partial N_S(x)}{\partial x} , \qquad \text{(VI.9)}$$

where $N_S(x)$ is the substrate impurity concentration. This varies from approximately $10^{14}/cm^3$ to a value on the order of 10^{17} to $10^{18}/cm^3$ (a value less than one-tenth of the effective density of states value for the substrate[*]). Note that if we desire a constant value of electric field, $\mathcal{E}$, then the substrate impurity concentration will be:

$$N_S(x) = N_S(o)exp\{q\mathcal{E}x/kT\} , \qquad \text{(VI.10)}$$

where $N_S(o)$ is the substrate impurity concentration at the junction and x is positive into the substrate[#].

Assume a substrate of 150 μm width. Then, for an electric field in the substrate of 16 volts/cm, the ratio $N_S(150)/N_S(o)$ is ten thousand. Given such a field, extending the width of the substrate, we can collect essentially all of the hole-electron pairs generated in the substrate and transport them to the edge of the depletion layer. In turn, the depletion layer separates the hole-electron pairs. As an added advantage, the graded substrate discussed here also serves to decouple the surface recombination velocity at the substrate contact [31-33].

Consider the photocurrent that may be expected in a solar cell of standard, inverted or vertical configuration. Suppose we have an inverted configuration solar cell, with a graded substrate, 95% efficient antireflection coating and 100% collection efficiency for all generated hole-electron pairs. The expected photocurrent density is that of Table VI.10.

In the case of a vertical configuration solar cell we must include additional losses from surface recombination owing to the fact that the

---

[*] Increasing the substrate impurity concentration above this value will decrease the minority carrier lifetime and mobility to such an extent as to decrease the collected photocurrent.

[#] Clearly, the same procedure can be followed when the substrate is n-type.

Table VI.10

Estimated photocurrent density (mA/cm^2) in an inverted configuration solar cell at 300° K

Semiconductor	Si	InP	GaAs	CdTe	AlSb	CdSe
AM0-conditions	44.65	41.7	37.2	35.8	28.6	26.0
AM1-conditions	36.1	31.8	28.7	27.2	21.9	20.5

surface which is illuminated has been cut from a wafer (see the discussion accompanying Figure VI.2). Assuming a reasonable surface recombination velocity of 10,000 cm/second, and using the data of Table VI.7 and Figure III.8, we can estimate that the maximum realizable photocurrent density will be some 5% or so below the photocurrents of the inverted solar cell, yielding the numbers of Table VI.11.

Table VI.11

The estimated photocurrent density (mA/cm^2) in a vertical configuration solar cell at 300°K

Semiconductor	Si	InP	GaAs	CdTe	AlSb	CdSe
AM0-conditions	42.7	39.5	35.3	33.9	27.1	24.7
AM1-conditions	34.2	30.1	27.2	25.7	20.8	19.5

The expected photocurrent from a standard configuration solar cell is still less. There are additional recombination losses due to the "dead layer" (which is a result of the heavily doped "front layer") in pn step junctions, from the interface effects in a heterojunction solar cell and from reflection effects in the case of mos and Schottky junction solar cells. In the following table the estimated photocurrent density for a standard configuration solar cell is provided. In the case of a pn step junction, the "front layer" is thin to minimize the "dead layer" thickness (keeping this layer from 0.3 to 0.6 μm thick). The "dead" layer does not completely fill the "front layer", but comprises the top third or so. Assuming that this "dead layer" is, indeed, totally dead, and utilizing Figures IV.10 and IV.11 the estimated maximum photocurrent density for standard configuration pn step junction solar cells is provided in Table VI.12. The estimated possible photocurrents for standard configuration heterojunctions are higher because the "dead layer" is not present--the semiconductor forming the "front layer" being transparent to the photons of interest. The photocurrent is also estimated for Schottky barrier standard configuration solar cells. The existence of a metal layer on the illuminated side of the solar cell drastically reduces the potential photocurrent and values indicated in Table VI.12 are, at best, estimates.

Table VI.12

The estimated photocurrent density (mA/cm^2) in a standard configuration solar cell, for pn step junction, heterojunction and Schottky junction devices at 300°K

Semiconductor	Si	InP	GaAs	CdTe	AlSb	CdSe
			heterojunction solar cells			
AM0-conditions	37.95	35.44	31.62	30.43	24.31	22.10
AM1-conditions	30.68	27.03	24.40	23.12	18.62	17.42
			pn step junction solar cells			
AM0-conditions	31.77	07.02	20.88	22.54	18.45	11.25
AM1-conditions	25.02	05.67	15.75	17.10	14.04	09.18
			Schottky junction solar cells			
AM1-conditions	10.59	02.34	06.96	07.53	06.15	03.75
AM1-conditions	8.34	01.89	05.25	05.70	04.68	03.06

In studying Table VI.12 it is clear that the high absorption coefficient of InP results in unusually high recombination losses in step pn junction standard configuration solar cells. Also note that, for every one of the example semiconductors, there is a decline in the expected photocurrent density from inverted configuration through vertical configuration and to standard configuration. This decline is minor when vertical and inverted configuration solar cells are compared, but of major proportions when standard configuration solar cells are considered. It cannot be overemphasized that the values of expected photocurrent density in Tables VI.10 through VI.12 are estimates and strongly dependent on the fabrication techniques employed in constructing the solar cells, upon surface crystal orientation, and on the semiconductor itself. The values listed above should be realizable, if sufficient care is exercised, but "mistakes" in fabrication technology and surface preparation can result in substantial reductions.

The overall purpose of this chapter is to provide an estimate of performance for several "realistic" situations. The heterojunction and pn step junction photocurrent density estimates of Table VI.12 are reasonable. The Schottky photocurrent density estimates are more problematic. To allow for photon penetration of the metal layer on top of a Schottky diode the layer must be very thin (<500 Å) [34]. Even so there is considerable loss due to photon reflection and the photocurrent density in such devices is small.

Device Design - Saturation Current

The saturation current densities for solar cells depend on the type of junction. From Chapter V and the discussions in Appendices B and C we have for the saturation current, J_S, in pn step junction solar cells:

$$J_S = \frac{qD_{pn}n_i^2}{L_{pn}N_D} + \frac{qD_{np}n_i^2}{L_{np}N_A} , \qquad (VI.11)$$

where D_{pn} and D_{np} are the diffusion constants for holes on the n-side of the junction and electrons on the p-side of the junction, L_{pn} and L_{np} are the diffusion lengths for holes on the n-side of the junction and electrons on the p-side of the junction. N_D and N_A are the net donor impurity concentration on the n-side and acceptor concentration on the p-side of the junction.

The saturation current for a heterojunction is given by:

$$J_S = J_{SH} = \frac{qD_S n_i^2}{L_S N_S} X_T , \qquad (VI.12)$$

where D_S, L_S and N_S are the minority carrier diffusion constants and diffusion length in the substrate semiconductor (the semiconductor with the smaller energy gap) and the impurity concentration in the substrate. The quantity, X_T, is used to account for internal effects at the interface between the two semiconductors (see Chapter V).

The saturation current density for a Schottky junction is:

$$J_S = J_{SS} = A^* T^2 \exp\{-\lambda \phi_{Bn}\} , \qquad (VI.13)$$

where λ is q/kT and the Richardson constant, A^*, is given by:

$$A^* = 120 m^* , \qquad (VI.14)$$

where m^* is the effective mass of the majority charge carrier in the semiconductor (see Appendix B). The quantity, ϕ_{Bn}, is the barrier voltage adjusted for image force lowering (see Table VI.4).

As discussed earlier, mos junctions are imperfectly understood at present. This being the case, further discussions of mos cells will not be conducted in this work. The reader who wishes to estimate the performance of the mos solar cell is referred to, as an introduction, [16-20, 35 and 36].

Using Equations VI.11 through VI.14 in conjunction with a substrate impurity concentration level of $3 \times 10^{15}/cm^3$, the barrier energies for Schottky barriers of Table VI.4, the carrier mobility and lifetime values of Table VI.7, the diffusion lengths of Table VI.8, the intrinsic carrier concentrations from Table III.2 and the effective mass values from Appendix B, it is possible to estimate the saturation currents for pn step junctions, Schottky junctions and heterojunctions. In the case of heterojunctions, we will assume that X_T, the interface factor, is unity. In practice, it is probable that X_T is less than unity [37] so that the estimate used here is conservative. Because the impurity concentration of the illuminated or "front" layer in a pn step junction is much larger than the impurity concentration of the substrate (by approximately four orders of magnitude), Equation VI.11 is reduced to a single term, dependent on the substrate. The estimated values of saturation current are presented in Table VI.13.

Examination of Table VI.13 in conjunction with the minimum values of saturation current density used in the preceding chapter indicates that the values of saturation current density that we may reasonably expect to encounter in practice are one to two orders of magnitude

Table VI.13

Typical values of saturation current density for heterojunctions, pn step junctions and Schottky junctions on the example semiconductors (A/cm^2). The junction temperature is 300°K. (Note that the junction value for CdSe is valid for heterojunctions only, as p-type CdSe is not available)

Semi-conductor	For pn step junctions and heterojunctions		For Schottky Barriers	
	p+n	n+p	on n	on p
Si	4.50×10^{-12}	7.80×10^{-12}	9.15×10^{-9}	6.89×10^{-9}
InP	1.27×10^{-16}	2.16×10^{-16}	6.43×10^{-5}	7.58×10^{-7}
GaAs	7.20×10^{-19}	2.19×10^{-18}	6.10×10^{-5}	3.13×10^{-5}
CdTe	1.31×10^{-20}	9.53×10^{-21}	3.02×10^{-8}	9.76×10^{-7}
AlSb	1.22×10^{-21}	8.20×10^{-22}	1.01×10^{-4}	7.56×10^{-4}
CdSe	1.89×10^{-22}	--------------	8.38×10^{-3}	-------------

higher than the ideal minimum values for pn and heterojunctions. The saturation current densities for Schottky barrier junctions are much higher. It is possible that breakthroughs in Schottky (or mos) technology will permit lower saturation current densities and that improvements in material preparation and junction fabrication technology will, by increasing minority carrier lifetime, lower the saturation current densities for both types of pn junctions. For the present, however, Table VI.13 is a reasonable approximation.

Device Design - Series Resistance

We have arrived at estimates for photocurrent density (Tables VI.10 through VI.12) and junction saturation current density (Table VI.13). These values can now be used in Equations VI.16 - VI.19 to determine the operating junction voltage for maximum power output density, V_p', the maximum power density which is transferable to the load, P_{max}/A, the efficiency for maximum power transference, η_s, and the internal loss factor, K':

$$(\lambda V_D' + 1 + K')\exp\{\lambda V_D'\} = (J_{ph} + J_S)(1 + K')/J_S \;, \tag{VI.16}$$

$$K' = 2(r_D A_D)J_S \lambda \exp\{\lambda V_D'\} \;, \tag{VI.17}$$

$$P_{max}/A_D = \frac{(J_{ph} + J_S)(\lambda V_D')V_D'}{(\lambda V_D' + 1 + K')}\;(1 - (K'/2)/(1 + K')) \;, \tag{VI.18}$$

and

$$\eta_S = (P_{max}/A_D)(I_{ns}) \;, \tag{VI.19}$$

where $\lambda = q/kT$, A_D is the junction area of the solar cell (assumed to be equal to the light collecting area of the cell), J_{pn} is the photocurrent, J_S is the saturation current density, r_D is the solar cell series resistance, and I_{ns} is 135.3 mW/cm^2 under AMO conditions and 107 mW/cm^2 under AM1 conditions.

In Chapter V we assumed values of the loss factor, K' and then computed efficiencies and power output densities for solar cells. We are now in a position to make estimates of the series resistance of a solar cell, and to calculate K' directly. To do this, let us assume the geometries for solar cells of the three configurations. Probable geometries for standard and vertical configuration cells are shown in Figure VI.4.

The standard cell is assumed to be round with a light sensitive area of πa^2; a "front layer" thickness, d, of 0.4 μm for pn step junctions, 0.5 μm for heterojunctions and 0.05 μm for Schottky junctions; and a substrate thickness, T_S, of 150 μm. Vertical configuration solar cells are rectangular parallelepipeds with light collecting areas of $T_S W$, where T_S is 150 μm and W is arbitrary[*]; a substrate width (distance in the direction of photon travel, Y, of 150 μm; and a total thickness of barrier plus substrate contact (d + T_S + Δ) of 150.4 μm. The inverted solar cell we will take to be equivalent to the standard cell turned upside down (see Figure VI.1). Following Figure VI.1, the light sensitive area for inverted

---

[*] If the procedure for making vertical cells discussed in Figure VI.2 is followed, reasonable values of W might be as large as 10 cm. Since this dimension drops out of our calculations, the precise value is unimportant.

solar cells is taken to be twice the area of the junction (assuming that both contacts for the inverted solar cell are on the down or non-illuminated side. The actual junction, then, occupies only about half of the down side).

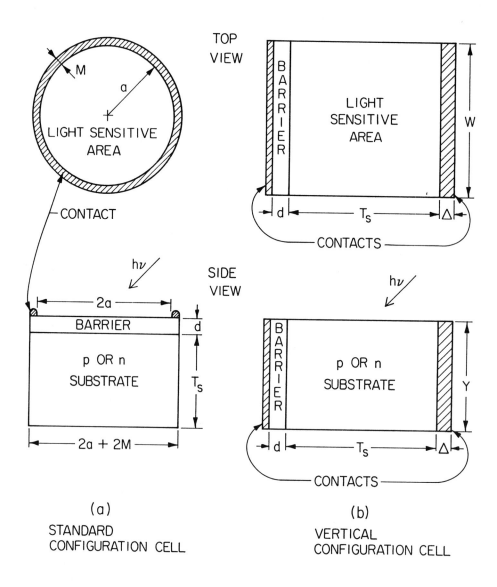

Figure VI.4. Probable standard and vertical solar cell configurations.

Note that in both the standard and vertical configurations some of the illuminated surface is obscured by the contacts on the illuminated surface. For the standard configuration solar cells the ratio of optically absorbing surface to total surface area, Φ, is given by:

$$\Phi_{standard} = a^2/(a + M)^2 , \qquad\qquad (VI.20a)$$

where M is the width of the metal contact ring (see Figure VI.4). Even with a heavily doped barrier region, the metal contact ring is much better at carrying current. A practical ratio of contact area to light absorbing area is 1:10.

For vertical configuration solar cells, Φ is:

$$\Phi_{vertical} = T_S/(T_S + d + \Delta) , \qquad\qquad (VI.20b)$$

where, from the preceding discussion, $\Phi_{vertical} = .997$.

For inverted configuration solar cells, the entire illuminated surface can be light absorbing; there being no contacts on the illuminated surface.

Note that, for standard configuration Schottky junction solar cells, the illuminated surface must be coated with a thin metal contact layer. This layer must be thick enough to conduct current to the metal contacts without excessive electrical resistance and thin enough to permit light to pass through it and reach the semiconductor substrate. The estimated photocurrent for this case is provided in Table VI.12.

The junction area, the area through which the current flows, is YW for vertical configuration solar cells, πa^2 for standard configuration solar cells and $\pi a^2/2$ for inverted configuration devices[*]. Now the series resistance of a solar cell consists of contact and lead resistance (small

---

[*] Note that the majority of the absorbed photons in a vertical configuration solar cell are absorbed close to the illuminated surface. Thus the photocurrent is generated close to the upper surface. The fact that the hole-electron pairs do spread via diffusion and the effect of electric fields does spread the hole-electron population, but the fact remains that the photocurrent does flow through a somewhat smaller area than that indicated. Lacking sufficient experimental data to predict this value exactly, we make the conservative assumption provided here.

since these are metal parts); resistance due to the substrate and resistance due to the "front layer". We will ignore the contact and lead resistance.

The substrate impurity concentration is often graded in order to improve the collection efficiency of the generated electron-hole pairs. From the discussion on this topic earlier in this chapter and using Equation VI.10, we can write for the series resistance times area product of the substrate:

$$(r_D A_D)_S = z \int_o^{T_S} [\exp\{-\lambda \mathscr{E}_S x\}/q\mu_{SM} N_{Sj}]dx \; . \tag{VI.21}$$

In Equation VI.21, A_D is the area of the junction[*], z is unity for vertical and standard configuration devices and is 0.96 for inverted configuration solar cells[#]. $\mathscr{E}_S$ is the electric field in the substrate region of the solar cell, μ_{SM} is the majority carrier mobility in the substrate and N_{Sj} is the impurity concentration in the substrate at the junction. Equation VI.21 is further complicated by the fact that the mobility of the carrier is a function of the impurity concentration (see data in Appendix B). Assume that the substrate thickness is 150 μm, N_{Sj} is 10^{15}/cm^3 and $\mathscr{E}_S$ is 16 volts/cm[**]. From the data of Chapter III, Appendix B and using Equation VI.21, we can obtain for the approximate substrate series resistance times area product the results in Table VI.14.

The series resistance times junction area product of the heavily doped "front layer" for a vertical configuration solar cell is determined by using Equation VI.22:

---

[*] Recall that the junction and optically absorbing surfaces are approximately the same in vertical and standard configuration devices, while the optically absorbing surface is some twice the area of the junction for inverted configuration devices.
[#] This factor reflects the efficiency of collection of hole electron pairs as a result of recombination on the lateral surfaces.
[**] For this value of substrate electric field we have seen that the substrate impurity concentration varies from 10^{15}/cm^3 at the junction to 10^{19}/cm^3 at the far side of the substrate. Recall that this range results from balancing the need for low saturation current (and hence high impurity concentration in the substrate) and a large extent for the electric fields produced by the junction in the substrate (and hence a low impurity concentration in the substrate).

Table VI.14

The approximate substrate series resistance times junction area $(r_D A_D)_S$ product for solar cells with a 150 μm thick substrate with an exponentially graded substrate impurity concentration ranging from $10^{15}/cm^3$ at the junction to $10^{19}/cm^3$ at the far substrate surface, at a temperature of 300°K (in ohm-cm^2)

Semiconductor	Si	InP	GaAs	CdTe	AlSb	CdSe
		vertical and standard configurations				
n-type substrate	.0069	.0032	.0016	.0110	.0746	.0176
p-type substrate	.0192	.1033	.0294	.1127	.0258	*
		inverted configuration				
n-type substrate	.0066	.0030	.0015	.0106	.0695	.0169
p-type substrate	.0183	.0953	.0282	.1078	.0250	*
* p-type single crystal CdSe is not available.						

$$(r_D A_D)_L = (\Delta + d)/(q\mu_{LM}N_L) \; . \tag{VI.22}$$

The resistance area product of the "front layer" can be easily computed using the approximation of a constant impurity concentration and the mobility values from Table VI.7. For Schottky barriers we can use the resistivity of a typical metal (approximately 5 μ ohm-cm) and find that $(r_D A_D)_L \cong 2 \times 10^{-10}$ ohm-cm^2. For devices utilizing heterojunctions the resistance time junction area product for the "front" layer is on the order of 10^{-7} ohm-cm^2. The product for devices with pn junctions can similarly be determined, again using Table VI.7 and Equation VI.22, along with a "front layer" thickness of 0.4×10^{-4} centimeters.

Upon investigation of the "front layer" series resistance times junction area product for inverted configuration solar cells, we discover that the values are essentially of the same order of magnitude as those for the vertical configuration. These values are present in Table VI.15. Note that they are far smaller than the resistance times junction area product of the substrate region--so small as to enable us to ignore them in future calculations of solar cell performance.

In a standard configuration solar cell the "front layer" resistance times junction area product is not necessarily less than the substrate

Table VI.15

The series resistance times junction area product for the heavily doped "front layer" of vertical and inverted configuration solar cells as a function of the type of junction and at 300°K (ohm-cm^2)

Semiconductor	Si	InP	GaAs	CdTe	AlSb	CdSe
			pn junctions			
n-type "front layer"	3.12 x 10^{-7}	1.67 x 10^{-7}	2.50 x 10^{-8}	3.57 x 10^{-8}	1.79 x 10^{-7}	8.33 x 10^{-7}
p-type "front layer"	1.85 x 10^{-7}	5.56 x 10^{-8}	2.48 x 10^{-7}	4.99 x 10^{-7}	1.37 x 10^{-7}	*
			heterojunctions*			
	10^{-7}	--				10^{-7}
			Schottky junctions*			
	2 x 10^{-10}	--------------------------------			2 x 10^{-10}	

* p-type single crystal CdSe is not available

area times junction product. Here, the current flows through the substrate and "front layer" to the metal contacts (See Figure VI.4). If we assume that the current flowing at the junction proper is uniformly distributed, then, from Ohm's law:

$$(R_D A_D)_L = (\rho_L a^2)/4d .$$ (VI.23)

Equation VI.23 has been used to compute the "front layer" series resistance times junction area product for standard configuration solar cells. The data of Table VI.7 and Appendix B have been used, together with a value of d of 0.4 μm for pn junctions, ρ_L equal to 5 x 10^{-6} ohm-cm for Schottky metals and d equal to 0.05 μm for Schottky barriers. For a heterojunction we assume ρ_L equal to 5 μm and a resistivity of 0.01 ohm-cm [38]. Under these conditions, we have the results outlined in Table VI.16 for the "front layer" series resistance of standard configuration solar cells.

We are now in a position to determine the potential performance of the three configurations. Note that the results which we are about to determine are not intended to be the maximum potential. Rather, these

Table VI.16

The series resistance times junction area product for the "front layer" of a standard configuration solar cell as a function of the junction type and at 300°K (ohm-cm^2)

Semiconductor	Si	InP	GaAs	CdTe	AlSb	CdSe
			Schottky junctions*			
	0.25a^2	-----------	-----------	-----------	-----------	--0.25a^2
			heterojunctions*			
	5a^2--	-----------	-----------	-----------	-----------	--5a^2
substrate			pn junctions			
n-type	28.94a^2	08.68a^2	03.91a^2	05.58a^2	27.9a^2	*
p-type	48.82a^2	26.04a^2	39.06a^2	78.12a^2	21.7a^2	*

* p-type CdSe is not available

results are meant to indicate the "practical" output values achievable with more or less "normal" fabrication technologies. We will discuss potential, all-out, maximum output potentials later. The expected solar cell delivered power density is given by Equation V.33 where the junction voltage (the photovoltage) and loss factor for maximum power delivered (under the assumed conditions) must mutually satisfy Equations V.31 and V.32. The series resistance times junction area products for the "front layer" and substrate are provided in Tables VI.14 and VI.16. Note that for inverted and vertical configuration solar cells only the substrate series resistance is of importance. For the standard configuration solar cell, the "front layer" resistance times area product is of importance. The larger the area of the solar cell the larger the "front layer" resistance and the less expensive (within reason) it is to construct the device. For our purposes we will assume that the standard configuration solar cells are sufficiently large to exhibit a "front layer" series resistance times area product equal to one-tenth of the substrate resistance times area product. Thus, the effective resistance times junction area product for our example standard configuration solar cells is given by the substrate value from Table VI.14 times one point one. The saturation current densities for the various junction types are provided in Table VI.13. The photocurrent density for inverted configuration solar cells is taken to be 1.95 times the values in

Table VI.10 (the physical structure assumed in the discussion of this chapter leads to a degree of optical concentration equal to this value). The photocurrent density for vertical configuration solar cells is provided in Table VI.11 and the photocurrent for standard configuration solar cells lies between 0.3 and 0.9 times that of the values furnished in Table VI.12[*]. The solar spectra employed are AMO and AM1.

The loss factor, which depends directly on the series resistance times junction area product, for these devices is considerably less than unity. Selected values for K' are presented in Table VI.17 for standard configuration solar cells. Table VI.18 provides the loss factors for vertical configuration solar cells as a function of junction and substrate type. Table VI.19 presents the same data for inverted configuration solar cells. Note that p-type CdSe is not available. Comparison of these values of loss factor with those used in computing Tables V.2 and V.3 indicates that the assumptions used in those tables were conservative. Also note that the loss factor is higher for inverted configuration solar cells than for vertical configuration devices, which are, in turn, higher than K' values for standard configuration solar cells.

Figures VI.5 through VI.10 present values of the photovoltage, V_D', under conditions of maximum delivered power density, the maximum solar cell delivered power density, P_{max}/A_D, and η_S, the solar cell efficiency under these conditions, for standard configuration solar cells under AMO and AM1 illumination. Note that the realtively low barrier energies of the Schottky barrier solar cells, lead to high saturation current denisty and poor overall performance of Schottky barrier solar cells.

---

[*] The physical configuration of an inverted solar cell, as described here, leads to a light collected area which is twice that of the junction. The collection of holes and electrons from the furthest distances within the solar cell is not perfect--hence the photocurrent density is 1.95 times Table VI.13. In the case of standard configuration solar cells, the necessity to make electrical contact to the illuminated surface results in a loss of at least 10% of the incoming light, owing to refection from the metal contacts. In the case of a Schottky barrier standard configuration solar cell, the requirement for a metal surface (to complete the Schottky junction) results in further decreases in the photocurrent density. Thus to determine the photocurrent density for pn and heterojunction standard configuration solar cells we multiply the values of Table VI.12 by 0.9, while for Schottky barrier diodes, we multiply by 0.3.

Table VI.17

The loss factor, K', as a function of the illumination conditions, barrier type, and substrate for six semiconductors used in standard configuration solar cells (at 300°K)

| | AMO light input | | | | | |
| | n-type substrate | | | p-type substrate | | |
Junction	pn	hetero	Schottky	pn	hetero	Schottky
Si	.000904	.001072	.000500	.002581	.003058	.001361
InP	.000065	.000311	.000191	.002148	.010363	.002910
GaAs	.000082	.000123	.000228	.001524	.002281	.003620
CdTe	.000536	.000719	.000639	.005480	.007357	.009218
AlSb	.002833	.003712	.010524	.000969	.001269	.006453
CdSe	*	.000763	.015264	*	*	*
	AM1 light input					
Si	.000720	.000875	.000401	.002055	.002496	.001091
InP	.000053	.000239	.000163	.001747	.007955	.002413
GaAs	.000062	.000096	.000182	.001158	.001773	.002871
CdTe	.000410	.000550	.000495	.004181	.005618	.007190
AlSb	.002168	.002859	.008512	.000742	.000978	.005461
CdSe	*	.000605	.014785	* p-type CdSe is not available		

Table VI.18

The loss factor, K', as a function of the illumination conditions, barrier type, and substrate for six semiconductors used in vertical configuration solar cells (at 300°K)

| | AMO light input | | | | | |
| | n-type substrate | | | p-type substrate | | |
Junction	pn	hetero	Schottky	pn	hetero	Schottky
Si	.001089	.001089	.001660	.003114	.003114	.004544
InP	.000316	.000316	.001724	.010468	.010468	.033751
GaAs	.000122	.000122	.000777	.002312	.002312	.013064
CdTe	.000726	.000726	.002327	.007430	.007430	.032825
AlSb	.003750	.003750	.032108	.001282	.001282	.016549
CdSe	*	.000772	.025312	*	*	*

Table VI.18, continued

| Junction | AM1 light input | | | | | |
| | n-type substrate | | | p-type substrate | | |
	pn	hetero	Schottky	pn	hetero	Schottky
Si	.000881	.000881	.001350	.002519	.002519	.003691
InP	.000243	.000243	.001369	.008025	.008025	.026237
GaAs	.000095	.000095	.000623	.001794	.001794	.010425
CdTe	.000554	.000554	.001800	.005662	.005662	.025427
AlSb	.002894	.002894	.025687	.000990	.000990	.013570
CdSe	*	.000613	.022677	* p-type CdSe is not available		

Table VI.19

The loss factor, K', as a function of the illumination conditions, barrier type, and substrate for six semiconductors used in inverted configuration solar cells (at 300°K)

| Junction | AMO light input | | | | | |
| | n-type substrate | | | p-type substrate | | |
	pn	hetero	Schottky	pn	hetero	Schottky
Si	.002059	.002059	.003092	.005871	.005871	.008459
InP	.000596	.000596	.003001	.019598	.019598	.061330
GaAs	.000230	.000230	.001348	.004477	.004477	.023573
CdTe	.001417	.001417	.004389	.014478	.014478	.061249
AlSb	.006915	.006915	.054453	.002454	.002454	.002759
CdSe	*	.001499	.035937	*	*	*
AM1 light input						
Si	.001680	.001680	.002534	.004789	.004789	.006924
InP	.000458	.000458	.002376	.015009	.015009	.047348
GaAs	.000179	.000179	.001079	.003475	.003475	.018742
CdTe	.001083	.001083	.003398	.011037	.011037	.047620
AlSb	.005457	.005457	.044116	.001939	.001939	.002226
CdSe	*	.001188	.031298	* p-type CdSe is not available		

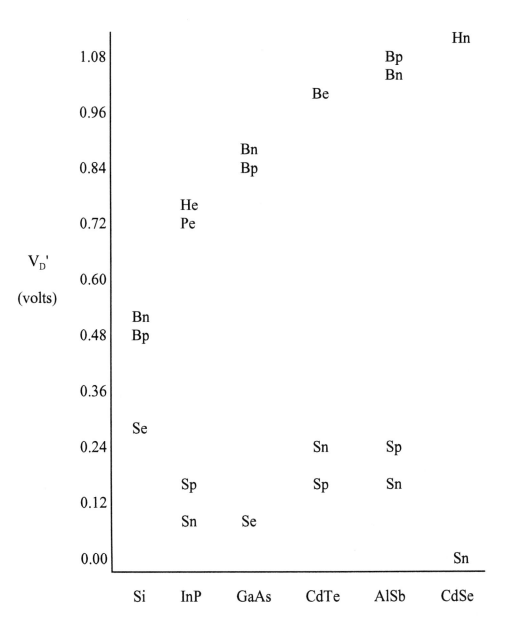

Figure VI.5. Photovoltage when the delivered solar cell power is a maximum, V_D', for standard configuration solar cells, as a function of barrier and substrate, under AMO light, at 300°K and for six example semiconductors.

Junction symbols: H for heterojunction, P for pn junction, S for Schottky barrier and B for both pn and heterojunctions.

Substrate symbols: n for n-type, p for p-type and e for either type.

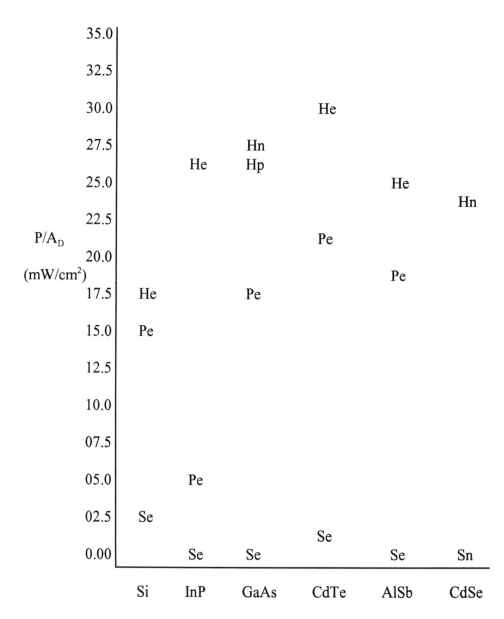

Figure VI.6. The maximum delivered power density, P/A_D, for standard configuration solar cells, as a function of barrier and substrate, for six example semiconductors under AMO illumination and at 300°K.
Junction symbols: H for heterojunction, P for pn junction, S for Schottky barrier and B for both pn and heterojunctions.
Substrate symbols: n for n-type, p for p-type and e for either type.

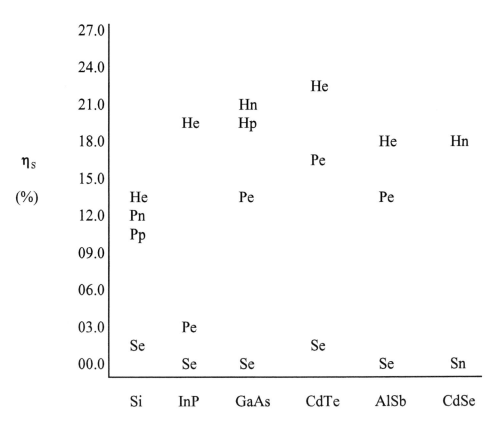

Figure VI.7. Solar cell efficiency, η_S, under the conditions of maximum delivered power density, for standard configuration solar cells, as a function of barrier and substrate, for six example semiconductors under AMO illumination and at 300°K.
Junction symbols: H for heterojunction, P for pn junction, S for Schottky barrier and B for both pn and heterojunctions.
Substrate symbols: n for n-type, p for p-type and e for either type.

Figures VI.11 through VI.16 present values of photovoltage, maximum delivered power density and solar cell efficiency at 300°K for vertical configuration solar cells of various junction and substrate types. Note that the performance of these vertical configuration solar cells is higher than for standard configuration solar cells, partly as a result of higher effective (potentially higher!) photocurrents and partly as a result of the lower series resistance times junction area product of these solar cells. Further note that the low Schottky barrier heights result in higher saturation current density and poorer solar cell performance.

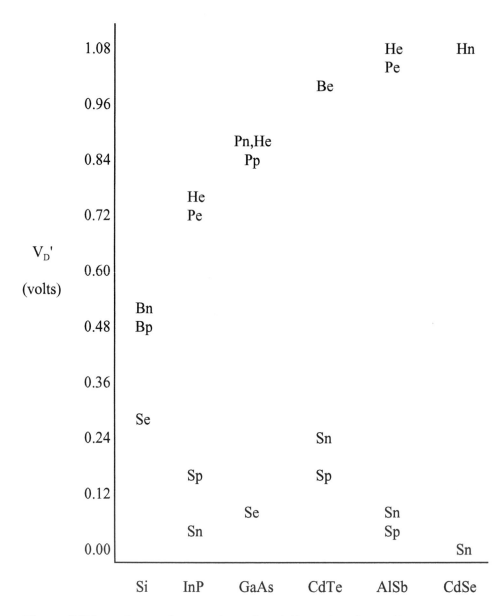

Figure VI.8. Photovoltage when the delivered solar cell power is a maximum, V_D', for standard configuration solar cells, as a function of barrier and substrate, under AM1 light, at 300°K and for six example semiconductors.

Junction symbols: H for heterojunction, P for pn junction, S for Schottky barrier and B for both pn and heterojunctions.

Substrate symbols: n for n-type, p for p-type and e for either type.

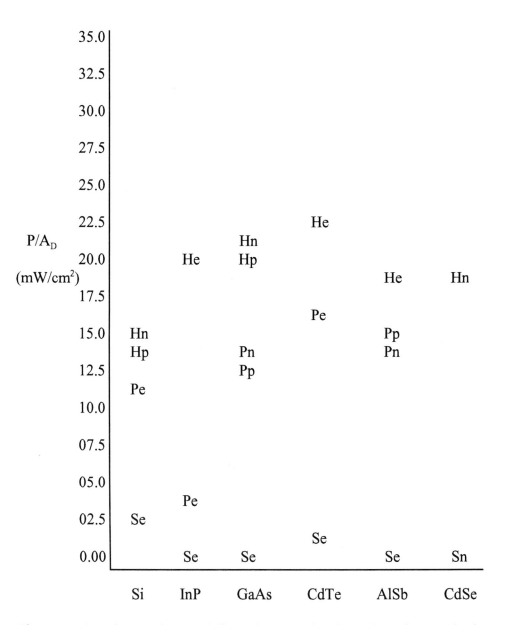

Figure VI.9. The maximum delivered power density, P/A_D, for standard configuration solar cells, as a function of barrier and substrate, for six example semiconductors under AM1 illumination and at 300°K.
Junction symbols: H for heterojunction, P for pn junction, S for Schottky barrier and B for both pn and heterojunctions.
Substrate symbols: n for n-type, p for p-type and e for either type.

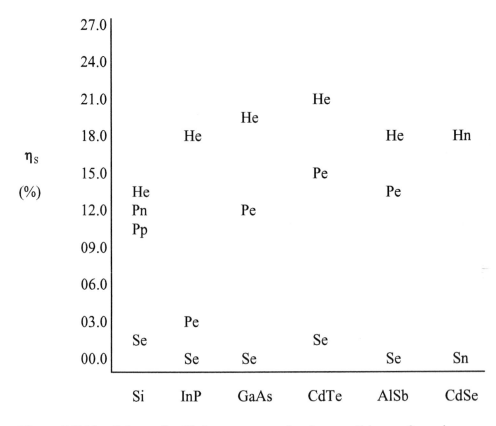

Figure VI.10. Solar cell efficiency, η_s, under the conditions of maximum delivered power density, for standard configuration solar cells, as a function of barrier and substrate, for six example semiconductors under AM1 illumination and at 300°K.
Junction symbols: H for heterojunction, P for pn junction, S for Schottky barrier and B for both pn and heterojunctions.
Substrate symbols: n for n-type, p for p-type and e for either type.

Figures VI.17 through VI.22 present values of photovoltage, maximum delivered power density and solar cell efficiency at 300°K for inverted configuration solar cells of various junction and substrate types. Note that the performance of these inverted configuration solar cells is higher than for standard configuration solar cells, partly as a result of higher effective (potentially higher!) photocurrents and partly as a result of the lower series resistance times junction area product of these solar cells. Further note that the low Schottky barrier heights result in higher saturation current density and poorer solar cell performance.

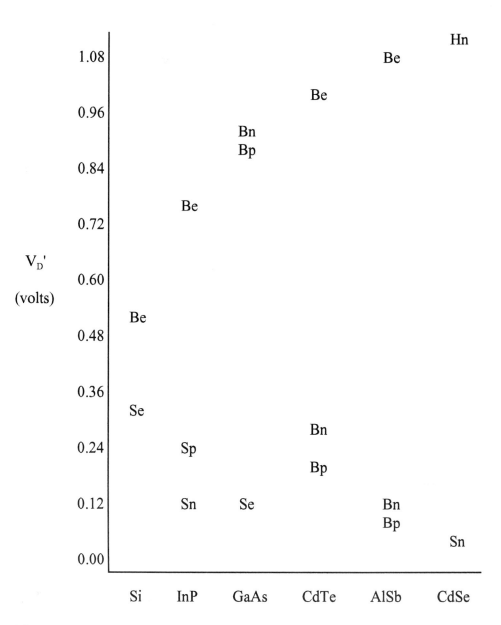

Figure VI.11. Photovoltage when the delivered solar cell power is a maximum, V_D', for vertical configuration solar cells, as a function of barrier and substrate, under AMO light, at 300°K and for six example semiconductors.

Junction symbols: H for heterojunction, P for pn junction, S for Schottky barrier and B for both pn and heterojunctions.

Substrate symbols: n for n-type, p for p-type and e for either type.

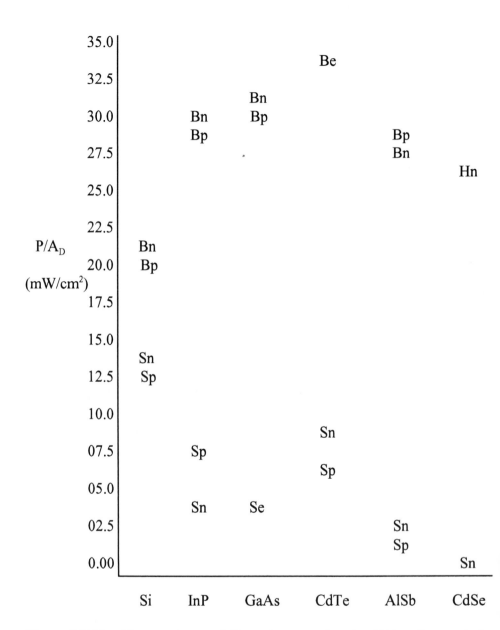

Figure VI.12. The maximum delivered power density, P/A_D, for vertical configuration solar cells, as a function of barrier and substrate, for six example semiconductors under AMO illumination and at 300°K.
Junction symbols: H for heterojunction, P for pn junction, S for Schottky barrier and B for both pn and heterojunctions.
Substrate symbols: n for n-type, p for p-type and e for either type.

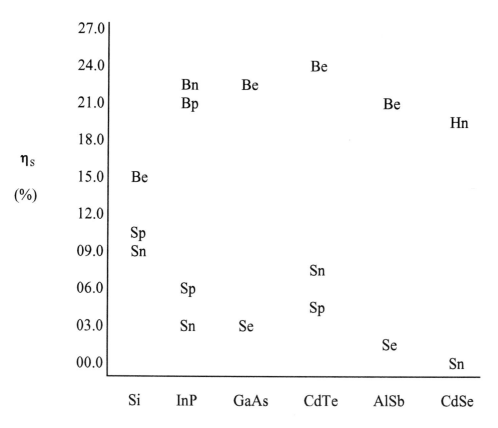

Figure VI.13. Solar cell efficiency, η_s, under the conditions of maximum delivered power density, for vertical configuration solar cells, as a function of barrier and substrate, for six example semiconductors under AMO illumination and at 300°K.
Junction symbols: H for heterojunction, P for pn junction, S for Schottky barrier and B for both pn and heterojunctions.
Substrate symbols: n for n-type, p for p-type and e for either type.

Comparing the inverted solar cell types of Figures VI.17 through VI.22 with the vertical solar cell performance estimated in Figures VI. 11 through VI.16 note that the performance of the inverted solar cells is estimated to be significantly higher than that of the vetical solar cells. The principal reason for this is the higher photocurrent of the inverted type of solar cell. For inverted cells the effective junction area is approximately one half of the illuminated area. Thus there is significantly higher photocurrent in the inverted solar cells as these devices exhibit a moderate amount of built-in optical concentration.

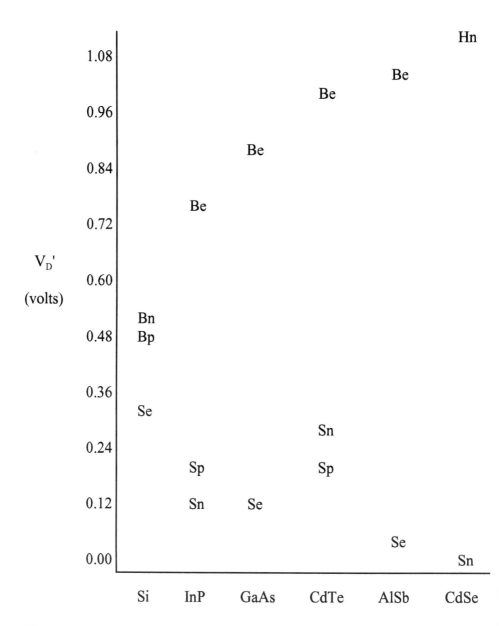

Figure VI.14. Photovoltage when the delivered solar cell power is a maximum, V_D', for vertical configuration solar cells, as a function of barrier and substrate, under AM1 light, at 300°K and for six example semiconductors.

Junction symbols: H for heterojunction, P for pn junction, S for Schottky barrier and B for both pn and heterojunctions.

Substrate symbols: n for n-type, p for p-type and e for either type.

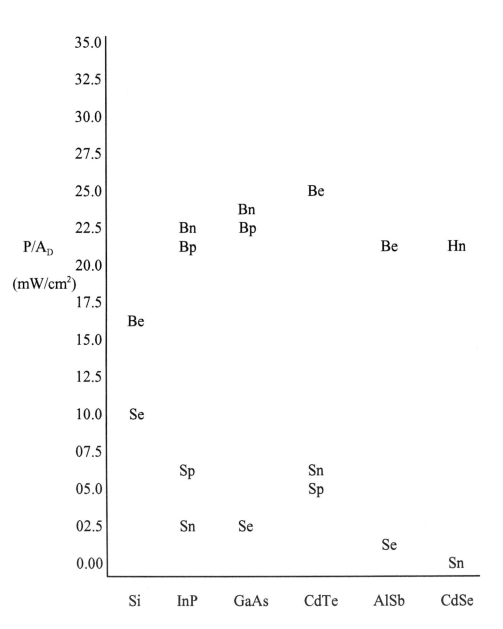

Figure VI.15. The maximum delivered power density, P/A_D, for vertical configuration solar cells, as a function of barrier and substrate, for six example semiconductors under AM1 illumination and at 300°K.
Junction symbols: H for heterojunction, P for pn junction, S for Schottky barrier and B for both pn and heterojunctions.
Substrate symbols: n for n-type, p for p-type and e for either type.

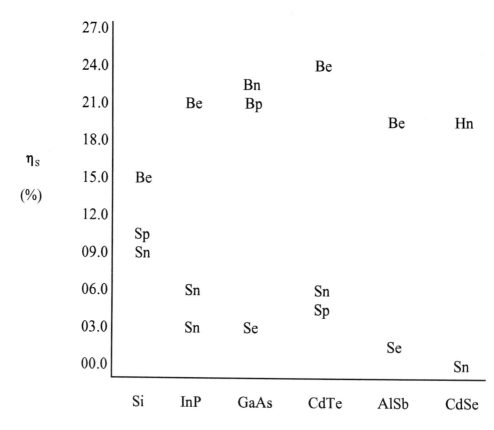

Figure VI.16. Solar cell efficiency, η_s, under the conditions of maximum delivered power density, for vertical configuration solar cells, as a function of barrier and substrate, for six example semiconductors under AM1 illumination and at 300°K.
Junction symbols: H for heterojunction, P for pn junction, S for Schottky barrier and B for both pn and heterojunctions.
Substrate symbols: n for n-type, p for p-type and e for either type.

Solar Cell Performance - Discussion

Let us conclude this chapter with some discussion regarding Figures VI.5 through VI.22. Let it be emphasized that the devices whose photovoltage, maximum delivered power density and efficiency are shown in these figures are not considered to be the "best there can be" at performing as solar cells. Instead, these solar cells are constructed to represent practical devices. In other words, they are taken to be

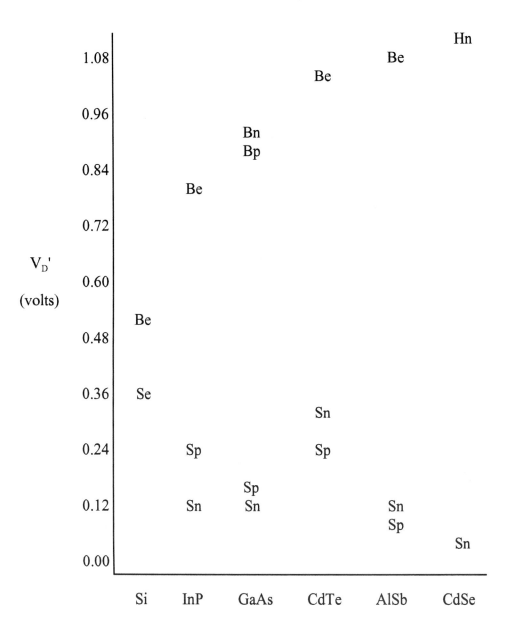

Figure VI.17. Photovoltage when the delivered solar cell power is a maximum, V_D', for inverted configuration solar cells, as a function of barrier and substrate, under AMO light, at 300°K and for six example semiconductors.
Junction symbols: H for heterojunction, P for pn junction, S for Schottky barrier and B for both pn and heterojunctions.
Substrate symbols: n for n-type, p for p-type and e for either type.

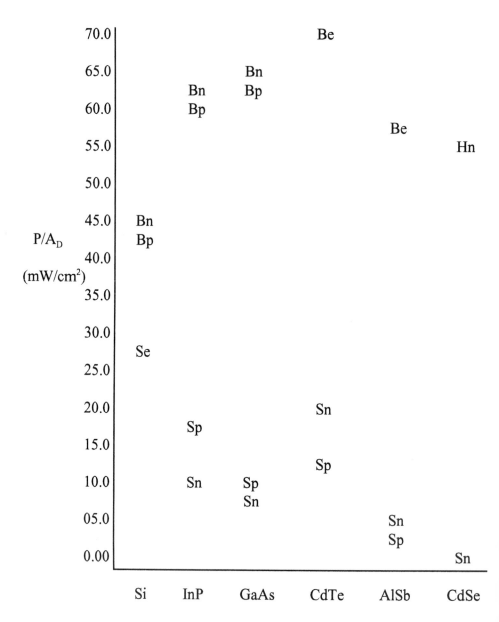

Figure VI.18. The maximum delivered power density, P/A$_D$, for inverted configuration solar cells, as a function of barrier and substrate, for six example semiconductors under AMO illumination and at 300°K.

Junction symbols: H for heterojunction, P for pn junction, S for Schottky barrier and B for both pn and heterojunctions.

Substrate symbols: n for n-type, p for p-type and e for either type.

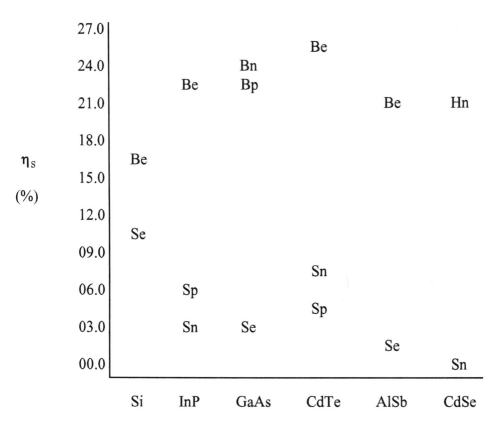

Figure VI.19. Solar cell efficiency, η_s, under the conditions of maximum delivered power density, for inverted configuration solar cells, as a function of barrier and substrate, for six example semiconductors under AMO illumination and at 300°K.
Junction symbols: H for heterojunction, P for pn junction, S for Schottky barrier and B for both pn and heterojunctions.
Substrate symbols: n for n-type, p for p-type and e for either type.

representative of devices which can be reliably and repeatedly constructed to operate at the performance level indicated. It is possible to better the performance levels indicated in Figures VI.5 through VI.22 by spending a considerable amount of effort at maximizing the photocurrent, while minimizing the saturation current density and series resistance. For example, silicon devices with efficiencies in excess of 20% have been reported [39]. It is also, unfortunately, possible to "snatch defeat from the jaws of victory" and do much worse. The solar cells considered in this chapter are achievable, and allow us to draw some design conclusions.

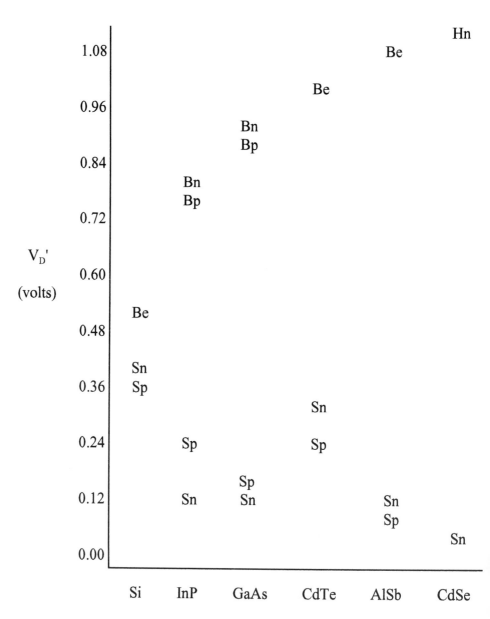

Figure VI.20. Photovoltage when the delivered solar cell power is a maximum, V_D', for inverted configuration solar cells, as a function of barrier and substrate, under AM1 light, at 300°K and for six example semiconductors.

Junction symbols: H for heterojunction, P for pn junction, S for Schottky barrier and B for both pn and heterojunctions.

Substrate symbols: n for n-type, p for p-type and e for either type.

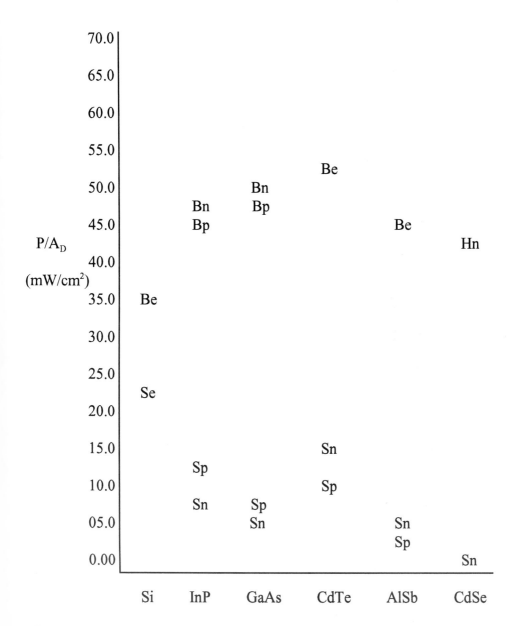

Figure VI.21. The maximum delivered power density, P/A_D, for inverted configuration solar cells, as a function of barrier and substrate, for six example semiconductors under AM1 illumination and at 300°K.
Junction symbols: H for heterojunction, P for pn junction, S for Schottky barrier and B for both pn and heterojunctions.
Substrate symbols: n for n-type, p for p-type and e for either type.

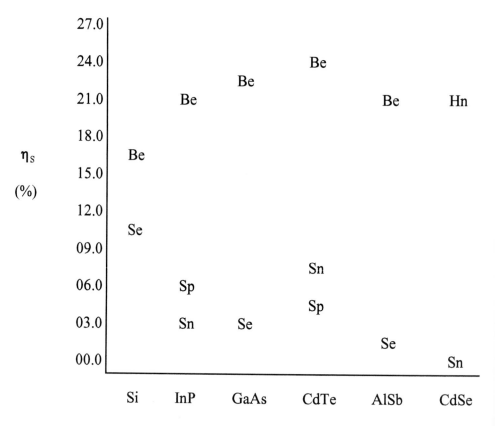

Figure VI.22. Solar cell efficiency, η_s, under the conditions of maximum delivered power density, for inverted configuration solar cells, as a function of barrier and substrate, for six example semiconductors under AM1 illumination and at 300°K.
Junction symbols: H for heterojunction, P for pn junction, S for Schottky barrier and B for both pn and heterojunctions.
Substrate symbols: n for n-type, p for p-type and e for either type.

Overall, pn junction and heterojunction solar cells are potentially capable of generating larger amounts of photopower than are Schottky barrier solar cells. This is the result of lower saturation current densities, particularly for non-silicon semiconductors. In all three optical configurations; standard, vertical and inverted; the predicted performance for solar cells appears to be at a maximum for CdTe. Also note that, in general, the substrate type is relatively unimportant.

From our discussion in Chapter V, the values for photovoltage, given in Figures VI.5, VI.8, VI.11, VI.14, VI.17 and VI.20, should be in excess of V', the voltage at which the diode current is contributed equally by diffusion and by generation-recombination components. We can solve Equation V.11 to determine V' for the devices considered in this chapter. The results of this operation, along with the effective built-in voltage for pn junctions are provided in Table VI.20.

Table VI.20

Values of V_B and V' (in volts) for pn junctions made from the example semiconductors with the material parameters indicated in this chapter

Semiconductor	Si	InP	GaAs	CdTe	AlSb
		for p-type substrates			
V_B	0.815	1.14	1.23	1.29	1.46
V'	0.063	0.56	0.66	0.71	0.86
		for n-type substrates			
V_B	0.815	1.14	1.23	1.29	1.46
V'	0.036	0.52	0.59	0.65	0.87

Examination of Figures VI.5, VI.8, VI.11, VI.14, VI.17 and VI.20 shows that the experienced photovoltages for heterojunction and pn junction solar cells, for standard, inverted and vertical configurations, are in excess of V'. The photovoltages for Schottky diode solar cells are less than V' in Table VI.20 but the above mentioned limitation does not apply, because the overall performance is so low as to make Schottky barrier devices of little, or no, interest. Note that the optimum output solar cell voltage is less than V_B for all pn and heterojunction devices and is less than ϕ_{Bo} (see Table VI.4) for Schottky junction solar cells.

The performance of standard configuration solar cells as displayed in Figures VI.5 through VI.10 involves an approximation. The power density obtainable depends on the actual physical dimensions of the solar cell. The series resistance times area product contains a term due to the substrate which is independent of the illuminated area radius, a, and which is provided in Table VI.14. However, it also includes a term due to the series resistance of the "front layer". This term was presented in Table VI.16, assuming a circular area for the solar cells, and it was indicated in Table VI.16 that the series resistance times area product for standard configuration solar cells varies as a^2. The performance displayed in Figures VI.5 through VI.10 is based on the assumption that the series resistance times area product for standard configuration solar cells is that of the substrate plus a contribution from the "front layer"--a contribution that amounts to 10% of the substrate value. Note that the greater the radius of the standard configuration solar cell the greater the light collecting area and the greater the experienced photocurrent. However, the resistance times area product will also increase, leading to a larger loss factor and a lower output power density and operating efficiency. The exact size which is most efficient is a function of the type of junction, the intensity of the light illuminating the solar cell and the economics of the fabrication process employed in constructing the solar cells. Clearly, if output power density is the sole important item, then the ideal solar cell has no internal resistance what-so-ever--a topic which will be addressed in the following chapter.

We will not pursue this topic further at present. For now, the reader will be interested in Table VI.21 which presents the effective radius of standard configuration solar cells for the case in which the "front layer" contributes an amount equal to 10% of the substrate series

Table VI.21

The illuminated region radius, a, (in μm) for standard configuration solar cells for a "front layer" series resistance times area product which is 10% of that of the substrate product

Semiconductor	Si	InP	GaAs	CdTe	AlSb	CdSe
Junction Type						
			n-type substrates			
Schottky	525	358	253	663	1727	839
heterojunction	152	103	073	191	0499	242
pn	038	035	020	038	0185	*
			p-type substrates			
Schottky	876	2032	1084	2123	1015	*
heterojunction	253	0567	0313	0613	0294	*
pn	081	0345	0027	1412	0096	*

* p-type CdSe is not available

resistance times area product--the situation leading to Figures VI.5 through VI.10. Note that the dimensions, a, are small, leading the reader to conclude that a larger device would exhibit a significantly higher series resistance area product, and hence would have a substantially lower performance.

Careful study of the theoretical predictions contained in Figures VI.5 through VI.22 and of the rest of this chapter indicates that inverted configuration solar cells are potentially capable of the highest level of performance, followed by vertical configuration solar cells and, followed, at the lowest level, by standard configuration solar cells. Considering the type of junction, pn or heterojunction (given the availability of suitable matching semiconductors) solar cells appear to be preferable, as compared to the poorly performing Schottky barriers. Substrate type, while it does exhibit some effect on solar cell performance, does not play a major role in device output. Finally, it would seem that the performance of solar cells (in terms of output power density and efficiency) is highest for those situations in which the semiconductor employed has an energy gap in the vicinity of 1.5 eV.

What we have not done, as yet, is consider the economics of the fabrication processes employed and the effects of increased illumination levels and solar cell junction temperatures. These topics will be covered in the succeeding chapters. In addition, we will consider additional techniques for improving solar cell performance. Even without these improvements, the potential performance for solar cells, as indicated in this chapter, is encouraging when we consider solar cells as a major energy source for our civilization.

References

1 Electronics, 29 May, 1975, p. 41.
2 J. A. Hitchby and R. L. Fudurich, in Journal of Applied Physics, Vol. 47, 1976, p. 3152.
3 A. M. Sekela, D. L. Feucht and A. G. Milnes, in IEEE Transactions on Electron Devices, Vol. 24, 1977, p. 373.
4 J. L. Shay, et. al., in Journal of Applied Physics, Vol. 47, 1976, p. 614.
5 R. L. Soukap, in Journal of Applied Physics, Vol. 47, 1976, p. 555.

6 A. K. Sreedhar, B. L. Sharma and R. K. Purchet, in IEEE Transactions on Electron Devices, Vol. 16, 1969, p. 309.

7 W. G. Thompson, et. al., in IEEE Transactions on Electron Devices, Vol. 24, 1977, p. 463.

8 R. H. Bube, et. al., in IEEE Transactions on Electron Devices, Vol. 24, 1977, p. 487.

9 M. D. Lammert and R. J. Schwartz, "A Silicon Solar Cell for Use at High Illumination Intensities, Technical Report TR-EE-75-37, 1975. See also IEEE Transactions on Electron Devices, 1977, p. 337.

10 H. J. Hovel, Semiconductors and Semimetals, Vol.11, Academic Press, New York, 1975, p. 14.

11 S. M. Sze, Physics of Semiconductors, 2nd Edition, John Wiley and Sons, New York, p. 291.

12 R. C. Neville, unpublished data on CdTe, AlSb and InP.

13 A. N. Saxena, et. al., personal communication on hafmium-silicon surface barriers.

14 C. A. Mead, in Solid-State Electronics, Vol. 9, 1966, p. 1023.

15 M. F. Millea, in Physics Review, Vol. 177, 1969, p. 1164.

16 W. A. Anderson, et. al., in Journal of Vacuum Science and Technology, Vol. 13, 1976, p. 1158.

17 J. P. Ponpon and P. Siffer, in Journal of Applied Physics, Vol. 47, 1976, p. 89.

18 R. J. Stirm, Y. C. M. Yeh and S. J. Fonash, in Journal of Vacuum Science and Technology, Vol. 13, 1976, p. 894.

19 R. Singh and J. Sewchun, in Journal Vacuum Science and Technology, Vol. 14, 1977, p. 89.

20 W. A. Anderson, J. K Kim and A. E. DeLahoy, in IEEE Transactions on Electron Devices, Vol. 24, 1977, p. 453.

21 A. K. Sreedhar, B. L. Shavna and R. K. Purchit, in IEEE Transactions on Electron Devices, Vol. 16, 1969, p. 309.

22 R. Rahi and A. G. Milnes, in Solid State Electronics, Vol. 13, 1970, p. 1289.

23 A. S. Grove, Physics and Technology of Semiconductor Devices, John Wiley and Sons, New York, 1967, Chap. 3.

24 S. M. Sze, Semiconductor Devices, Physics and Technology, John Wiley and Sons, New York, 1985, Chap. 9.

25 The upper limit on impurity densities in a semiconductor is the solid solubility of the impurity species in the semiconductor. For typical examples see, F. A. Trumbore, in The Bell System Technical Journal, Vol. 39, 1960, p. 205.

26 A full discussion of ion implantation and diffusion as techniques for introduction of impurities into solar cells is too lengthy for inclusion in this work. The scientific literature is replete with references to the technique. For those who wish to further pursue these subjects, suggested titles are: D. C. Gupta, editor, Silicon Processing, ASTM Technical Publications, Philadelphia, 1983 and Reference [23], pages 40 and 210.

27 J. Singh, Physics of Semiconductors and Their Heterostructures, McGraw-Hill, New York, 1993, p. 47-49.

28 The topic of molecular beam epitaxy (MBE) requires considerably more space to adequately cover than is available here. It is suggested that the interested reader begin study of MBE by reading S. M. Sze, Semiconductor Device Physics and Technology, John Wiley and Sons, New York, 1985, p. 333-338. There is volumnious literature on this topic, reaching back to 1974, particularly A. Y. Cho, et. al., in Applied Physics Letters, Vol. 25, 1974, p. 224.

29 K. B. Wolfstirn, in Physics Chemistry Solids, Vol. 16, 1960, p. 279.

30 Unpublished data obtained by the author.

31 M. A. Green, in IEEE Transactions on Electron Devices, Vol. 23, 1976, p.11.

32 J. Mandelkorn, J. H. Lamneck and L. R. Scudder, in the Conference Proceedings of the 10th IEEE Photovoltaic Specialists Conference, Nov. 1973, p. 20.

33 M. Wolf, in Proceedings of IEEE, Vol. 51, 1963, p. 674.

34 A. Adibi, "Schottky Barrier Solar Cells", Thesis, University of California, Santa Barbara, 1977.

35 E. J. Charlson and J. C. Lein, in Journal of Applied Physics, Vol. 47, 1975, p. 3982.

36 J. Schewchun, R. Singh and M. A. Green, in Journal of Applied Physics, Vol. 48, 1977, p. 765.

37 H. J. Hovel, Semiconductors and Semimetals, Volume III, Solar Cells, Academic Press, New York, 1975, p. 765.

38 N.C. Wyeth, in Solid State Electronics, Vol. 20, 1977, p. 629. This article treats the "front layer" resistance in greater detail than space permits here--for both circular and rectangular solar cell geometries.

39 M. Green, University of New South Wales, reported in the Photovoltaic Insider's Report, 30 April, 1993.

CHAPTER VII: ADVANCED APPROACHES

Introduction

In Chapters IV through VI we considered the performance of solar cells both in the abstract and utilizing concrete examples wherein the semiconductor, optical orientation, and construction technology were specified. These examinations of performance took place under AMO and AM1 conditions. We may consider a solar cell system operating under simple AMO or AM1 conditions to be a stage one solar cell system.

Let us define a stage two solar cell system as one which concentrates sunlight and allows this concentrated light to impact the solar cells. There exist several powerful reasons for using concentrated sunlight as the energy source. The principal reason is cost. Note that the single crystal semiconductor solar cells which we have been considering are expensive (both in terms of "dollars" and in terms of time) to fabricate and that they exhibit a number of loss mechanisms (hole-electron pair recombination, series resistance, etc.). It is, generally, less expensive to fabricate mirrors and/or lenses which can then be used to collect sunlight over a large area and to focus this light on a relatively small area of semiconducting solar cells. Not only do we replace large amounts of expensive solar cells with inexpensive lenses (or mirrors), but the intense, focused sunlight which illuminates the solar cells tends to saturate a number of solar cell loss mechanisms and so improve overall solar cell efficiency. It should be noted that when we concentrate sunlight we need to have some tracking mechanism so that the solar power system is aligned with the sun. However, even with the additional cost and complication of a tracking mechanism, the overall system cost per provided KWH of a concentrating solar energy system can be significantly lower than for a non-concentrating, non-tracking system [1-4].

Another reason for the use of concentrated sunlight arises from the availability of thermal as well electrical energy from the solar power system. The solar cells are exposed to a high concentration of photons.

Though the efficiency of the solar cells increases, a significant fraction of the solar energy is still converted to heat in the semiconductor. This is undesirable since, as we shall see in this chapter, increased solar cell temperature results in reduced efficiency in converting solar energy to electrical energy. Thus, this heat must be removed from the solar cells and if the energy density is sufficient, then the thermal energy removed from the solar cells while cooling them can be used to heat water, or living spaces, or for some industrial purpose We will spend most of this chapter on these stage two solar cell systems.

There is a stage three type of solar power system. From our earlier discussions it is evident that no semiconductor is an ideal match for the solar spectrum (either AMO or AM1). There are too many high energy photons in sunlight, which, when absorbed, yield the standard E_g of electrical energy with the rest of the photon energy ending up as heat. There are also too many low energy photons in the solar spectrum-- photons that pass completely through the semiconductor. Third stage solar energy systems initially alter the spectral distribution of sunlight to a form which is a closer match to the ideal spectrum for a particular semiconductor. These third stage systems also furnish the energy consumer with both electrical and thermal energy. We will consider the theoretical aspects of these systems in the following chapter.

If we consider the effects of optical concentration on solar cells, the observed phenomena are of three kinds. First, with increased photon flux density the hole-electron pair generation increases. From the discussions in Chapter III, the greater density of holes and electrons will increase the recombination rate, however, the recombination mechanisms will begin to saturate as the carrier density increases. Thus a smaller percentage of the hole-electron pairs will be lost via recombination, and the photocurrent and device output power density will increase, yielding a higher operating efficiency. Second, the increased number of electron-hole pairs in the relatively lightly doped substrate layer of the solar cells tends to reduce the resistance of the substrate (an effect improving device efficiency) while depressing the collecting electric field. This results, unfortunately, in an increased surface recombination rate. Finally, the principal effects of the use of concentrated sunlight are thermal. The incoming light energy is converted to both electrical and thermal forms of energy. The thermal energy will act to raise the junction temperature of the solar cells, in turn decreasing the efficiency with which they convert optical to electrical energy.

As in our earlier studies, our examination of the effects of optical concentration on solar cells will be based on studying the example semiconductors and design philosophies addressed for the first stage systems in Chapter VI. We begin with an examination of temperature.

Temperature Effects

From Chapter V, the saturation current densities for the various types of junctions under consideration are:

(a) for pn junctions

$$J_S = \frac{qD_{pn}n_i^2}{L_{pn}N_D} + \frac{qD_{np}n_i^2}{L_{np}N_A} \ , \qquad \text{(VII.1)}$$

where the first term on the right hand side refers to the n-type region of the junction and the second term to the p-type region. Recall from the discussions in Chapter VI, that pn junction solar cells are constructed from a moderately doped substrate and a heavily doped, "front layer". From the typical impurity concentration levels of the examples in Chapter VI it is clear that the contribution of the "front layer" to the saturation current density is minimal, reducing Equation VII.1 to but a single term, that due to the substrate:

(b) for heterojunction devices

$$J_S = J_{SH} = (qD_S n_i^2 / L_S N_S) X_T \ , \qquad \text{(VII.2)}$$

where the narrow energy gap substrate is primarily responsible for the saturation current density and the only effect from the wide energy gap "front layer" is the X_T which accounts for carrier transport effects across the junction. This factor is characteristically less than unity [5]. In this analysis, we will make the conservative estimate that X_T is unity and, furthermore, is not a function of the temperature [6]:

(c) for a Schottky barrier the saturation current is

$$J_S = J_{SS} = A^*T^2 exp\{-q\phi_{Bo}/kT\} \ , \qquad\qquad (VII.3)$$

and (d) for mos type Schottky barriers

$$J_S = J_{SS} = A^{**}T^2 exp\{-q\phi_{Bo}/kT\} \ , \qquad\qquad (VII.4)$$

where A^* is the Richardson constant and A^{**} is the modified Richardson constant. As indicated in Chapters V and VI, the mos-Schottky device is imperfectly understood at present and sufficient data for a full engineering analysis is not available. We will not consider mos-Schottky barriers further.

Beginning with the pn junction let us consider the saturation current densities for pn, heterojunction and Schottky barriers in detail. From Chapters III and VI we have, for the diffusion constant and diffusion length of the lightly doped substrate region:

$$D_S = \{kT/q\}\mu_S \ and \ L_S = \sqrt{\{D_S\tau_S\}} \ . \qquad\qquad (VII.5)$$

From Chapter III it is clear that the mobility and lifetime in the substrate are temperature dependent. Assuming that the carrier capture cross section and trap density are temperature independent, then lifetime for trap recombination will vary roughly inversely with the square root of the temperature. Thus, the principal variation with temperature of the saturation current in pn junctions is due to the temperature variation of the intrinsic carrier concentration which has a dependence of the form:

$$n_i^2 \propto T^3 exp\{E_g(T)/kT\} \ . \qquad\qquad (VII.6)$$

To first order, we may write for pn junctions:

$$J_S(T)/J_S(300) = (T/300)^3 exp\{E_g(300)/k300 - E_g(T)/kT\} \ . \qquad (VII.7)$$

The variations in energy gap with temperature for the example semiconductors used in this work are provided in Appendix B. Specific values of the intrinsic carrier concentration squared are furnished in Table III.2 and the ratio of $J_S(T)$ to $J_S(300)$ for Si, InP, GaAs, CdTe and AlSb are displayed in Figure VII.1. Note that the range in temperature is from 300°K to 500°K. Study of Figure VII.1 combined with observation of Figures V.8 through V.15 indicates that the output power density of pn junction solar cells at 500°K lies between 10 and 40% of the value at 300°K. Even at 400°K the performance is much degraded.

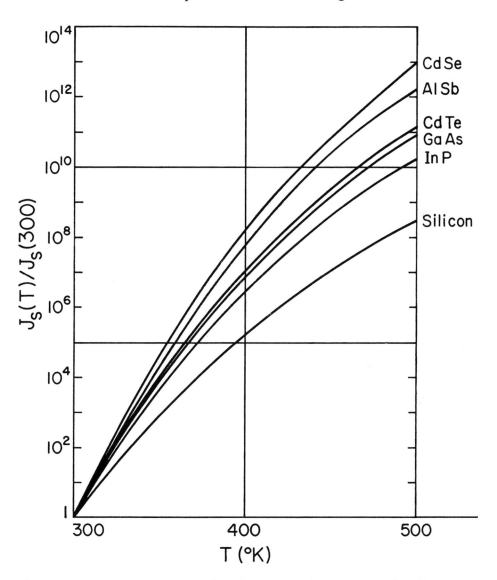

Figure VII.1.   PN and heterojunction saturation current densities as a function of the temperature, normalized to the saturation current density at 300°K with the substrate semiconductor as a parameter.

Examination of Equation VII.2, which provides the saturation current density for heterojunctions as a function of temperature, indicates that the temperature dependence of the saturation current for these devices exhibits the same characteristics as the pn junctions of Figure VII.1. The curve in this figure for cadmium selenide is seen to be that for a heterojunction on n-type CdSe.

Note that we can maintain a low junction temperature in a solar cell by cooling it. If we do this, we can use a portion of this thermal energy to provide space and/or material (e.g., water) heating. In such a situation, overall system energy requirements may actually encourage some decrease in electric output from the solar cells while the resulting increase in junction temperature yields a more efficient extraction of thermal energy (see Equation I.4). At this juncture, without full system and economic details, we can do no more than to remark that operation for solar cell junction temperatures in excess of approximately 400°K is inefficient.

Consider the changes in Schottky barrier saturation current induced by shifts in temperature. Assuming that the effective mass is temperature independent, then, from Equation VII.3:

$$J_S(T)/J_S(300) = (T/300)^2 \exp\{q\phi_{Bo}(300)/k300 - q\phi_{Bo}(T)/kT\} .\qquad (VII.8)$$

In Table VI.4, a number of Schottky barrier energy values (at 300°K) for the six example semiconductors are presented. As the temperature increases, we expect the barrier energy to vary as [8]:

$$\phi_{Bo}(T) = \phi_{Bo}(300) + A_1(T) .\qquad (VII.9)$$

In covalent semiconductors, the presence of surface states pins the Fermi level, at the surface, relative to the conduction band edge. In such instances, it has been found that the barrier energy varies in temperature as does the energy gap [9]. This is the situation for silicon. The other five example semiconductors are ionic, or nearly ionic. In an ionic semiconductor, the Fermi level, at the semiconductor surface, can move freely between the conduction and valence band edges. This variation of surface barrier energy with temperature is expected to be proportional to that fraction of the energy gap occupied by the Schottky barrier. Therefore:

$$A_1 = \partial\phi_{Bo}/\partial T = (\phi_{Bo}/E_g)(\partial E_g/\partial T) .\qquad (VII.10)$$

The temperature dependence of the saturation current density for selected Schottky barriers is presented in Figure VII.2. Because of the

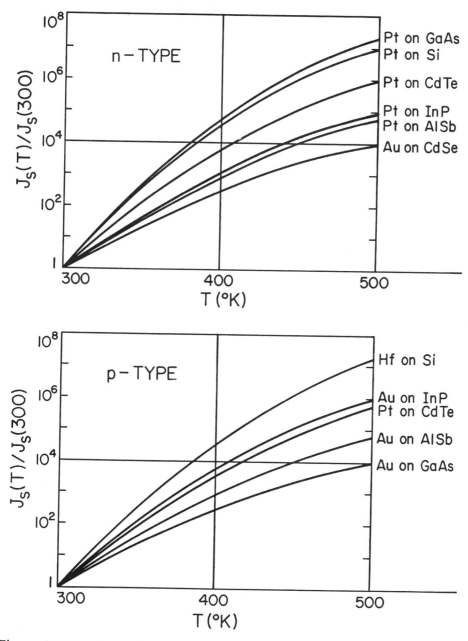

Figure VII.2.   Saturation current density for selected Schottky barriers versus temperature, normalized to the saturation current density at 300°K.

smaller barrier heights, the increase in current with temperature is less for Schottky barriers than for pn or heterojunctions. However, our studies in the preceding chapters have indicated that, with present day technology, we start with a higher level of saturation current density, and hence, with a lower solar cell efficiency.

Study of Figures VII.1 and VII.2 makes it all-too-plain that the allowable degree of temperature rise is limited. The exact degree of permissible temperature rise is the result of trading off photovoltaic efficiency and produced electric power with increasing utilization of the heat rejected by the solar cells to some cooling medium [10].

Heat Flow within a Solar Cell

Figure VII.3 displays thermal diagrams for standard and vertical configuration solar cells, assuming heat flow into the sink located on the down or non-illuminated surface only. Note that heat is generated in two ways in these solar cells. First, there are ohmic losses (I^2r_D) as a result of the current flowing through the resistance, r_D, of the solar cell. Second, there is the thermal energy which represents the difference between the energy of the absorbed solar photons and the realized electrical energy of the generated electron-hole pairs. For air-mass-one conditions these two components of heat energy will range in size from $107N$ mW/cm^2 (a solar cell with zero percent efficiency for conversion to electrical energy where N is the degree of optical concentration) to approximately $80N$ mW/cm^2 for a solar cell of 25% electrical conversion efficiency.

This heat energy is distributed over the entire volume of the solar cell. However, as can be seen from study of the absorption curves of Chapter IV and the series resistance discussion of Chapter VI, the bulk of the heat is generated near the illuminated surface[*]. The conservative approach to thermal analysis that we will follow here, is to assume that

---

[*] Figure VII.3 depicts standard and vertical configuration solar cells. An inverted configuration resembles an upside down standard configuration cell. For such a device the thermal energy derived from the excess photon energy is generated near the illuminated surface. However, the bulk of the heat energy derived from the flow of current is generated close to the heat sink.

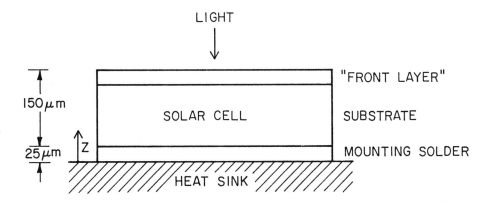

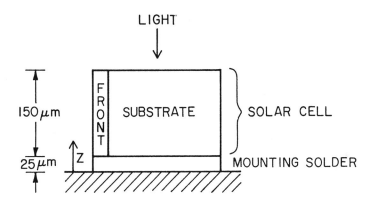

Figure VII.3. Thermal configurations for standard (the upper figure) and vertical (the lower figure) configuration solar cells.

the thermal energy is generated at the illuminated surface and then flows through the semiconductor to reach the heat sink.

    Let T_S be the temperature of the heat sink, and R_T be the thermal resistance of the solar cell and other material between the heat sink and the solar cell junction. Then, for a heat energy flow, Θ_T, between the junction and the heat sink, the temperature of the junction, T_J, is:

$$T_J = T_S + R_T \Theta_T .$$
(VII.11)

where R_T is given by:

$$R_T = \int \{\rho_T(z)/A(z)\}dz \ , \tag{VII.12}$$

where the integral is from the heat sink to the solar cell junction (in the direction z); A(z) is the cross sectional area through which the heat is flowing; and $\rho_T(z)$ is the thermal resistivity of the semiconductor, the mounting solder and any other materials between the heat sink and the junction.

The thermal resistivity of mounting solder is difficult to accurately determine. From sales literature, a value of 4°C-cm per watt is obtainable, if the solar cell is correctly mounted on the heat sink. If it is not, the solder thermal resistivity may be one to ten orders of magnitude higher. The thermal resistivity of the example semiconductors is presented in Table VII.1.

Table VII.1

The thermal resistivity (°C-cm/watt) at 300°K

Semiconductor	Si	InP	GaAs	CdTe	AlSb	CdSe
ρ_T	1	1.43	1.13	14.29	1.11	15.87
Reference	[11]	[12]	[13]	[14]	[15]	[16]

It should be noted that the thermal resistivity is a function of temperature for most materials and certainly is so for semiconductors [17]. However, it is clear, from our earlier study of saturation current density, that large excursions in temperature are undesirable. Therefore, for this first order study of thermal effects, we will assume a constant value of thermal resistivity for our semiconductors.

The junction temperature of a standard, vertical or inverted configuration solar cell can now be determined if we know the degree of optical concentration, N; the thermal resistance or the solder and semiconductor; and the operating temperature of the heat sink. The heat flow power density under the conditions of maximum electrical power output power density is:

$$\Theta_T = NI_{ns}\{1 - \eta_s/100\} \ , \tag{VII.13}$$

where $I_{ns} = 135.3$ mW/cm^2 for AMO conditions and equals 107 mW/cm^2 for AM1 conditions, and:

$$\eta_s = P_{max}/\{A_D N I_{ns}\} \ , \tag{VII.14}$$

where P_{max}/A_D is determined using Equation VI.18 subject to the conditions that the photocurrent density, J_{ph}, is N times its unconcentrated value and that the operating voltage for maximum power output, V_D', and the diode loss factor, K', must mutually satisfy Equations VI.16 and VI.17.

For vertical configuration solar cells the situation is somewhat more complicated than for the standard and inverted configuration devices. The saturation current depends upon the temperature of the junction and from Figure VII.3 it is clear that the junction temperature for vertical configuration cells varies across the device. Returning to Equation VII.11, we can write for vertical configuration solar cells:

$$T_j = T_S + \Delta T_s \ (z), \tag{VII.15}$$

where ΔT_s (z) is the temperature rise across the semiconductor and mounting solder, assuming z = 0 to be the heat sink location.

This complication makes it necessary to integrate from the heat sink to the illuminated surface in order to determine the saturation current density for a vertical configuration solar cell. As discussed in Chapter VI, the vertical configuration solar cell has a tendency to exhibit a higher photovoltage at the illuminated surface--due to the nature of the absorption coefficient. The combination of temperature and photovoltage variations makes it difficult to analyze the efficiency of these devices in detail, unless the exact structure of the cell is known. Rough calculations suggest a saturation current density of from 20 to 60 percent of the value calculated assuming that the entire junction is at a temperature equivalent to the illuminated surface.

The inverted solar cell has its junction close to the heat sink which means that its operating junction temperature is somewhat lower than the other devices; again depending on construction details.

It is clear, at this point, that further pursuit of the thermal analysis of solar cell systems would involve much detailed analysis of materials, fabrication techniques, thermal energy exchange between solar cell and heat sink, and of the general ambient conditions. Still, it is important that we consider the effects of temperature on the performance of solar cells. Either deliberately (owing to the desire for extra thermal energy) or accidentally (because of faulty construction techniques) it is possible to encounter high operating temperatures. Study of Figures VII.1 and VII.2 and the general performance data of Chapter V indicates that it is unwise to consider solar cell operation at junction temperatures much in excess of 400°K. For the purposes of our initial study of the effects of junction temperature, let us assume that the junction temperature is either 300, 350 or 400°K. Earlier, in Chapter VI, we considered the 300°K case under single sun (N=1) illumination. We now commence our broader study by determining the saturation current density for our example semiconductors as a function of temperature using Figures VII.1 and VII.2.

Table VII.2

$\underline{M(T)} = J_S(T)/J_S(300°K)$ as a function of barrier type, temperature, substrate and semiconductor

Semiconductor	Si	InP	GaAs	CdTe	AlSb	CdSe
			350°K			
PN & Heterojunction	1.1×10^3	4.2×10^3	7.8×10^3	9.3×10^3	2.5×10^4	5.0×10^4
Schottky on n-type	290	053	430	130	48	25
Schottky on p-type	381	140	026	120	48	--*
			400°K			
PN & Heterojunction	1.8×10^5	2.4×10^6	7.5×10^6	1.0×10^7	5.6×10^7	1.3×10^8
Schottky on n-type	3.0×10^4	1000	5.6×10^4	5620	1000	237
Schottky on p-type	3.2×10^4	4730	237	1540	0750	---*

*p-type CdSe is not commercially available

We will be examining the temperature variations by using specific examples. In turn, this will indicate that some designs of solar cells are more temperature sensitive than others and some semiconductors are more sensitive than others. In this work, it will merely be noted that the design of solar cells for high junction temperature operation may be beneficial in that it allows more efficient use of the heat rejected to the cooling system. The scientific literature has been relatively silent on such designs, but some work is being done [18-21]. As we treat optical concentration in this chapter, the reader will be able to draw conclusions as to the amount of effort which may prudently be expended on heat sinking and thermal matters in general.

Optical Concentration - Photocurrent

The hole-electron pairs generated by incoming photons serve to reduce the series resistance in the substrate as well as to provide the photocurrent. In Chapter VI an estimate was made of the photocurrent generated under single sun conditions for the various device configurations and junctions. For convenience, these values are reproduced in Table VII.3. Note that these values are not the ideal current densities, but have been reduced to account for reflection and transmission losses of incoming photons, bulk and surface recombination effects and surface metallization effects. In particular note that the metallization required to form Schottky barrier standard configuration cells drastically reduces the available photocurrent. Also note that the expected photocurrent for inverted configuration solar cells is quite high. This is a direct result of the construction employed for these cells--a construction which effectively results in a light collecting area which is double the junction area. This, in turn, results in an effective optical concentration of ˜1.95 for an inverted configuration solar cell.

The photocurrent densities listed in Table VII.3 are conservative estimates based upon the discussions in Chapters III through VI and upon work elsewhere upon various collection models [19, 22]. The distribution of the hole-electron pairs which gives rise to this current, is not uniform over the entire thickness of the solar cell, but is skewed towards the illuminated side. The exact nature of the generated carrier distribution may be inferred from Figures IV.6 and IV.7. Let us currently assume that the distribution of generated carriers is roughly even over the entire solar cell. (Note--a photocurrent density of 32 mA/cm^2 such as that predicted for GaAs standard configuration heterojunctions under AMO conditions in Table VII.3 implies an electron-hole pair generation rate of 2 x 10^{17}/cm^2-sec.)

From the discussion on solar cell design in Chapter VI, the optimum solar cell design has its lowest substrate impurity

Table VII.3

The photocurrent density (mA/cm^2) for standard, inverted and vertical configuration solar cells under AMO and AM1 illumination for pn, heterojunction and Schottky barrier solar cells

Semiconductor	Si	InP	GaAs	CdTe	AlSb	CdSe
Inverted configuration (all junctions)						
AMO	87.07	81.31	72.54	69.81	54.34	50.70*
AM1	70.40	62.01	55.97	53.04	42.71	39.98*
Vertical configuration (all junctions)						
AMO	42.70	39.50	35.30	33.90	27.10	24.70*
AM1	34.20	30.10	27.20	25.70	20.80	19.50*
Standard configuration (Schottky barrier)						
AMO	10.59	02.34	06.96	07.53	06.15	03.75*
AM1	08.34	01.89	05.25	05.70	04.68	03.06*
(pn junctions)						
AMO	31.77	07.02	20.88	22.54	18.45	*
AM1	25.02	05.67	15.75	17.10	14.04	*
(heterojunctions)						
AMO	37.97	35.44	31.62	30.43	24.31	22.10*
AM1	30.68	27.03	24.40	23.12	18.62	17.42*

*Recall that p-type CdSe is not commercially available

concentration close to the junction. Here, N_S is on the order of 10^{15}/cm^3. In the "front layer" the impurity concentration is close to 10^{19}/cm^3 in both pn and heterojunction solar cells. Near the substrate contact, when the substrate is graded to facilitate carrier collection, the impurity concentration approaches 10^{19}/cm^3. Thus, solely near the junction in the substrate region of the solar cell, is the addition of hole-electron pairs liable to make a major contribution to the reduction of the device series resistance. In a well designed solar cell it is just this region that is subject to the greatest density of optically generated charge carriers. The actual contribution to the conductivity of these optically generated charge carriers is the result of specific device parameters. The expected resistance times area product of the substrate for solar cells made from our six example semiconductors was presented in Chapter VI. Call this value $(A_D r_D)_{so}$. In keeping with the first order approximations made elsewhere in this chapter, and utilizing a non-uniform generated carrier concentration, peaking near the junction, we have, from Figures IV.6 and IV.7, the following useful approximation for the effects of optical concentration on the series resistance of the substrate:

$$(A_D r_D)_s = (A_D r_D)_{so}/\{1 + .1N\} .$$
(VII.16)

Consider the effect on recombination and the reflecting back contact of the increasing electron-hole pair density as the optical concentration increases. In standard configuration solar cells the generation for optically driven hole-electron pairs is primarily near the "front layer" surface, and does not affect the reflecting nature of the substrate contact. In vertical configuration solar cells this is not necessarily so, particularly in the direct gap semiconductors where the bulk of carrier generation is close to the illuminated surface. In inverted configuration solar cells the optical generation of hole-electron pairs is, in direct semiconductors, close to the reflecting surface and can have major effects. Once again, exact analysis, for all three configurations, depends on the surface recombination velocity of the substrate contact, on the precise nature of the impurity concentration in the substrate, and upon the absorption coefficient of the semiconductor. For our purposes, in this first order analysis, let us assume that the substrate contact remains reflecting.

From Chapter III, the recombination rate, U, for hole-electron pairs increases in proportion to:

$$U \propto \{pn - n_i^2\}/\{p + n\} .$$
(VII.17)

As p and n increase above n_i^2, the recombination rate becomes proportional to the electron (or hole) concentration. Thus, recombination, for high optical concentration levels, is dependent on the generated carrier density. At the same time, the lifetime, which is a measure of the time available to collect these optically generated carriers, increases to a maximum, τ_m, equal to [23]:

$$\tau_m = \tau_{po} + \tau_{no} .$$

The mechanism responsible for recombination becomes saturated at high carrier concentrations and the result is an increase in the photocurrent. In our first order analysis of solar cells this effect will be neglected and the photocurrent will be taken to be the unconcentrated photocurrent (see Table VII.3) times the optical concentration, N.

Performance Under Concentration

We are now in a position to make estimates of solar cell performance as a function of the level of optical concentration using Equations VI.16 through VI.18 and VII.14 to compute output power density, efficiency, operating voltage and loss factor with temperature as a parameter.

In the following figures, we examine solar cell performance for optical concentrations, N, varying from unity to 1000. At each concentration level we will assume junction temperatures of 300°K, 350°K and 400°K. The saturation current density is obtained from Tables VII.2 and VI.13 for pn, heterojunction and Schottky barrier solar cells. The photocurrent is obtained from Table VII.3 using the appropriate value of N. The resistance times area product is that determined in Chapter VI[*]. Figures VII.4 through VII.9 present computed output power density and efficiency for standard configuration solar cells exposed to an AMO type of spectrum (N = 1 corresponds to AMO light, while N = 10 corresponds to light of the same AMO spectral distribution, but contains 10 times as many photons). Data are presented for solar cells made from silicon, indium phosphide, gallium arsenide, cadmium telluride, aluminum antimonide and cadmium selenide with n-type substrates. Results for solar cells with p-type substrates are similar to these values, with variations primarily due to Schottky barrier height variations and minor resistance and lifetime changes.

Study of Figures VII.4 through VII.9 reveals that the output power density increases with increasing optical concentration level for all of

---

[*] Recall that this product depends upon both the substrate and "front layers" as demonstrated in Tables VI.14 through VI.16 and, for standard configuration solar cells, the "front layer" area times resistance product is a function of the geometry.

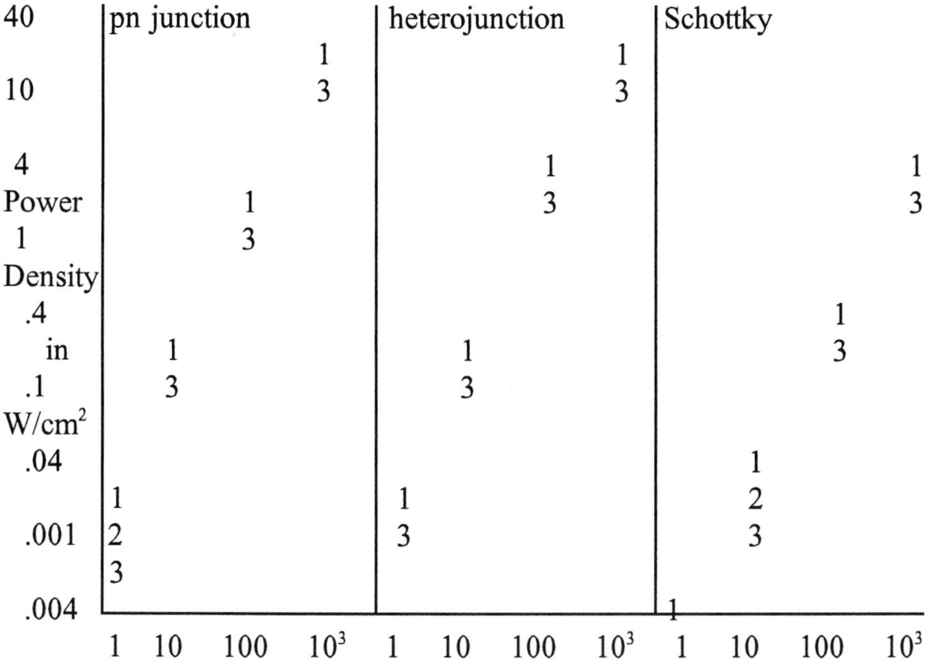

```
40      pn junction        heterojunction       Schottky
                      1                      1
10                    3                      3

4                               1                            1
Power           1               3                            3
1               3
Density
.4                                                      1
 in            1                1                       3
.1             3                3
W/cm²
.04                                                    1
              1                1                        2
.001          2                3                        3
              3
.004    ┗━━━━━━━━━━━━━━━━━━━━━━━━━━━━━━━━━━━━━━━━━━━━━━━━━━━━
        1  10  100  10³    1  10  100  10³    1  10  100  10³
                        Solar Concentration
        Key: 1--> T = 300°K, 2--> T = 350°K, 3--> T = 400°K
```

Efficiency (%)

Temperature (°K)	Solar Concentration Level						
	1	3.16	10	31.6	100	316	1000
			pn junctions				
300	11.4	12.0	12.6	13.2	13.5	13.6	10.3
350	08.6	09.3	10.1	10.7	11.1	10.8	08.4
400	06.0	06.9	07.7	08.4	09.0	08.8	06.8
			heterojunctions				
300	13.7	14.5	15.2	15.9	16.2	15.4	11.4
350	10.4	11.3	12.1	12.9	13.3	12.7	09.2
400	07.4	08.4	09.3	10.2	10.8	10.3	07.3
			Schottky barriers				
300	2.13	2.34	2.56	2.76	2.94	3.03	2.88
350	1.28	1.52	1.76	1.99	2.20	2.33	2.23
400	0.48	0.70	0.94	1.18	1.42	1.58	1.54

Figure VII.4. Output electrical power density and efficiency as functions of solar concentration with junction temperature as a parameter for standard configuration silicon solar cells with n-type substrates and under AMO-like spectral distribution.

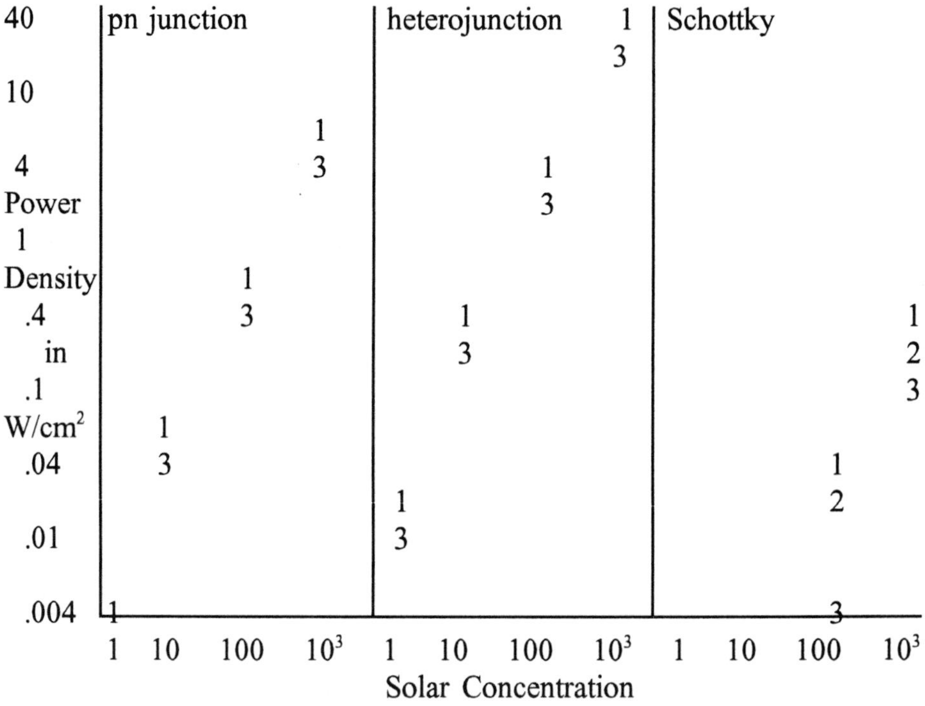

```
40  | pn junction            | heterojunction    1 | Schottky
    |                        |                   3 |
10  |                        |                     |
    |              1         |                     |
 4  |              3         |         1           |
Power|                       |         3           |
 1  |                        |                     |
Density|          1          |                     |
 .4 |              3         |     1               |                    1
 in |                        |     3               |                    2
 .1 |                        |                     |                    3
W/cm²|   1                   |                     |
 .04|   3                   |                     |                1
    |                        |     1               |                2
 .01|                        |     3               |
    |                        |                     |
.004 |_____|_____|_____3___
      1  10  100  10³    1   10   100   10³    1   10   100   10³
                        Solar Concentration
     Key: 1--> T = 300°K, 2--> T = 350°K, 3--> T = 400°K
```

Temperature	Efficiency (%) Solar Concentration Level						
(°K)	1	3.16	10	31.6	100	316	1000
	pn junctions						
300	3.67	3.81	3.96	4.11	4.25	4.37	4.44
350	3.02	3.19	3.36	3.54	3.70	3.85	3.94
400	2.41	2.57	2.77	2.96	3.15	3.32	3.44
	heterojunctions						
300	19.6	20.3	21.0	21.7	22.3	22.4	21.0
350	16.5	17.3	18.2	19.0	19.7	19.9	18.7
400	13.4	14.4	15.3	16.2	17.0	17.4	16.4
	Schottky barriers						
300	.078	.114	.154	.196	.239	.281	.319
350	.007	.002	.042	.073	.113	.156	.197
400	.001	.002	.005	.014	.032	.061	.097

Figure VII.5. Output electrical power density and efficiency as functions of solar concentration with junction temperature as a parameter for standard configuration indium phosphide solar cells with n-type substrates and under AMO-like spectral distribution.

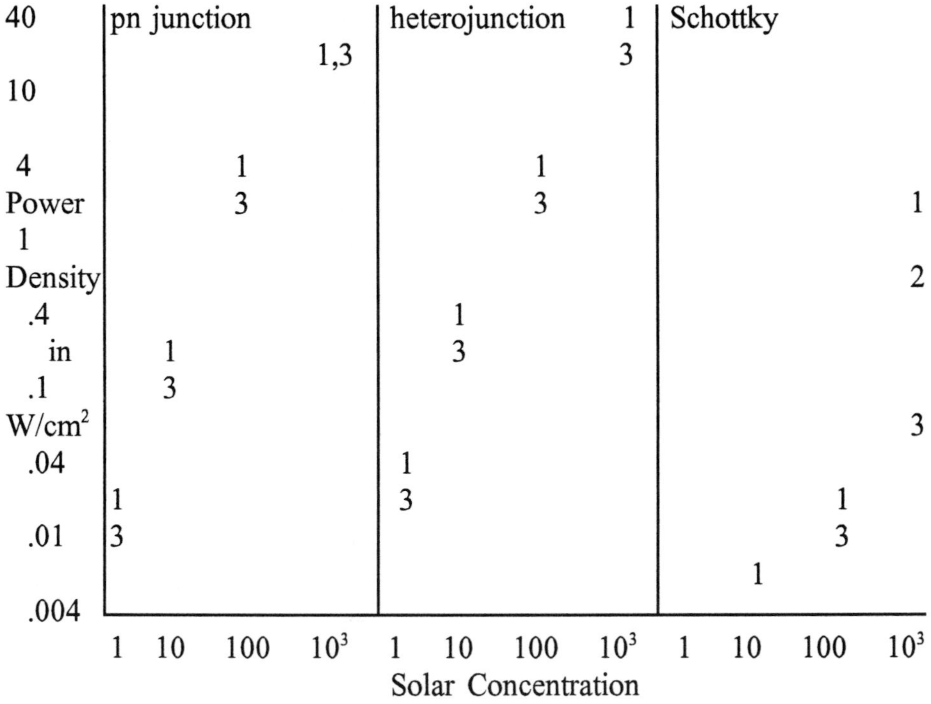

```
40  | pn junction           | heterojunction    1 | Schottky
    |                 1,3   |                   3 |
10  |                       |                     |
    |                       |                     |
 4  |          1            |          1          |
Power|         3            |          3          |                    1
 1  |                       |                     |
Density|                    |                     |                    2
 .4 |                       |          1          |
 in |          1            |          3          |
 .1 |          3            |                     |
W/cm²|                      |                     |                    3
 .04|                       |          1          |
    | 1                     |          3          |                1
 .01| 3                     |                     |                3
    |                       |                     |           1
.004|_____|_____|_____
     1   10   100   10³   1   10   100   10³   1   10   100   10³
                         Solar Concentration
    Key: 1--> T = 300°K, 2--> T = 350°K, 3--> T = 400°K
```

Efficiency (%)

Temperature (°K)	1	3.16	10	31.6	100	316	1000
			pn junctions				
300	13.3	13.8	14.2	14.7	15.1	15.4	15.5
350	11.5	12.0	12.5	13.0	13.5	13.9	14.1
400	09.6	10.2	10.8	11.4	11.9	12.4	12.6
			heterojunctions				
300	20.4	21.1	21.8	22.4	23.0	23.4	23.2
350	17.7	18.5	19.3	20.0	20.7	21.2	21.2
400	14.9	15.8	16.7	17.6	18.4	19.0	19.0
			Schottky barriers				
300	.339	.457	.582	.711	.841	.966	1.07
350	.010	.027	.067	.137	.235	.348	.455
400	.000	.000	.001	.002	.007	.020	.047

The column subheader "Solar Concentration Level" spans the concentration columns.

Figure VII.6. Output electrical power density and efficiency as functions of solar concentration with junction temperature as a parameter for standard configuration gallium arsenide solar cells with n-type substrates and under AMO-like spectral distribution.

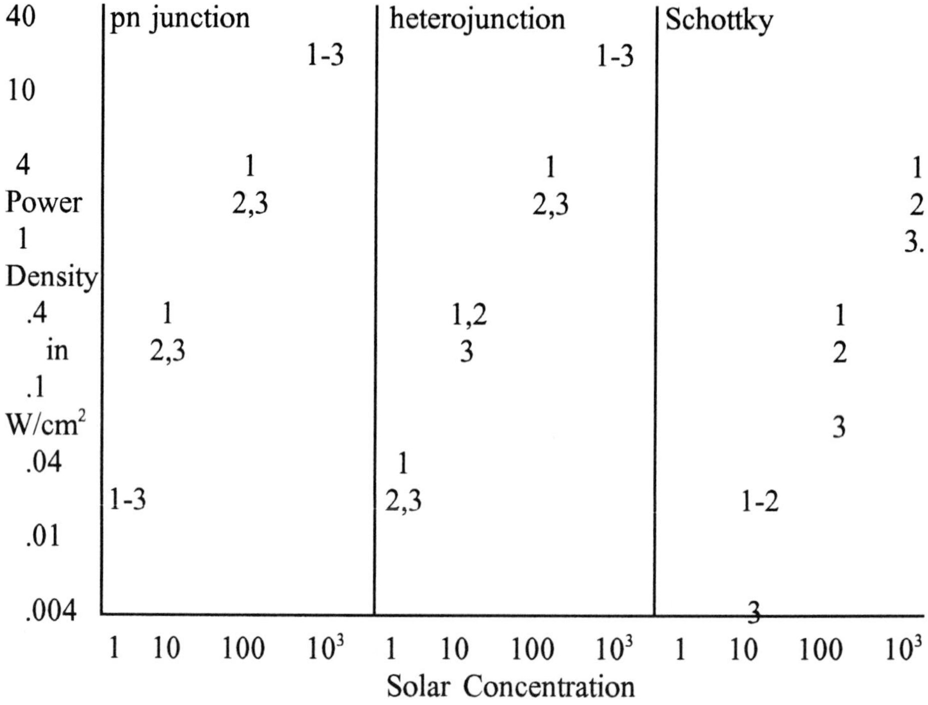

	pn junction		heterojunction		Schottky
40		1-3		1-3	
10					
4	1		1		1
Power	2,3		2,3		2
1					3.
Density					
.4	1		1,2		1
in	2,3		3		2
.1					
W/cm²					3
.04			1		
	1-3		2,3		1-2
.01					
.004					3

```
     1  10  100  10³   1  10  100  10³   1  10  100  10³
```

Solar Concentration

Key: 1--> T = 300°K, 2--> T = 350°K, 3--> T = 400°K

Efficiency (%)

Temperature (°K)	Solar Concentration Level						
	1	3.16	10	31.6	100	316	1000
			pn junctions				
300	16.1	16.6	17.0	17.4	17.6	17.2	14.7
350	14.3	14.9	15.4	15.9	16.1	15.8	13.5
400	12.5	13.1	13.7	14.3	14.6	14.4	12.2
			heterojunctions				
300	21.9	22.5	23.1	23.6	23.7	22.7	18.1
350	19.5	20.3	21.0	21.6	21.8	20.9	16.5
400	17.1	17.9	18.7	19.4	19.8	19.0	14.8
			Schottky barriers				
300	1.31	1.46	1.61	1.76	1.88	1.94	1.80
350	0.81	0.97	1.14	1.30	1.45	1.54	1.43
400	0.12	0.22	0.36	0.51	0.66	0.76	0.73

Figure VII.7. Output electrical power density and efficiency as functions of solar concentration with junction temperature as a parameter for standard configuration cadmium telluride solar cells with n-type substrates and under AMO-like spectral distribution.

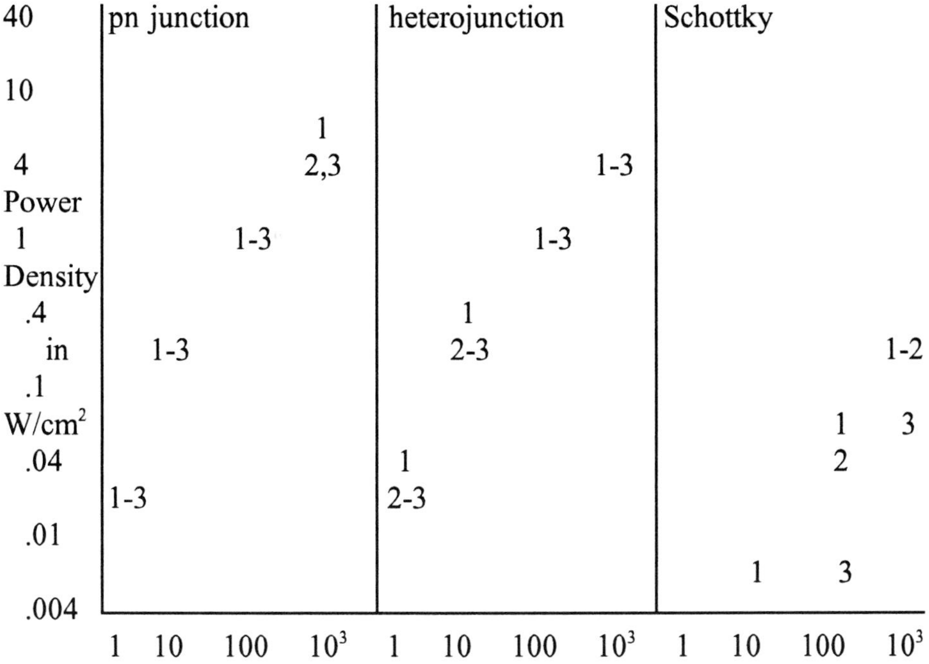

```
40   | pn junction            | heterojunction          | Schottky

10
                1
 4             2,3                      1-3
Power
 1         1-3                        1-3
Density
 .4                              1
  in      1-3                   2-3                              1-2
 .1
W/cm²                                                      1    3
 .04                           1                           2
         1-3                   2-3
 .01
                                                          1    3
.004 |_____|_____|_____
        1   10  100  10³    1   10  100  10³    1   10  100  10³
                          Solar Concentration
```

Key: 1--> T = 300°K, 2--> T = 350°K, 3--> T = 400°K

Temperature	Efficiency (%) Solar Concentration Level						
(°K)	1	3.16	10	31.6	100	316	1000
				pn junctions			
300	13.9	14.3	14.5	14.5	13.6	09.9	03.8
350	12.2	12.6	12.9	13.0	12.1	08.6	03.4
400	10.4	10.9	11.3	11.4	10.6	07.4	02.9
				heterojunctions			
300	18.4	18.9	19.2	19.0	17.2	10.9	03.9
350	16.2	16.7	17.1	17.0	15.3	09.4	03.4
400	13.9	14.5	15.0	14.9	13.4	08.0	02.9
				Schottky barriers			
300	.245	.341	.438	.516	.522	.370	.173
350	.033	.076	.141	.212	.243	.175	.098
400	.002	.006	.016	.037	.060	.062	.043

Figure VII.8. Output electrical power density and efficiency as functions of solar concentration with junction temperature as a parameter for standard configuration aluminum antimonide solar cells with n-type substrates and under AMO-like spectral distribution.

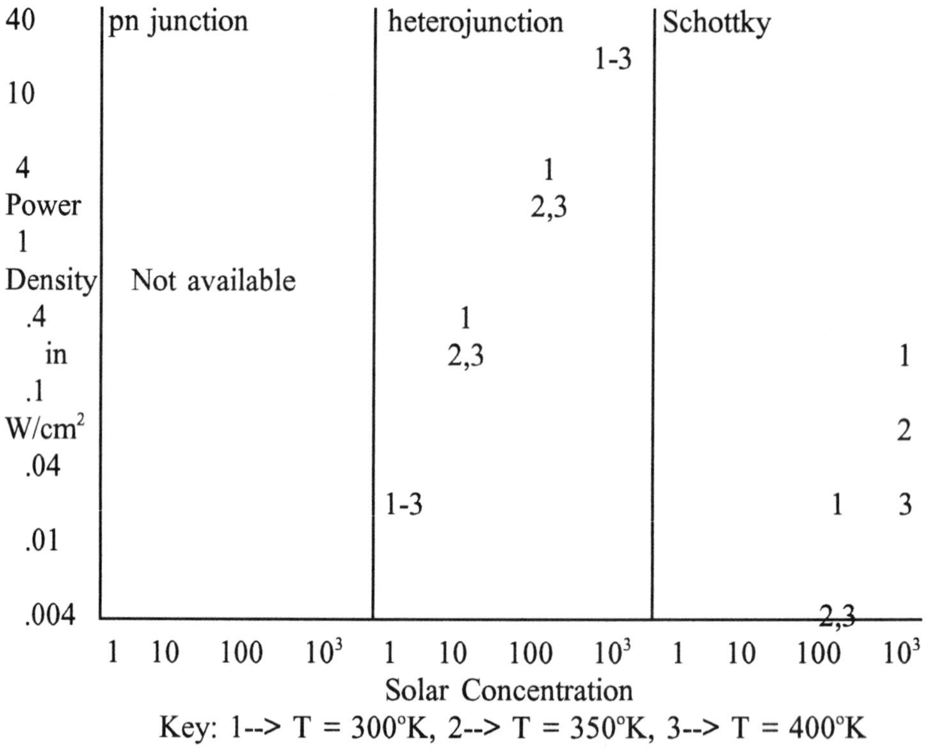

	pn junction	heterojunction	Schottky
40			
		1-3	
10			
4		1	
Power		2,3	
1			
Density	Not available		
.4		1	
in		2,3	1
.1			
W/cm^2			2
.04			
		1-3	1 3
.01			
.004			2,3

Solar Concentration axis: 1 10 100 10^3 1 10 100 10^3 1 10 100 10^3

Solar Concentration

Key: 1--> T = 300°K, 2--> T = 350°K, 3--> T = 400°K

Efficiency (%)

Temperature (°K)	Solar Concentration Level						
	1	3.16	10	31.6	100	316	1000
				pn junctions			
300							
350		Not available, owing to lack of p-type CdSe					
400							
				heterojunctions			
300	17.5	18.0	18.9	18.7	18.8	17.8	13.8
350	15.2	15.8	16.3	16.7	16.8	15.9	12.1
400	13.1	13.7	14.3	14.8	15.0	14.2	10.6
				Schottky barriers			
300	.007	.019	.043	.081	.124	.158	.151
350	.000	.001	.003	.009	.022	.040	.050
400	.000	.000	.000	.001	.002	.005	.010

Figure VII.9. Output electrical power density and efficiency as functions of solar concentration with junction temperature as a parameter for standard configuration cadmium selenide solar cells with n-type substrates and under AMO-like spectral distribution.

the example semiconductors irrespective of the junction type, while an increase in junction temperature reduces the output power density. Note, however, that as the optical concentration level increases the operating efficiency of the solar cells increases rapidly, slows and eventually begins to decrease with increasing optical concentration levels. Also note that the heterojunction devices perform at a higher level with greater output power densities and higher efficiencies than do the pn junction devices. This is a consequence of a superior collection efficiency with a smaller effective "dead layer". Also, the standard configuration Schottky barrier devices are seen to perform very poorly--owing to smaller effective photocurrent and higher saturation current density.

Figures VII.10 through VII.15 present computed output power density and solar cell efficiency for vertical configuration solar cells exposed to an AMO-like spectrum as the concentration level and junction temperature are varied. Data are presented for silicon, indium phosphide, gallium arsenide, cadmium telluride, aluminum antimonide and cadmium selenide solar cells constructed on n-type substrates. Results for solar cells constructed on p-type substrates are similar to these values, with variations primarily due to changes in Schottky barrier height, and minor resistance and lifetime changes.

Study of Figures VII.10 through VII.15 reveals that as the optical concentration level increases, the output power density for vertical configuration solar cells increases, while an increase in operating junction temperature results in a decrease in output power density. Note that the Schottky barrier solar cells have a significantly lower performance than either the pn or heterojunctions--a result of their higher saturation current density. Also, note that the vertical configuration solar cells have an improved performance (output power density and efficiency) over the standard configuration solar cells of Figures VII.4 through VII.9.

Figures VII.16 through VII.21 complete our study of the performance of the various solar cell optical configurations by presenting the theoretical output power density and efficiency for inverted configuration solar cells exposed to AMO-like spectral inputs of varying concentration levels, while the junction temperatures are varied. Data are presented for silicon, indium phosphide, gallium arsenide, cadmium telluride, aluminum antimonide and cadmium selenide solar cells constructed on n-type substrates. Results for solar cells constructed on p-type substrates are essentially similar with variations due to changes in Schottky barrier height and minor changes in resistance and lifetime.

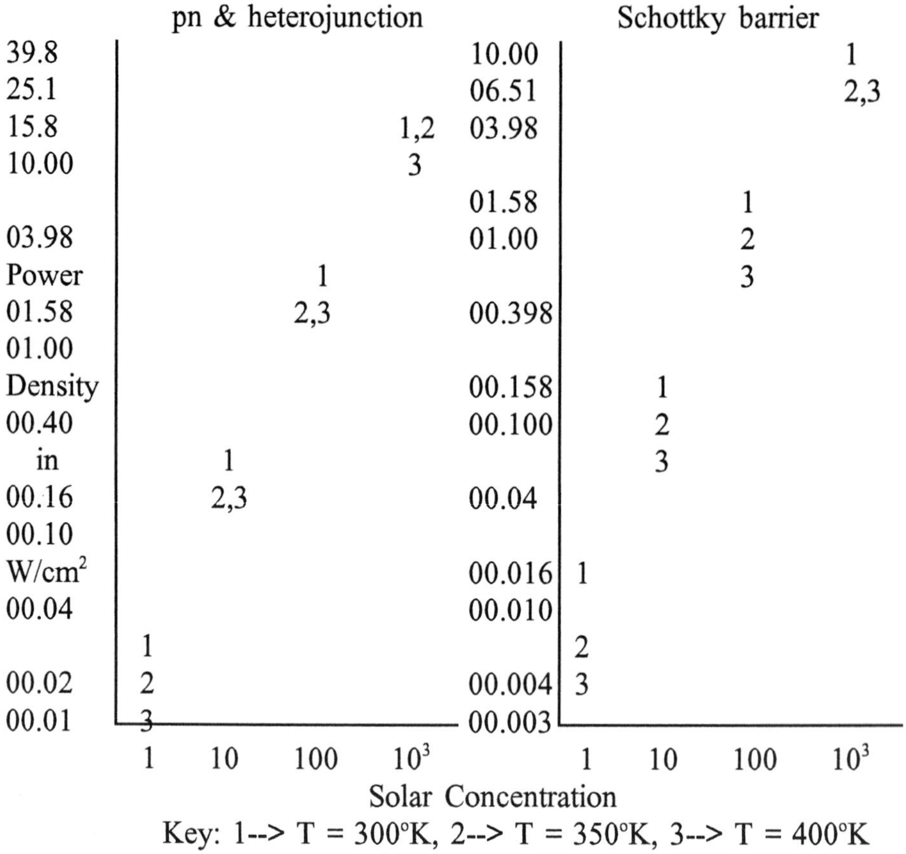

pn & heterojunction   Schottky barrier

```
39.8 |                          10.00 |                        1
25.1 |                          06.51 |                        2,3
15.8 |                    1,2   03.98 |
10.00|                    3           |
                                01.58 |                   1
03.98 |                         01.00 |                   2
Power |           1                   |                   3
01.58 |         2,3           00.398 |
01.00 |
Density|                       00.158 |              1
00.40 |                        00.100 |              2
 in   |        1                      |              3
00.16 |       2,3             00.04   |
00.10 |
W/cm² |                        00.016 | 1
00.04 |                        00.010 |
      | 1                             |              2
00.02 | 2                    00.004 | 3
00.01 |_3_____          00.003 |_____
       1  10  100  10³          1    10   100   10³
                Solar Concentration
```
Key: 1--> T = 300°K, 2--> T = 350°K, 3--> T = 400°K

Temperature	Efficiency (%) Solar Concentration Level						
(°K)	1	3.16	10	31.6	100	316	1000
	pn junctions and heterojunctions						
300	15.5	16.4	17.2	18.0	18.2	17.3	12.8
350	11.8	12.8	13.8	14.6	15.1	14.4	10.3
400	08.4	09.5	10.6	11.6	12.2	11.7	08.2
	Schottky barriers						
300	09.6	10.5	11.3	12.0	12.3	11.5	07.5
350	06.3	07.3	08.2	09.1	09.5	08.9	05.7
400	03.0	04.0	0.50	0.59	06.5	06.1	03.9

Figure VII.10. Output electrical power density and efficiency as functions of solar concentration with junction temperature as a parameter for vertical configuration silicon solar cells with n-type substrates and under AMO-like spectral distributions.

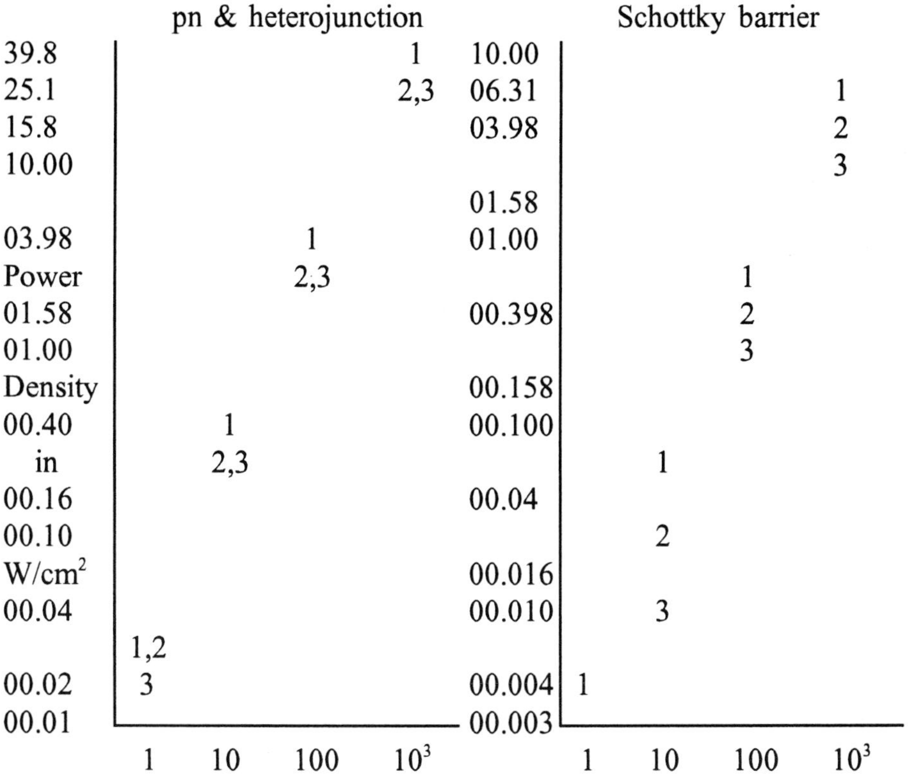

```
         pn & heterojunction              Schottky barrier
39.8   │                        1   10.00 │
25.1   │                      2,3  06.31  │                          1
15.8   │                           03.98  │                          2
10.00  │                                  │                          3
       │                           01.58  │
03.98  │              1            01.00  │
Power  │            2,3                   │              1
01.58  │                           00.398 │              2
01.00  │                                  │              3
Density│                           00.158 │
00.40  │          1                00.100 │
  in   │          2,3                     │           1
00.16  │                           00.04  │
00.10  │                                  │           2
W/cm²  │                           00.016 │
00.04  │                           00.010 │           3
       │      1,2                          │
00.02  │      3                    00.004 │ 1
00.01  └──────────────────────── 00.003 └──────────────────────────
          1    10   100   10³           1    10   100   10³
                        Solar Concentration
        Key: 1--> T = 300°K, 2--> T = 350°K, 3--> T = 400°K
```

Temperature (°K)	Efficiency (%) Solar Concentration Level						
	1	3.16	10	31.6	100	316	1000
	pn junctions and heterojunctions						
300	21.9	22.7	23.5	24.3	24.9	25.0	23.5
350	18.4	19.4	20.4	21.3	22.0	22.2	20.9
400	15.0	16.1	17.2	18.2	19.1	19.5	18.3
	Schottky barriers						
300	2.91	3.64	4.37	5.06	5.64	5.78	4.72
350	0.93	1.53	2.22	2.96	3.62	3.91	3.22
400	0.13	0.34	0.74	1.30	1.91	2.31	2.00

Figure VII.11. Output electrical power density and efficiency as functions of solar concentration with junction temperature as a parameter for vertical configuration indium phosphide solar cells with n-type substrates and under AMO-like spectral distributions.

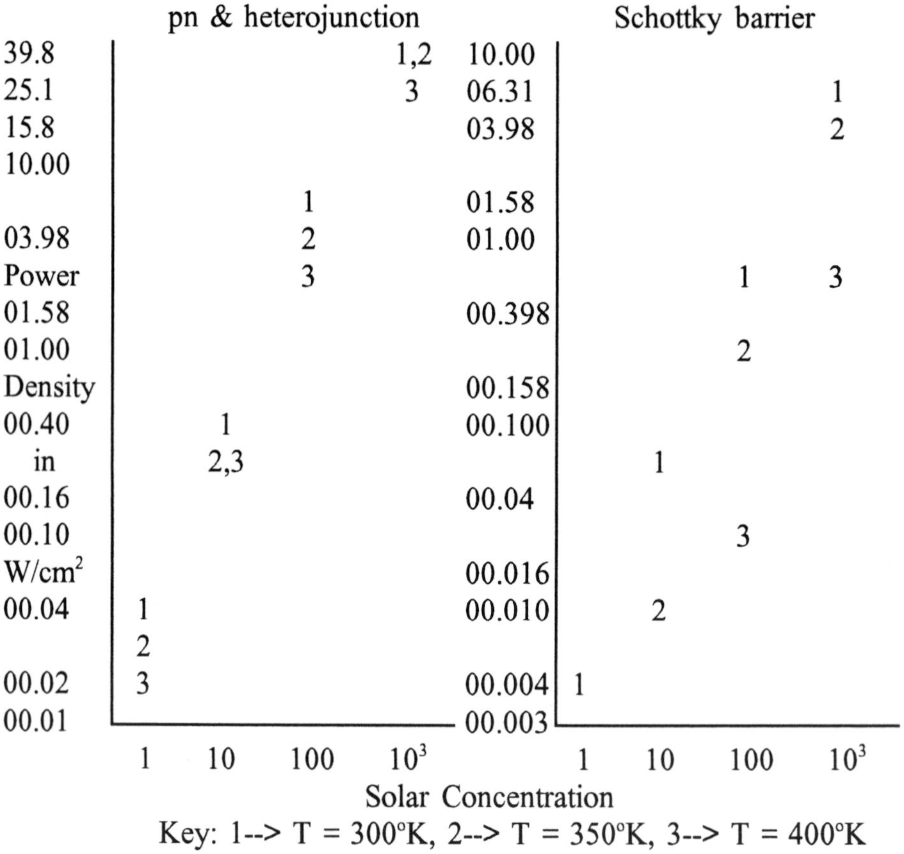

pn & heterojunction   Schottky barrier

```
39.8    |                    1,2  10.00 |
25.1    |                      3  06.31 |                        1
15.8    |                         03.98 |                        2
10.00   |
        |              1          01.58 |
03.98   |              2          01.00 |
Power   |              3                |              1         3
01.58   |                         00.398|
01.00   |                                |              2
Density |                         00.158|
00.40   |        1                00.100|
  in    |        2,3                     |        1
00.16   |                         00.04  |
00.10   |                                |              3
W/cm²   |                         00.016 |
00.04   |  1                      00.010 |        2
        |  2                             |
00.02   |  3                      00.004 | 1
00.01   |_____          00.003|_____
           1   10   100   10³         1   10   100   10³
                    Solar Concentration
        Key: 1--> T = 300°K, 2--> T = 350°K, 3--> T = 400°K
```

Temperature	Efficiency (%) Solar Concentration Level						
(°K)	1	3.16	10	31.6	100	316	1000
	pn junctions and heterojunctions						
300	22.9	23.6	24.4	25.1	25.8	27.3	26.0
350	19.9	20.7	21.6	22.4	23.2	23.8	23.8
400	16.8	17.8	18.8	19.7	20.6	21.3	21.4
	Schottky barriers						
300	2.58	3.22	3.88	4.54	5.15	5.59	5.48
350	0.20	0.47	0.89	1.43	2.01	2.51	2.61
400	0.00	0.01	0.02	0.06	0.15	0.32	0.47

Figure VII.12. Output electrical power density and efficiency as functions of solar concentration with junction temperature as a parameter for vertical configuration gallium arsenide solar cells with n-type substrates and under AMO-like spectral distributions.

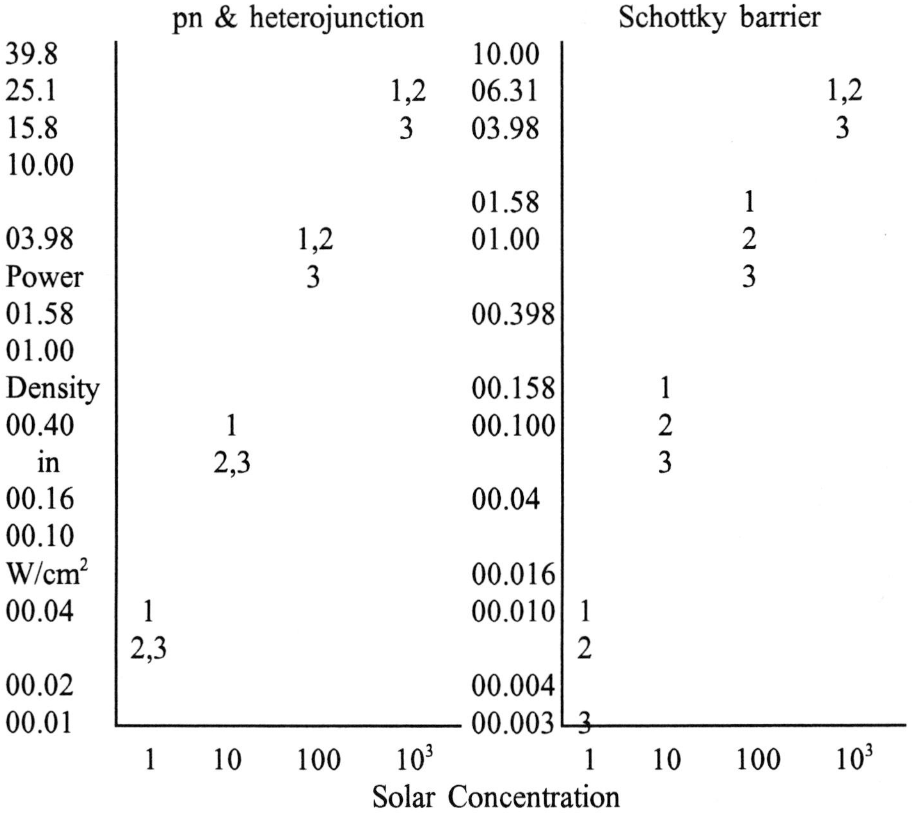

pn & heterojunction   Schottky barrier

```
39.8  |                          10.00  |
25.1  |               1,2        06.31  |                    1,2
15.8  |               3          03.98  |                    3
10.00 |
                                 01.58  |              1
03.98 |          1,2             01.00  |              2
Power |          3                                     3
01.58 |                          00.398 |
01.00 |
Density|                         00.158 |        1
00.40 |        1                 00.100 |        2
 in   |        2,3                               3
00.16 |                          00.04  |
00.10 |
W/cm² |                          00.016 |
00.04 |    1                     00.010 | 1
      |    2,3                           | 2
00.02 |                          00.004 |
00.01 |_____  00.003 |_3_____
         1   10   100   10³              1   10   100   10³
```

Solar Concentration

Key: 1--> T = 300°K, 2--> T = 350°K, 3--> T = 400°K

Efficiency (%)

Temperature (°K)	Solar Concentration Level						
	1	3.16	10	31.6	100	316	1000
	pn junctions and heterojunctions						
300	24.5	25.2	25.8	26.4	26.5	25.3	20.1
350	21.8	22.6	23.4	24.1	24.3	23.3	18.3
400	19.2	20.1	21.0	21.7	22.1	21.2	16.5
	Schottky barriers						
300	6.80	7.47	8.10	8.62	8.77	7.74	4.40
350	4.62	5.37	6.09	6.72	6.98	6.18	3.49
400	2.56	3.34	4.12	4.83	5.21	4.64	2.64

Figure VII.13. Output electrical power density and efficiency as functions of solar concentration with junction temperature as a parameter for vertical configuration cadmium telluride solar cells with n-type substrates and under AMO-like spectral distributions.

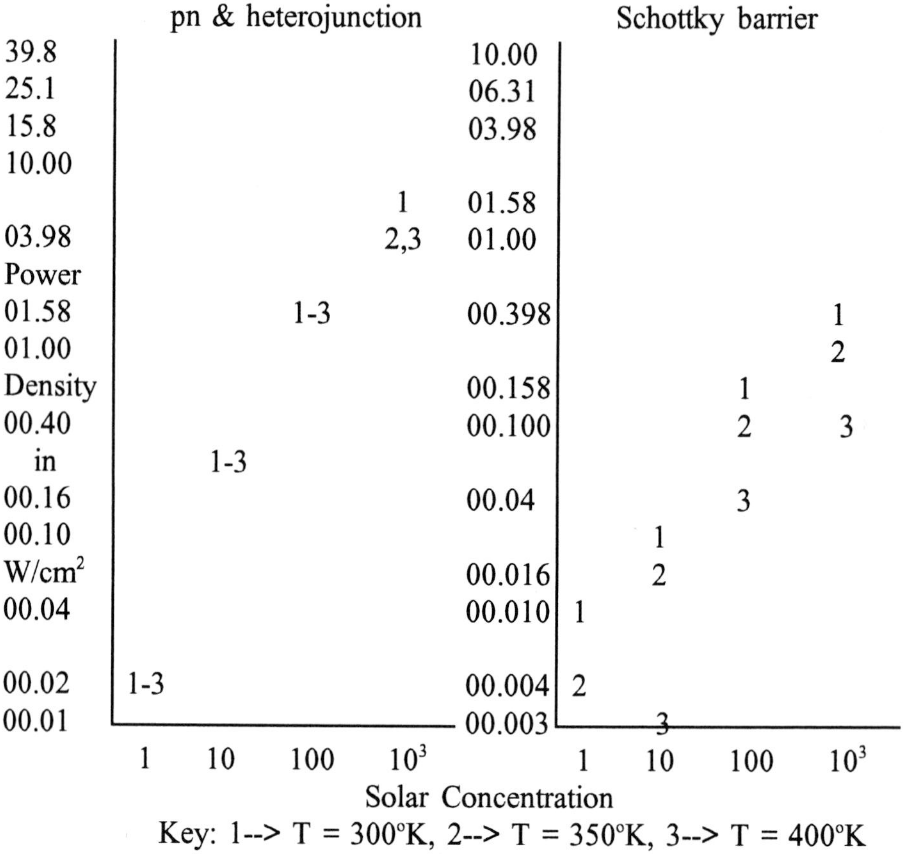

```
            pn & heterojunction                    Schottky barrier
39.8    |                          10.00 |
25.1    |                          06.31 |
15.8    |                          03.98 |
10.00   |
                          1        01.58 |
03.98   |                 2,3      01.00 |
Power   |
01.58   |        1-3               00.398|                          1
01.00   |                                |                          2
Density |                          00.158|                  1
00.40   |                          00.100|                  2       3
  in    |        1-3                      |
00.16   |                          00.04 |                          3
00.10   |                                |                  1
W/cm²   |                          00.016|                  2
00.04   |                          00.010| 1

00.02   | 1-3                      00.004| 2
00.01   |_____        00.003|_____3_____
         1    10    100   10³             1    10    100   10³
                      Solar Concentration
       Key: 1--> T = 300°K, 2--> T = 350°K, 3--> T = 400°K
```

Temperature	Efficiency (%) Solar Concentration Level						
(°K)	1	3.16	10	31.6	100	316	1000
	pn junctions and heterojunctions						
300	20.6	21.1	21.4	21.2	19.2	12.1	04.3
350	18.1	18.7	19.1	19.0	17.1	10.5	03.8
400	15.5	16.2	16.7	16.7	14.9	08.9	03.3
	Schottky barriers						
300	1.63	2.05	2.36	2.27	1.46	0.65	0.26
350	0.41	0.72	1.02	1.09	0.76	0.38	0.17
400	0.03	0.09	0.20	0.30	0.27	0.17	0.09

Figure VII.14. Output electrical power density and efficiency as functions of solar concentration with junction temperature as a parameter for vertical configuration aluminum antimonide solar cells with n-type substrates and under AMO-like spectral distributions.

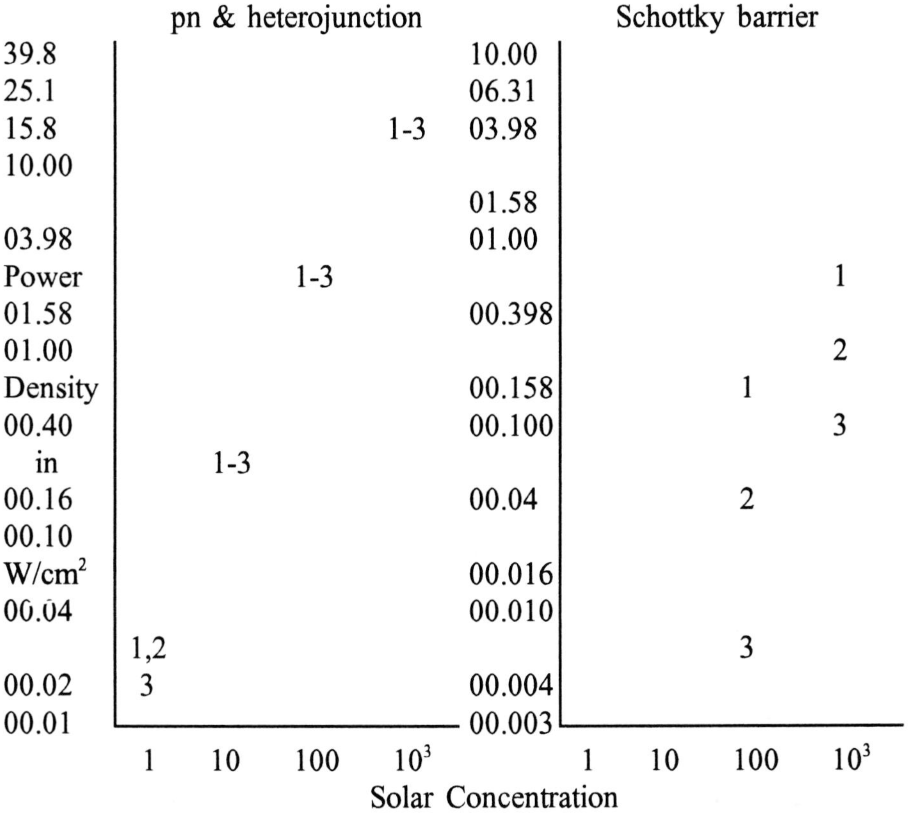

pn & heterojunction   Schottky barrier

```
39.8   |                                10.00  |
25.1   |                                06.31  |
15.8   |                        1-3     03.98  |
10.00  |                                       |
       |                                01.58  |
03.98  |                                01.00  |                              1
Power  |               1-3                     |
01.58  |                                00.398 |                                    2
01.00  |                                       |
Density|                                00.158 |                         1
00.40  |                                00.100 |                                    3
  in   |         1-3                            |
00.16  |                                00.04  |                         2
00.10  |                                       |
W/cm²  |                                00.016 |
00.04  |                                00.010 |
       |  1,2                                  |                         3
00.02  |  3                                    |
00.01  |_____        00.004 |
                                        00.003 |_____
        1    10    100    10³             1    10    100    10³
                      Solar Concentration
        Key: 1--> T = 300°K, 2--> T = 350°K, 3--> T = 400°K
```

Efficiency (%)

Temperature (°K)	Solar Concentration Level						
	1	3.16	10	31.6	100	316	1000
	heterojunctions						
300	19.6	20.1	20.6	21.0	21.0	19.9	15.4
350	17.1	17.7	18.3	18.7	18.8	17.8	13.5
400	14.7	15.4	16.1	16.6	16.8	15.9	11.8
	Schottky barriers						
300	0.22	0.44	0.72	0.99	1.10	0.82	0.40
350	0.01	0.04	0.11	0.23	0.34	0.32	0.20
400	0.00	0.00	0.01	0.03	0.06	0.08	0.08

Figure VII.15. Output electrical power density and efficiency as functions of solar concentration with junction temperature as a parameter for vertical configuration cadmium selenide solar cells with n-type substrates and under AMO-like spectral distributions.

Study of Figures VII.16 through VII.21 indicates that inverted configuration Schottky barrier solar cells are less efficient than inverted configuration pn and heterojunction solar cells. As was the case for standard and vertical configuration solar cells, inverted configuration solar cell performance increases with increasing optical concentration level and decreases with increasing junction temperature.

Note that the pn and heterojunction solar cells of all three optical configurations are more efficient and provide higher output power density than the Schottky barrier solar cells. It is, of course, possible that, at sometime in the future a Schottky junction will be discovered that possesses a sufficiently high junction barrier height, or a mos barrier will be devised that allows for improved solar cell performance.

Before discussing these optical configurations under AM1 spectral illumination, let us spend a few moments discussing the operating voltage and loss factors for some of the example solar cells discussed in Figures VII.4 through VII.21.

Study of Table VII.4 reveals that the operating voltage for maximum delivered output power density increases slowly as the optical concentration level is increased. Note also that the operating voltage decreases with increasing junction temperature. The operating voltage for vertical configuration devices lies between that for the standard configuration design and the inverted optical configuration devices. Recall that the higher the operating voltage, the fewer devices which must be electrically stacked to provide a given system voltage. The reader will observe that solar cells made from indium phosphide, aluminum antimonide and cadmium selenide have been ignored in Table VII.4. This is because the data contained in this chapter, in Chapter VI and the following chapters indicate that the semiconductors listed in Table VII.4 are the most likely to be useful in future semiconductors. The operating voltages for pn junctions are essentially the same as those given for heterojunctions, while the operating voltages for Schottky solar cells are somewhat less. The overall performance of the Schottky solar cells from Figures VII.4 through VII.21 are so poor, compared to the heterojunction and pn junction devices that we will not comment on them further.

The loss factors considered in Chapters V and VI were quite small, normally less than unity. The theoretical loss factors for heterojunction devices of Table VII.4 are presented in Table VII.5. From Equation VII.15 the loss factor depends on the series resistance times area product of the solar cell. Owing to the high carrier mobility of GaAs this

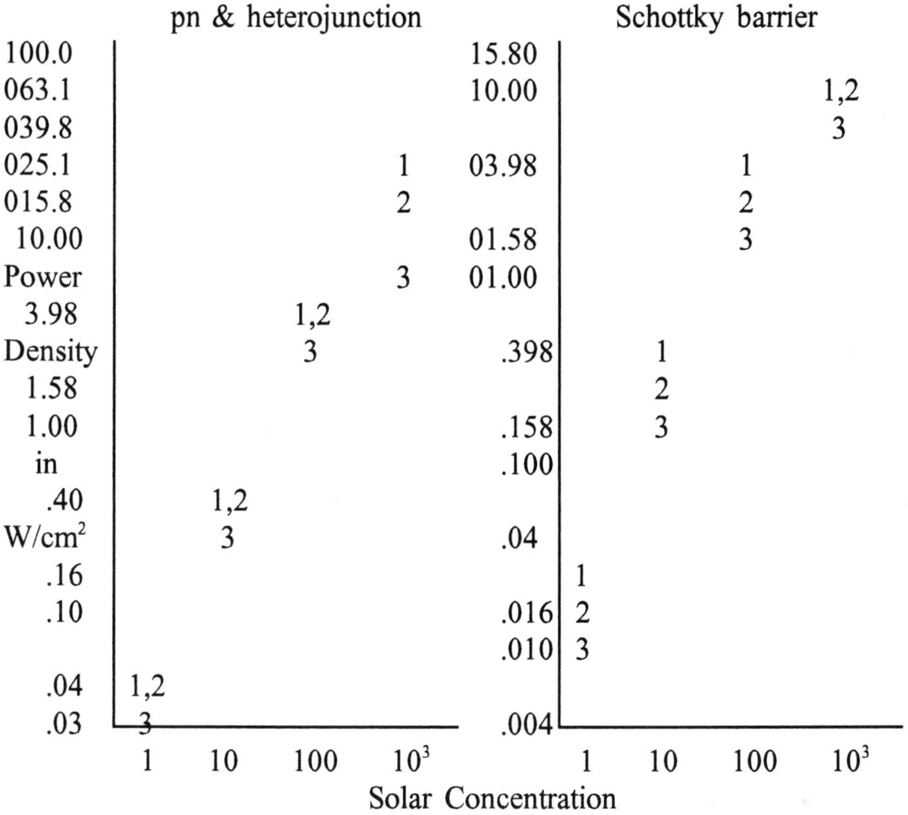

```
        pn & heterojunction              Schottky barrier
100.0 |                        15.80 |
063.1 |                        10.00 |                          1,2
039.8 |                                                         3
025.1 |                  1     03.98 |                1
015.8 |                  2                            2
10.00 |                        01.58 |                3
Power |               3        01.00 |
 3.98 |            1,2
Density|            3            .398 |        1
 1.58 |                                        2
 1.00 |                         .158 |         3
  in  |                         .100 |
  .40 |         1,2
W/cm² |         3                .04 |
  .16 |                               |  1
  .10 |                         .016 |  2
                                .010 |  3
  .04 | 1,2
  .03 |_3_____       .004 |_____
        1   10  100  10³              1    10   100   10³
                      Solar Concentration
        Key: 1--> T = 300°K, 2--> T = 350°K, 3--> T = 400°K
```

Temperature	Efficiency (%) Solar Concentration Level						
(°K)	1	3.16	10	31.6	100	316	1000
	pn junctions and heterojunctions						
300	16.4	17.2	18.0	18.6	18.3	15.7	08.1
350	12.6	13.7	14.6	15.3	15.2	12.9	06.6
400	09.3	10.4	11.5	12.3	12.4	10.4	05.3
	Schottky barriers						
300	10.4	11.2	12.0	12.6	12.4	09.9	04.7
350	07.1	08.0	08.9	09.6	09.6	07.6	03.6
400	03.7	04.7	05.7	06.5	06.7	05.2	02.5

Figure VII.16. Output electrical power density and efficiency as functions of solar concentration with junction temperature as a parameter for inverted configuration silicon solar cells with n-type substrates and under AMO-like spectral distributions.

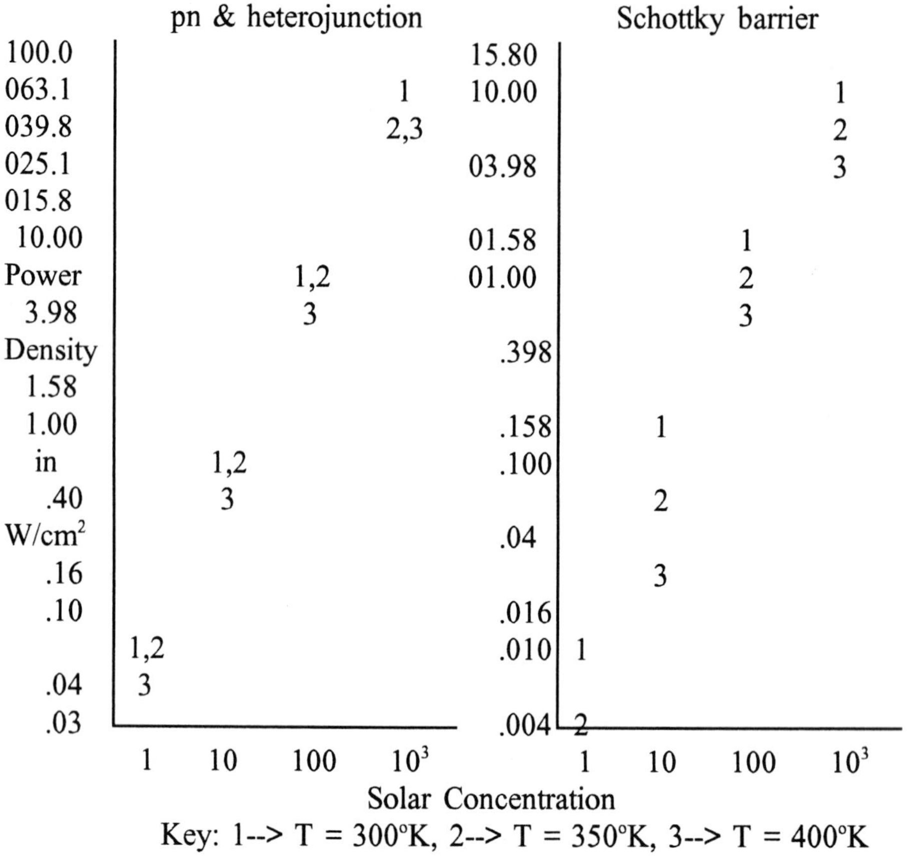

pn & heterojunction   Schottky barrier

```
100.0 |                        15.80 |
063.1 |               1        10.00 |                        1
039.8 |             2,3                                        2
025.1 |                        03.98 |                        3
015.8 |
10.00 |                        01.58 |            1
Power |          1,2           01.00 |            2
3.98  |          3                                 3
Density|                        .398 |
1.58  |
1.00  |                        .158 |   1
in    |         1,2             .100 |
.40   |         3                      |   2
W/cm² |                         .04 |
.16   |                                |   3
.10   |                         .016 |
      |        1,2              .010 | 1
.04   |        3
.03   |_____ .004 |_2_____
         1    10   100   10³          1   10   100   10³
                       Solar Concentration
         Key: 1--> T = 300°K, 2--> T = 350°K, 3--> T = 400°K
```

Temperature (°K)	Efficiency (%) Solar Concentration Level						
	1	3.16	10	31.6	100	316	1000
pn junctions and heterojunctions							
300	23.1	23.9	24.7	25.5	25.9	25.2	21.5
350	19.6	20.6	21.6	22.4	23.0	22.5	19.0
400	16.2	17.3	18.4	19.4	20.0	19.8	16.5
Schottky barriers							
300	3.47	4.22	4.96	5.62	6.02	5.60	3.49
350	1.33	2.03	2.77	3.50	4.02	3.84	2.43
400	0.25	0.58	1.11	1.75	2.30	2.31	1.58

Figure VII.17. Output electrical power density and efficiency as functions of solar concentration with junction temperature as a parameter for inverted configuration indium phosphide solar cells with n-type substrates and under AMO-like spectral distributions.

pn & heterojunction   Schottky barrier

```
100.0  |                      15.80 |                        1
063.1  |             1-3      10.00 |
039.8  |                            |                        2
025.1  |                      03.98 |
015.8  |                            |
10.00  |                      01.58 |          1             3
Power  |          1-3         01.00 |
3.98   |                            |          2
Density|                       .398 |
1.58   |                            |
1.00   |                       .158 |
in     |          1-3          .100 |     1
.40    |                            |
W/cm²  |                       .04  |     2
.16    |                            |
.10    |                       .016 |
       |  1,2                  .010 |
.04    |  3                         |     1                  3
.03    |_____   .004 |_____
          1   10   100   10³            1   10   100   10³
                    Solar Concentration
      Key: 1--> T = 300°K, 2--> T = 350°K, 3--> T = 400°K
```

Temperature	Efficiency (%) Solar Concentration Level						
(°K)	1	3.16	10	31.6	100	316	1000
	pn junctions and heterojunctions						
300	24.0	24.8	25.5	26.2	26.8	27.0	25.9
350	21.0	21.8	22.7	23.6	24.3	24.6	23.6
400	17.9	18.9	19.9	20.8	21.7	22.1	21.3
	Schottky barriers						
300	3.06	3.73	4.41	5.07	5.62	5.83	5.02
350	0.36	0.73	1.40	1.85	2.43	2.76	2.39
400	0.00	0.01	0.04	0.11	0.26	0.46	0.51

Figure VII.18. Output electrical power density and efficiency as functions of solar concentration with junction temperature as a parameter for inverted configuration gallium arsenide solar cells with n-type substrates and under AMO-like spectral distributions.

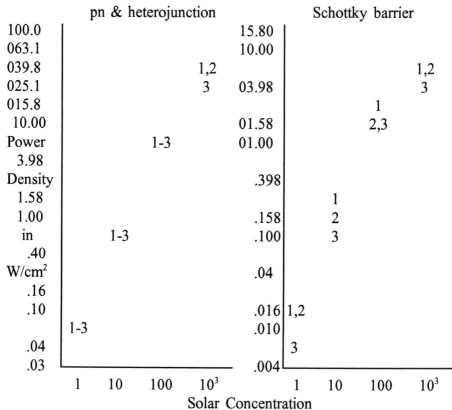

Key: 1--> T = 300°K, 2--> T = 350°K, 3--> T = 400°K

Temperature (°K)	Efficiency (%) Solar Concentration Level						
	1	3.16	10	31.6	100	316	1000
	pn junctions and heterojunctions						
300	25.6	26.4	27.0	27.4	26.9	23.7	13.6
350	23.0	23.8	24.6	25.1	24.6	21.7	12.3
400	20.3	21.2	22.1	22.7	22.5	19.7	11.0
	Schottky barriers						
300	7.43	8.11	8.71	9.07	8.66	6.18	2.60
350	5.24	6.00	6.70	7.18	6.93	4.88	2.11
400	3.14	3.95	4.72	5.30	5.23	3.66	1.64

Figure VII.19. Output electrical power density and efficiency as functions of solar concentration with junction temperature as a parameter for inverted configuration cadmium telluride solar cells with n-type substrates and under AMO-like spectral distributions.

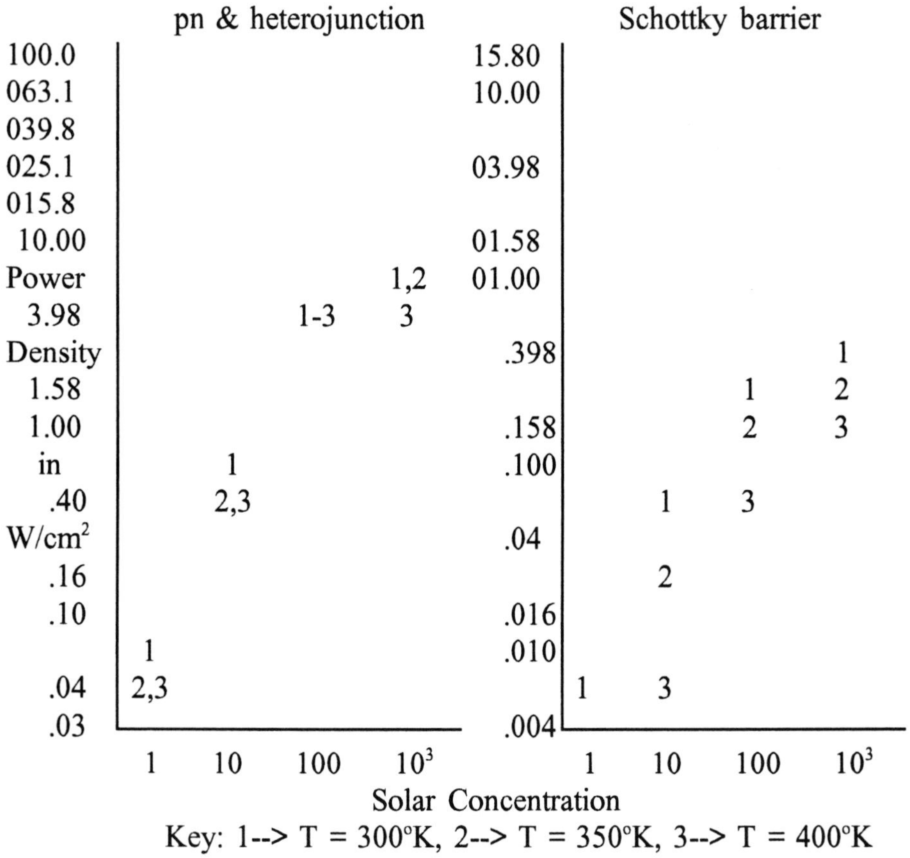

pn & heterojunction   Schottky barrier

```
100.0 |                        15.80 |
063.1 |                        10.00 |
039.8 |
025.1 |                        03.98 |
015.8 |
10.00 |                        01.58 |
Power |              1,2        01.00 |
3.98  |      1-3      3
Density|                         .398 |                   1
1.58  |                                        1      2
1.00  |                         .158 |          2      3
in    |        1                .100 |
.40   |       2,3                .04 |      1      3
W/cm² |
.16   |                                         2
.10   |                         .016 |
      |    1                    .010 |
.04   |   2,3                        |   1      3
.03   |_____.004 |_____
          1    10   100   10³          1    10   100   10³
                       Solar Concentration
       Key: 1--> T = 300°K, 2--> T = 350°K, 3--> T = 400°K
```

Temperature (°K)	Efficiency (%) Solar Concentration Level						
	1	3.16	10	31.6	100	316	1000
pn junctions and heterojunctions							
300	21.0	21.4	21.5	20.5	16.3	07.1	02.4
350	18.2	19.0	19.2	18.4	14.3	06.2	02.1
400	16.0	16.6	16.9	16.2	12.4	05.3	01.8
Schottky barriers							
300	1.90	2.28	2.42	1.93	0.97	0.40	0.15
350	0.60	0.93	1.14	0.97	0.55	0.25	0.10
400	0.07	0.16	0.28	0.32	0.23	0.13	0.06

Figure VII.20. Output electrical power density and efficiency as functions of solar concentration with junction temperature as a parameter for inverted configuration aluminum antimonide solar cells with n-type substrates and under AMO-like spectral distributions.

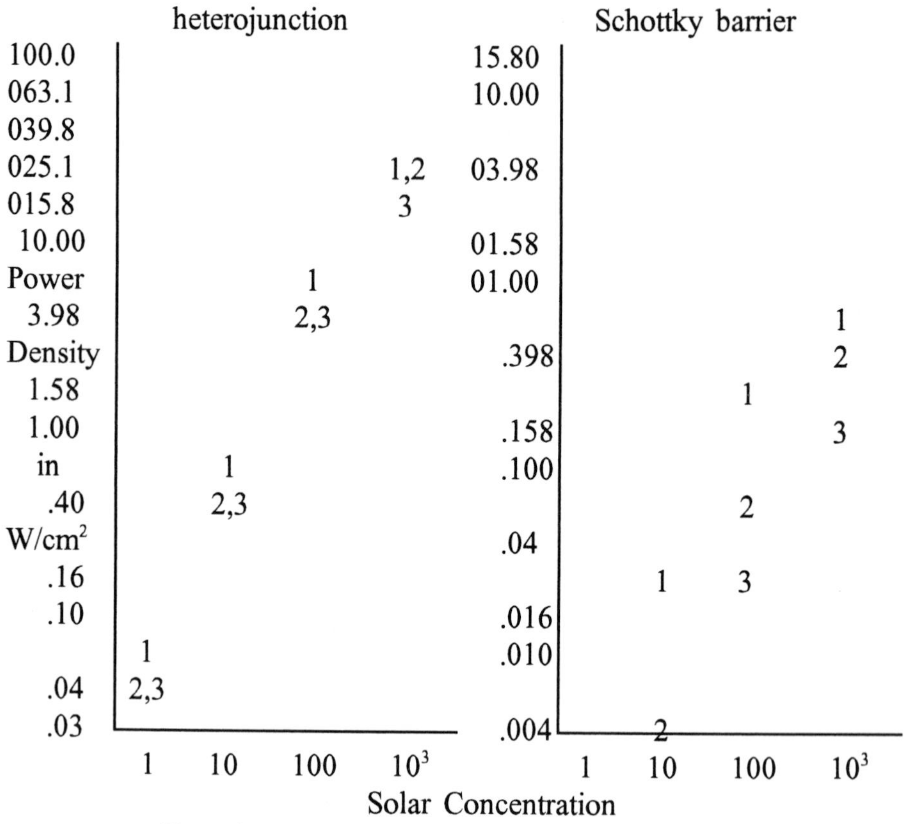

heterojunction   Schottky barrier

```
100.0                          15.80
063.1                          10.00
039.8
025.1              1,2         03.98
015.8               3
10.00                          01.58
Power         1                01.00                              1
 3.98        2,3                                                  2
Density                         .398
 1.58                                              1
 1.00                           .158                             3
  in          1                 .100
  .40        2,3                                    2
W/cm²                           .04
  .16                                          1   3
  .10                           .016
             1                  .010
  .04       2,3
  .03                           .004        2
       1   10   100   10³            1   10   100   10³
              Solar Concentration
   Key: 1--> T = 300°K, 2--> T = 350°K, 3--> T = 400°K
```

Temperature	Efficiency (%) Solar Concentration Level						
(°K)	1	3.16	10	31.6	100	316	1000
				heterojunctions			
300	20.5	21.0	21.4	21.6	21.1	18.4	10.1
350	17.9	18.5	19.1	19.4	19.0	16.4	08.7
400	15.6	16.2	16.9	17.3	17.0	14.5	07.6
				Schottky barriers			
300	0.35	0.62	0.92	1.13	1.02	0.59	0.27
350	0.03	0.08	0.18	0.32	0.36	0.27	0.14
400	0.00	0.01	0.02	0.05	0.08	0.09	0.07

Figure VII.21. Output electrical power density and efficiency as functions of solar concentration with junction temperature as a parameter for inverted configuration cadmium selenide solar cells with n-type substrates and under AMO-like spectral distributions.

Table VII.4

The solar cell operating voltage for maximum electrical output power density for heterojunction solar cells on n-type substrates as a function of optical configuration, solar concentration level and junction temperature

Semiconductor	Temperature (°K)	Solar Concentration			
		1	10	100	1000
		Standard Configuration			
Si	300	0.513	0.570	0.629	0.714
	350	0.399	0.464	0.532	0.633
	400	0.294	0.367	0.444	0.554
GaAs	300	0.899	0.957	1.015	1.076
	350	0.787	0.854	0.922	0.992
	400	0.672	0.748	0.824	0.904
CdTe	300	0.999	1.057	1.117	1.195
	350	0.897	0.965	1.034	1.127
	400	0.793	0.870	0.949	1.055
		Inverted Configuration			
Si	300	0.533	0.590	0.652	0.766
	350	0.422	0.488	0.559	0.687
	400	0.320	0.394	0.474	0.614
GaAs	300	0.920	0.978	1.037	1.100
	350	0.811	0.878	0.946	1.019
	400	0.699	0.775	0.852	0.936
CdTe	300	1.020	1.079	1.140	1.255
	350	0.922	0.990	1.061	1.193
	400	0.821	0.898	0.979	1.128

parameter is lower for GaAs than for the other semiconductors under consideration. Note that the loss factor for all of these semiconductors increases with increasing optical concentration and junction temperature. Similar computations for pn and Schottky barrier solar cells on both n- and p-type substrates yield similar results for Si, GaAs and CdTe as well as for InP, AlSb and CdSe.

Figures VII.4 through VII.21 employ concentrated light of an AMO-like spectrum. We could derive similar results for AM1-like spectral illumination. Because there is less energy in AM1 spectral

Table VII.5

The loss factors for standard and inverted configuration heterojunction solar cells on n-type substrates when exposed to AMO-like illumination of varying concentratiion as a function of junction temperatures

Semiconductor	Temperature (°K)	Solar Concentration			
		1	10	100	1000
		Standard Configuration			
Si	300	.001	.010	.096	2.53
	350	.001	.011	.113	3.17
	400	.002	.015	.135	3.85
GaAs	300	.000	.001	.011	0.115
	350	.000	.001	.012	0.125
	400	.000	.002	.013	0.137
CdTe	300	.001	.007	.069	1.41
	350	.001	.007	.074	1.57
	400	.001	.008	.081	1.78
		Inverted Configuration			
Si	300	.002	.019	.202	16.5
	350	.003	.023	.239	16.6
	400	.003	.028	.285	16.5
GaAs	300	.000	.002	.021	0.238
	350	.000	.002	.023	0.259
	400	.000	.003	.025	0.285
CdTe	300	.001	.014	.145	12.4
	350	.002	.015	.156	12.4
	400	.002	.016	.170	12.6

light, the ouput power density of solar cells under AM1-like illumination is less than that for AMO-like light. Rather than detail the performance of our example semiconductors as was done in Figures VII.4 through VII.21 and in Tables VII.4 and VII.5 for AMO, let us extract but a single set of values for each junction-configuration-semiconductor combination under concentrated AM1-like illumination. We will assume that the substrates are all n-type--the results for p-type substrates are similar. The values in Table VII.6 are those for the highest operating efficiencies while delivering the most power to an external load.

Table VII.6

The maximum delivered power operating conditions (V_D', K', P_{max}/A_D and η_s) under AM1-like concentrated sunlight as a function of optical concentration, junction type, junction temperature and semiconductor for standard configuration solar cells with n-type substrates

Semicon-ductor	Junction T(°K)	Type*	Concen-tration	V_D' (V)	K' --	P_{max}/A_D W/cm²	η_s %
Si	300	HET	100	0.62	.077	01.76	16.5
	350	HET	100	0.53	.091	01.45	13.6
	400	HET	100	0.44	.109	01.17	10.9
InP	300	HET	316	0.88	.068	07.32	21.6
	400	HET	316	0.73	.085	05.67	16.8
GaAs	300	HET	1000	1.07	.087	24.40	22.8
	350	HET	1000	0.98	.095	22.20	20.8
	400	HET	1000	0.90	.104	20.00	18.6
CdTe	300	HET	100	1.11	.052	02.44	22.8
	350	HET	100	1.03	.056	02.24	20.9
	400	HET	100	0.94	.061	02.03	19.0
AlSb	300	HET	10	1.10	.028	00.20	18.5
	400	HET	31.6	0.92	.112	00.49	14.5
CdSe	300	HET	100	1.21	.058	02.00	18.7
	400	HET	100	0.98	.071	01.60	14.9
Si	300	PN	100	0.62	.062	01.44	13.4
	400	PN	316	0.47	.304	03.01	08.9
InP	300	PN	1000	0.90	.045	04.85	04.5
GaAs	300	PN	1000	1.06	.055	15.80	14.8
	400	PN	1000	0.88	.066	12.90	12.0
CdTe	300	PN	100	1.10	.038	01.81	16.9
	400	PN	100	0.93	.045	01.50	14.0
AlSb	300	PN	31.6	1.13	.067	00.47	14.0
Si	300	SH	316	0.43	.095	01.02	03.0
InP	300	SH	1000	0.21	.059	00.34	00.3
GaAs	300	SH	1000	0.24	.077	01.07	01.0
CdTe	300	SH	316	0.39	.115	00.63	01.8
AlSb	300	SH	100	0.18	.555	00.05	00.5
CdSe	300	SH	1000	0.13	1.609	00.17	00.2

*HET denotes heterojunction; PN, pn junction; and SH, Schottky

Note that the performance of standard configuration Schottky barrier solar cells under AM1 conditions is very poor and that the heterojunction devices are superior to the types using other junctions. Table VII.7 provides similar data for vertical configuration solar cells and Table VII.8 does the same for inverted configuration solar cells.

Table VII.7

The maximum delivered power operating conditions (V_D', K', P_{max}/A_D and η_s) under AM1-like concentrated sunlight as a function of optical concentration, junction type, junction temperature and semiconductor for vertical configuration solar cells with n-type substrates

Semicon-ductor	\ T(°K)	Junction \ Type*	Concen-tration	V_D' (V)	K' --	P_{max}/A_D W/cm^2	η_s %
Si	300	PNH	100	0.63	.078	01.98	18.5
	350	PNH	100	0.53	.092	01.63	15.2
	400	PNH	100	0.44	.110	01.31	12.3
InP	300	PNH	316	0.91	.069	08.17	24.2
	400	PNH	316	0.73	.086	06.35	18.8
GaAs	300	PNH	1000	1.07	.086	27.30	25.5
	350	PNH	1000	0.99	.093	24.90	23.3
	400	PNH	1000	0.90	.102	22.40	20.9
CdTe	300	PNH	100	1.11	.052	02.72	25.4
	350	PNH	100	1.03	.056	02.50	23.3
	400	PNH	100	0.94	.061	02.26	21.2
AlSb	300	PNH	10	1.11	.028	00.22	20.8
	400	HET	31.6	0.92	.114	00.55	16.3
CdSe	300	HET	100	1.21	.059	02.25	21.0
	400	HET	100	0.99	.072	01.80	16.8
Si	300	SH	100	0.44	.112	01.34	12.5
InP	300	SH	316	0.25	.272	01.90	05.6
GaAs	300	SH	1000	0.28	.380	05.83	05.4
CdTe	300	SH	100	0.40	.150	00.90	08.4
AlSb	300	SH	31.6	0.18	.743	00.08	02.3
CdSe	300	SH	100	0.11	.791	00.12	01.1

*PNH denotes pn junction and heterojunction, HET is solely heterojunction, and SH is Schottky barrier

Table VII.8

The maximum delivered power operating conditions (V_D', K', P_{max}/A_D and η_s) under AM1-like concentrated sunlight as a function of optical concentration, junction type, junction temperature and semiconductor for inverted configuration solar cells with n-type substrates

Semicon-ductor	Junction T(°K)	Type*	Concen-tration	V_D' (V)	K' --	P_{max}/A_D W/cm²	η_s %
. Si	300	PNH	31.6	0.61	.048	01.28	18.9
	350	PNH	100	0.55	.188	03.36	15.7
	400	PNH	100	0.47	.224	02.73	12.8
InP	300	PNH	100	0.90	.042	05.33	24.9
	400	PNH	316	0.76	.173	13.00	19.3
GaAs	300	PNH	316	1.06	.051	17.80	26.4
	350	PNH	316	0.97	.056	16.20	23.9
	400	PNH	316	0.88	.061	14.60	21.5
CdTe	300	PNH	31.6	1.10	.032	01.77	26.2
	350	PNH	31.6	1.02	.035	01.62	24.0
	400	PNH	100	0.97	.126	04.66	21.8
AlSb	300	PNH	10	1.13	.054	00.46	21.4
	400	PNH	10	0.90	.067	00.36	16.8
CdSe	300	HET	31.6	1.20	.036	01.46	21.6
	400	HET	31.6	0.97	.044	01.16	17.2
Si	300	SH	100	0.46	.234	02.73	12.8
InP	300	SH	100	0.24	.160	01.23	05.8
GaAs	300	SH	316	0.26	.216	03.86	05.7
CdTe	300	SH	31.6	0.39	.092	00.59	08.7
AlSb	300	SH	10	0.17	.402	00.05	02.4
CdSe	300	SH	31.6	0.10	.466	00.08	01.1

*PNH denotes pn junction and heterojunction, HET is solely heterojunction, and SH is Schottky barrier

Note that the performance improves from standard, through vertical to inverted configuration solar cells, and that the performance of Schottky barriers is significantly the lowest. Solar cells with p-type substrates have roughly the same behavior under AM1-like illumination as the n-type solar cells surveyed in Tables VII.6 through VII.8.

Study of Tables VII.6 through VII.8 indicates that Si, GaAs and CdTe based heterojunction solar cells have better performance under AM1-like illumination than the other semiconductor and junction solar cells--a characteristic also seen under AMO-like illumination. The performance of these devices appears to be similar to, but, in general, not as "good" as under AMO conditions. Let us conclude this chapter by presenting the maximum efficiency operating characteristics for n-type substrate solar cells under AMO-like conditions in the same form as was done for AM1-like illumination, and drawing conclusions from them.

Table VII.9

The maximum delivered power operating conditions (V_D', K', P_{max}/A_D and η_s) under AM0-like concentrated sunlight as a function of optical concentration, junction type, junction temperature, optical orientation and semiconductor for solar cells with n-type substrates

Semiconductor	Junction T(°K)	Type*	Concentration	V_D' (V)	K' --	P_{max}/A_D W/cm^2	η_s %
				Standard Configuration			
Si	300	HET	100	0.63	.096	02.18	16.2
	350	HET	100	0.53	.113	01.80	13.3
	400	HET	100	0.44	.136	01.45	10.8
GaAs	300	HET	316	1.04	.035	10.00	23.4
	350	HET	316	0.96	.037	09.09	21.2
	400	HET	1000	0.90	.137	25.80	19.0
CdTe	300	HET	100	1.12	.069	03.21	23.7
	350	HET	100	1.03	.074	02.95	21.8
	400	HET	100	0.95	.081	02.67	19.8
				Vertical Configuration			
Si	300	PNH	100	.063	.098	02.47	18.2
	350	PNH	100	.054	.115	02.04	15.1
	400	PNH	100	.045	.138	01.65	12.2
GaAs	300	PNH	316	1.05	.036	11.70	27.3
	350	PNH	316	0.96	.037	10.20	23.8
	400	PNH	1000	0.90	.135	28.90	21.4
CdTe	300	PNH	100	1.12	.070	03.58	26.5
	350	PNH	100	1.04	.075	03.29	24.3
	400	PNH	100	0.95	.081	02.99	22.1

Table VII.9, continued

Semicon-ductor	Junction T(°K)	Type*	Concen-tration	V_D' (V)	K' --	P_{max}/A_D W/cm²	η_s %
					Inverted Configuration		
Si	300	PNH	31.6	0.62	.059	01.59	18.6
	350	PNH	31.6	0.52	.070	01.31	15.3
	400	PNH	100	0.53	1.440	08.87	12.4
GaAs	300	PNH	316	1.07	.067	23.10	27.0
	350	PNH	316	0.98	.073	21.00	24.6
	400	PNH	316	0.89	.079	18.90	22.1
CdTe	300	PNH	31.6	1.11	.043	02.34	27.3
	350	PNH	31.6	1.02	.046	02.14	25.1
	400	PNH	31.6	0.94	.051	01.94	22.7

*PNH denotes pn junction and heterojunction, HET is solely heterojunction

The second generation solar cells investigated in this chapter have exhibited improved theoretical performance over the first generation solar cells of Chapters V and VI. Considering the efficiency of optical energy to electrical energy conversion, it would seem that pn and heterojunctions are to be preferred to Schottky solar cells and that inverted configuration solar cells are slightly better than vertical configuration and much better than standard configuration solar cells.

We shall use the theoretical data derived in this chapter in the next chapter, as we consider additional important aspects of second generation solar systems; aspects dealing with thermal energy as well as electrical energy output from these second generation solar cell systems.

References

1 Electronics Review, 22 July 1976, p. 41.
2 Popular Science, March 1977, p. 14.
3 Electronics, 11 Nov. 1976, p. 86.
4 See numerous articles in the Proceedings of the Annual Meetings of the American and International Solar Energy Societies for the years 1976 through 1993.

5 H. J. Hovel, <u>Semiconductors and Semimetals, Vol. II</u>, Academic Press, New York, 1977, p. 133.

6 It is probable that X_T is caused by mechanisms similar to those responsible for tunnel current in Schottky barriers. If so, the temperature dependence will be small [7].

7 G. H. Parker, "Tunneling in Schottky Barriers", Thesis, California Institute of Technology, 1969, p. 8 and 19.

8 R. C. Neville and C. A. Mead, in Journal of Applied Physics, Vol. 41, 1970, p. 3795.

9 H. K. Henisch, <u>Rectifying Semiconductor Contacts</u>, Clarendon Press, Oxford, 1957.

10 R. C. Neville, "Study of Combined (Photovoltaic-Thermal) Solar Energy Systems", Proceedings of the 2nd Miami International Conference on Alternative Energy Sources, Miami Beach, FL, Dec. 1979, p. 1267.

11 T. Runyan, <u>Silicon Semiconductor Technology</u>, McGraw-Hill, New York, 1965.

12 T. Kadman and E. F. Steigmeer, in Physics Review, Vol. 133, 1964, p. A1665.

13 R. Keys, in Physics Review, Vol. 115, 1959, p. 564.

14 G. A. Slack and G. Galginatais, in Physics Review, Vol. 133, 1964, p. A253.

15 Unpublished data by author.

16 A. V. Ioffe and A. F. Ioffe, in Sov. Physics Solid State, Vol. 2, 1960, p. 719.

17 R. A. Smith, <u>Semiconductors</u>, Cambridge University Press, 1959, Chap. 6.

18 B. L. Slater, in Proceedings of the 11th Annual IEEE Photovoltaic Specialists Conference, 1975, IEEE Press, New York.

19 P. F. Ordung et al., Photovoltaic Solar Energy Conference, Luxembourg, Sept. 1977.

20 R. C. Neville, International Solar Energy Society Meeting, Brighton, UK, 1981.

21 R. C. Neville, International Solar Energy Society Meeting, Montreal, Canada, 1985.

22 H. C. Card, in Journal of Applied Physics, Vol. 47, 1976, p. 4964.

23 W. Schockley and G. T. Reed, Jr., in Physics Review, Vol. 87, 1952. p. 835.

CHAPTER VIII: ADVANCED APPROACHES-II

Introduction

This chapter follows up on the material introduced in Chapter VII; material concerning the use of concentrated sunlight as the photonic input to solar cells. As stated in Chapter VII, this approach implies a system which employs tracking mirrors (or lenses) to provide an intense photon input to the solar cells. As theoretically indicated in Chapter VII, this provides a more efficient energy conversion system (light energy to electrical energy) and a less expensive solar power system. The efficiency is improved as the large number of generated hole-electron pairs saturates the recombination mechanism resulting in an increased photocurrent-- increased even beyond that dictated by an increase in photon input*. The cost of the solar system comes down because a significant area of expensive solar cells is replaced by lenses (or mirrors). This reduction in system cost is counterbalanced, to some extent, by the necessity of providing for some mechanism for solar tracking [1 and 2] in order to follow the sun and enable the mirrors (or lenses) to perform their solar concentration function. In second generation solar systems the incoming photons are concentrated, but their spectral distribution is of a solar type (in our case this means AMO or AM1 type light). We shall later discuss third generation solar power systems in which the incoming light is also heavily concentrated but is so modified as to spectrum as to not resemble sunlight.

---

* The calculations in Chapter VII ignore this phenomenon, but still show an increase in solar cell efficiency for concentrated light systems as compared to non-concentrated systems. The reason for this phenomenon is that the ratio of the photocurrent to saturation current density plays an important role in solar cell performance and this ratio increases with increasing optical concentration.

In both second and third generation solar cell systems, the input is solar energy in the form of photons. The output is, in part, electrical energy--but only in part. Unlike first generation solar cell systems, there is enough thermal energy available in these concentrated light systems to permit consideration of the delivery of at least a part of this thermal energy to some end user. Thus, the overall system efficiency is a mixture of both electrical and thermal components.

We will not, in this chapter, consider the ideal mixture of thermal and electrical energy outputs for these solar systems. Rather, we will consider what variables affect this mix and which alterations in these variables have potential impact on overall system selection.

Second Stage Solar Power Systems

We begin this chapter by further study of second stage solar power systems. The general schematic for a second stage solar power system is indicated in Figure VIII.1:

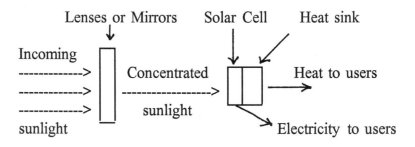

Figure VIII.1. The schematic of a second stage solar power system exclusive of the tracking mechanism.

Chapter VII was largely dedicated to this class of system; analyzing the output electrical power density and solar cell efficiency (at converting optical energy to electrical energy) for a number of example semiconductors as a function of the degree of concentration of either AMO or AM1 spectral type light, the type of junction and the optical orientation. This was done for typical, practical device parameters. In doing this, enough theoretical data was generated to permit us to eliminate Schottky barrier solar cells from the list of possible devices, to rule out indium phosphide, aluminum antimonide and cadmium selenide as potential semiconductors for solar cells and to suggest that inverted configuration solar cells appear to exhibit a performance level in excess of the vertical and standard configuration types[*]. In view of these comments, the reader is referred to Table VII.9 which presents the operating junction voltage, loss factor, output electrical power density and operating efficiency for stage two solar power systems using silicon, gallium arsenide and cadmium telluride as semiconductors operating under AMO-like concentrated sunlight. Tables VII.6 through VII.8 present similar data for a wider range of semiconductors operating under AM1-like concentrated sunlight. In concentrating both AMO and AM1 spectral sunlight, the degree to which the photon density was increased ranged from unity to a factor of one thousand[#]. Table VII.9 also presents the optical concentration level at which the solar cells (described in detail in Chapters VI and VII) exhibit maximum electrical efficiency.

One additional item of important information is included in Table VII.9. This is the junction temperature of the solar cells. Note that the operating efficiency of the solar cells decreases as the junction temperature is increased. Merely looking at this table would lead one to consider the use of a strong cooling system--a system holding the junction temperature to 300°K or lower. Such a cooling system requires the expenditure of energy to operate and adds significantly to the overall complexity of the solar energy converting system. A solution (at least partially) to this problem is to take the heat extracted (from the solar cells) in cooling the solar cells and to deliver it to some exterior

---

[*] As was done in earlier chapters, the author wishes to stress that the theoretical performance computed for various solar cell systems is based, in general, upon selecting practical values for various important semiconductor and solar cell characteristics. This led to ruling out various options. Even though certain materials, devices and optical configurations have been ruled out as candidates for high performance solar cell systems in this work, there is no fundamental law forbidding the discovery of new materials, fabrication techniques and operational methods which would lead to changes in which semiconductors, configurations and junctions are optimum for a given solar cell system.

[#] The reader is referred to the discussion in Chapter II concerning to the practical maximum degree of optical concentration--a maximum taking into consideration the optical focusing effects of mirrors and lenses and the degree to which the tracking system must accurately follow the sun.

customer. As in so many other situations encountered in this work, this delivery of thermal energy to some customer is not a simple procedure and involves considerable effort [3 and 4]. What we shall do is to consider the possible amount of thermal energy which can be moved from one place (the solar cell at a high temperature) to another (the customer at a lower temperature).

Following the approach of Chapter I we will use a Carnot engine to transfer the thermal energy from the high temperature to the low temperature. The efficiency of a Carnot cycle engine, η_c, is given by:

$$\eta_c = \{1 - T_c/T_h\} \times 100\% \ , \tag{VIII.1}$$

where T_c and T_h are the absolute temperatures of the high and low temperature thermal reservoirs.

Since we are discussing the delivery of thermal energy to locations on the earth's surface we are considering a cold reservoir temperature ranging from approximately 220°K (a Siberian location in January) to approximately 325°K (a Saharan location in July). To simplify our calculations let us assume the cold temperature reservoir to be at 300°K (this, moderately warm for the earth's surface, temperature allows for some inefficiency in the thermal extraction system). We will be transferring thermal energy from the junction of the solar cells to this hypothetical low temperature reservoir. Clearly, a solar cell junction temperature of 300°K will not permit any transfer of thermal energy. However, the solar cell junction temperatures of 350 and 400°K which we considered in Chapter VII will allow some energy transfer. Using Equation VIII.1 we can generate the thermal efficiency of the cooling system for the solar cells as:

Table VIII.1

Thermal efficiency of a second stage solar cell system

Solar cell junction temperature (°K)	Thermal Efficiency (η_c)
300	00.0
350	14.3
400	25.0

The efficiencies listed in Table VIII.1 are not impressive. However, the reader should recall that junction temperatures much in excess of 400°K will result in much reduced electrical output from a solar cell.

Let us return to Chapter VII and extract the maximum electrical efficiency for heterojunction solar cells made from silicon, gallium arsenide and cadmium telluride operating at maximum delivered power to the electrical load. This information is presented in Tables VIII.2 through VIII.5, along with other pertinent information.

Table VIII.2

The maximum electrical operating efficiency and corresponding optical concentration for heterojunction solar cells made from selected semiconductors, as a function of optical configuration, and junction temperatures under AMO-like light input

Semiconductor Substrate		Si	GaAs	CdTe
Temperature	Type	Efficiency, η_s, (%)/Optical concentration		
		(standard configuration)		
300°K	n-type	16.2/100	23.4/316	23.7/100
350°K	n-type	13.3/100	21.2/316	21.8/100
400°K	n-type	10.8/100	19.0/1000	19.8/100
		(vertical configuration)		
300°K	n-type	18.2/100	27.3/316	26.5/100
350°K	n-type	15.1/100	23.8/316	24.3/100
400°K	n-type	12.2/100	21.4/1000	22.1/100
		(inverted configuration)		
300°K	n-type	18.6/31.6	27.0/316	27.3/31.6
350°K	n-type	15.3/31.6	24.6/316	25.1/31.6
400°K	n-type	12.4/100	22.1/316	22.7/31.6

Recall that the values in Tables VIII.2 through VIII.4 were derived by applying "practical" values to the various parameters pertinent to predicting solar cell performance. It is possible to achieve higher levels of solar cell performance, but, as yet, not economical! Further, note that inverted configuration solar cells have a built-in optical concentration of approximately two. Thus, the actual optical concentration for

Table VIII.3

The maximum electrical operating efficiency and corresponding optical concentration for heterojunction solar cells made from selected semiconductors, as a funciton of optical configuration, and junction temperatures under AMO-like light input

Semiconductor	Substrate	Si	GaAs	CdTe
Temperature	Type	Efficiency, η_s, (%)/Optical concentration		
		(standard configuration)		
300°K	p-type	15.1/31.6	21.1/31.6	22.6/10
350°K	p-type	12.1/31.6	18.6/31.6	20.4/10
400°K	p-type	09.3/31.6	16.0/31.6	18.3/10
		(vertical configuration)		
300°K	p-type	17.0/31.6	23.6/31.6	25.2/10
350°K	p-type	13.7/31.6	20.8/31.6	22.8/10
400°K	p-type	10.7/31.6	18.0/31.6	20.4/10
		(inverted configuration)		
300°K	p-type	17.3/10	24.3/10	22.2/31.6
350°K	p-type	13.9/31.6	21.4/10	20.0/31.6
400°K	p-type	10.9/31.6	18.4/10	17.8/31.6

inverted configuration solar cells is some twice that listed in Tables VIII.2 through VIII.4. The optical concentration numbers provided in these tables, for all optical configurations, is the effective ratio of lens, or mirror, area to solar cell area. Clearly, assuming that the cost of a lens or mirror is significantly less than that of a single crystal solar cell on a per unit area basis, the larger the optical concentration number, the less costly, overall, will be the second stage solar cell system.

Study of Tables VIII.2 through VIII.4 does indicate that the performance of heterojunction solar cells made from the semiconductor examples is superior for n-type substrates, independent of the optical concentration. It is also clear that the electrical efficiency decreases rapidly with increasing temperature and that inverted configuration devices are superior.

Let us consider a second stage solar energy system receiving an input of 10 kW. (Note, the actual size in square meters, acres, or whatever is not specified--we are assuming that under a given set of

Table VIII.4

The maximum electrical operating efficiency and corresponding optical concentration for heterojunction solar cells made from selected semiconductors, as a function of optical configuration, and junction temperatures under AM1-like spectra light input

Semiconductor Substrate		Si	GaAs	CdTe
Temperature	Type	Efficiency, η_s, (%)/Optical concentration		
		(standard configuration)		
300°K	n-type	16.5/100	22.8/1000	22.8/100
350°K	n-type	13.6/100	20.8/1000	20.9/100
400°K	n-type	10.9/100	18.6/1000	19.0/100
		(vertical configuration)		
300°K	n-type	18.5/100	25.5/1000	25.4/100
350°K	n-type	15.2/100	23.3/1000	23.3/100
400°K	n-type	12.3/100	20.9/1000	21.2/100
		(inverted configuration)		
300°K	n-type	18.9/100	26.4/316	26.2/31.6
350°K	n-type	15.7/100	24.0/316	24.0/31.6
400°K	n-type	12.8/100	21.5/316	21.8/100
		(standard configuration)		
300°K	p-type	15.4/31.6	20.6/31.6	21.8/10
350°K	p-type	12.3/31.6	18.1/31.6	19.7/10
400°K	p-type	09.5/31.6	15.6/31.6	17.5/10
		(vertical configuration)		
300°K	p-type	17.2/31.6	23.0/31.6	24.2/10
350°K	p-type	13.8/31.6	20.3/31.6	21.9/10
400°K	p-type	10.6/31.6	17.5/31.6	19.6/10
		(inverted configuration)		
300°K	p-type	17.7/31.6	23.6/10	25.0/3.16
350°K	p-type	14.3/31.6	20.8/31.6	22.6/3.16
400°K	p-type	11.2/31.6	18.0/31.6	20.1/10

conditions--weather, lens/mirror, ratio of concentrator area to solar cell area, time of day, etc.--we have an input of 10 kW.) To simplify matters let us consider the input light to be of either an AMO or AM1 spectral nature and of a concentration level to permit maximum electrical efficiency. Using the operating efficiencies in Tables VIII.2 and VIII.4 (for n-type substrates) we can obtain the electrical power output from the solar cells. This represents energy which is removed from the solar cells and which is not available for thermal energy extraction. From our discussions in Chapters IV through VI, there is also a loss of approximately five percent of the optical energy due to reflection of photons from the semiconductor surface, irrespective of optical orientation. This loss is essentially all there is for inverted configuration solar cells. For vertical configuration solar cells, our discussion in Chapter VI suggests an additional loss of energy of some five percent owing to surface recombination effects and the "front layer" and substrate electrical contacts. Finally, for standard configuration solar cells we experience, for pn and heterojunction devices, a further loss of some five percent of the incoming optical energy owing to reflection from electrical contacts on the illuminated surface. For standard configuration Schottky barrier solar cells the metal layer on the surface, even if very thin, will still result in a loss of some two thirds of the incoming photons due to reflection. Defining the percentage of incoming light power which is lost to reflection as β, we have:

Table VIII.5

The percentage, β, of incoming light power which is lost to reflection as a function of optical configuration and junction

Optical configuration	Junction	β
Inverted	all	05 %
Vertical	all	10 %
Standard	pn	15 %
	heterojunction	15 %
	Schottky	67 %

Only after the delivered electrical energy and reflection losses have been deducted do we arrive at the amount of energy which is, theoretically, available for extraction by the heat sink/thermal energy source device. Defining the electrical power obtained from the solar cells under these conditions as P_E and the amount of power available for thermal extraction as P_{aT} , we have, where the incoming light power is

NI_{ns}, the following expression for P_{aT}^{*}:

$$P_{aT} = NI_{ns}\{1 - \beta/100\} - P_E .$$ (VIII.2)

Tables VIII.6 and VIII.7 provide values of the power available for thermal extraction assuming a 10 kW input of concentrated solar energy and the maximum electrical conversion efficiencies of Tables VIII.2 and VIII.4 (for n-type substrates) as a function of optical configuration. Since this analysis assumes that the cold temperature reservoir for the heat energy extraction engine is at 300°K, only junction temperatures of 350 and 400°K are considered.

Table VIII.6
Power available for thermal extraction, P_{aT}, under a AMO-like spectrum as a function of junction temperature and optical configuration for second stage solar power systems with n-type substrate heterojunctions. Initial optical input power is 10 kW with optical concentration levels as specified in Table VIII.2

Semiconductor Temperature (°K)	Si	GaAs	CdTe
		P_{aT} (watts)	
		(standard configuration)	
350	7,170	6,380	6,320
400	7,420	6,600	6,520
		(vertical configuration)	
350	7,490	6,620	6,570
400	7,780	6,860	6,790
		(inverted configuration)	
350	7,970	7,040	6,990
400	8,260	7,290	7,230

---

* Note, we are not attempting to specifiy the exact nature of the heat sink/thermal energy source device. This, like many other details and boundary conditions for the second stage solar cell system, must depend on knowing considerably more of the technical, economic and weather aspects of the system.

Table VIII.7
Power available for thermal extraction, P_{aT}, under a AM1-like spectrum as a function of junction temperature and optical configuration for second stage solar power systems with n-type substrate heterojunctions. Initial optical input power is 10 kW with optical concentration levels as specified in Table VIII.4

Semiconductor Temperature (°K)	Si	GaAs	CdTe
		P_{aT} (watts)	
		(standard configuration)	
350	7,140	6,420	6,410
400	7,410	6,640	6,600
		(vertical configuration)	
350	7,480	6,670	6,670
400	7,770	6,910	6,880
		(inverted configuration)	
350	7,930	7,100	7,100
400	8,220	7,350	7,320

Given the data of Tables VIII.6 and VIII.7 we can generate the amount of thermal power, P_{Th}, actually available to the end consumer using the thermal efficiencies of Table VIII.1. This is done in Figures VIII.2 and VIII.3 for junction temperatures of 350°K and 400°K under optimally concentrated (see Tables VIII.2 and VIII.4) AMO-like and AM1-like spectral inputs for silicon, gallium arsenide and cadmium telluride based heterojunction solar cells with n-type substrates. Note the considerable increase in thermal power for junction temperatures of 400°K over junction temperatures of 350°K and recall that the available thermal energy, P_T, for junction temperatures of 300°K is zero.

We conclude this section by considering the total useful energy, both thermal and electrical, delivered by typical second stage solar cell power systems to consumers. Again, we assume that we are using n-type substrate heterojunction solar cells made from silicon, gallium arsenide or cadmium telluride, and Carnot cycle heat extraction engines. Here, we consider semiconductor junction temperatures of 300°K, 350°K and 400°K as well as standard, vertical and inverted optical concentrations. We continue to assume that the optical concentrations are those of Tables VIII.2 and VIII.4 under AMO and AM1-like spectral inputs of 10 kW.

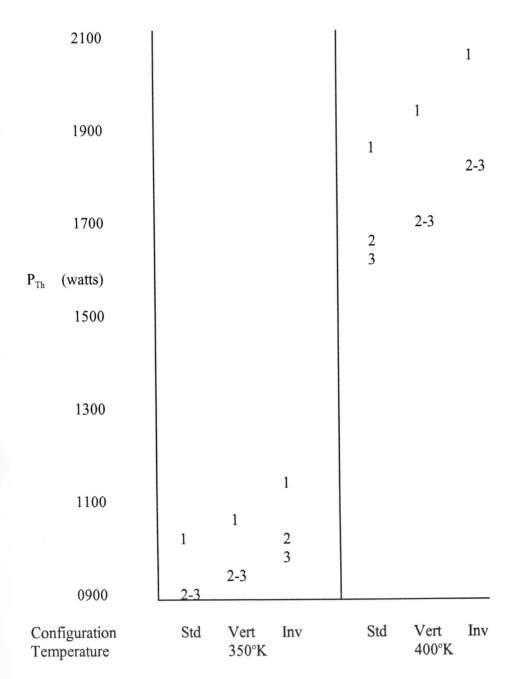

Figure VIII.2. The thermal power, P_{Th}, supplied to the consumer as a function of junction temperature and optical configuration for Si (1), GaAs (2), and CdTe (3) n-type substrate heterojunction second stage solar cell systems. The input is 10 kW of optimally concentrated AMO light.

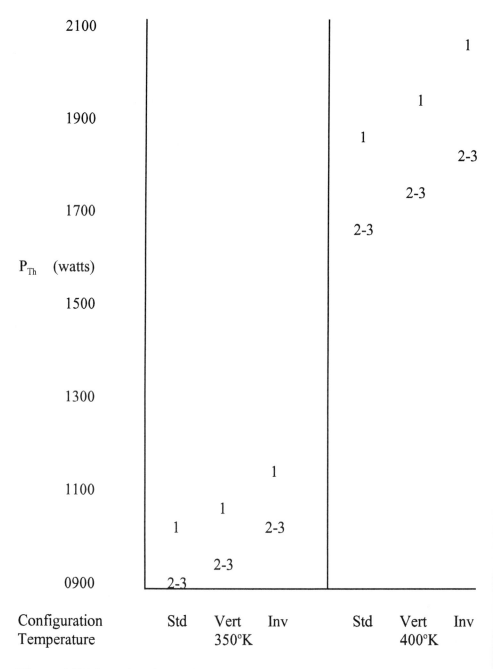

Figure VIII.3. The thermal power, P_{Th}, supplied to the consumer as a function of junction temperature and optical configuration for Si (1), GaAs (2), and CdTe (3) n-type substrate heterojunction second stage solar cell systems. The input is 10 kW of optimally concentrated AM1 light.

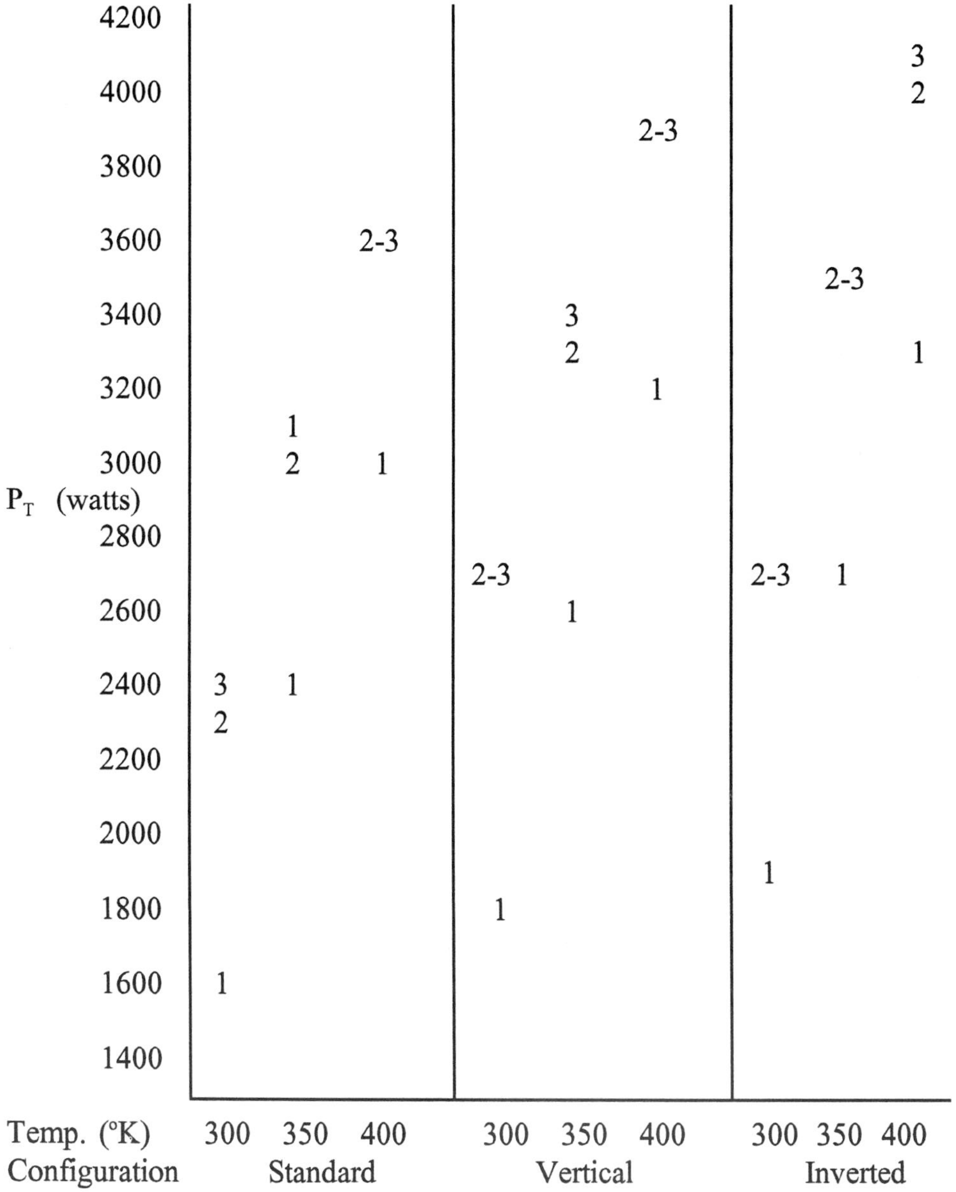

Figure VIII.4. The total power, P_T, supplied (in theory) to customers by a second stage solar cell system utilizing n-type substrate heterojunction solar cells of Si (1), GaAs (2) or CdTe (3) with a cold temperature heat engine reservoir at 300°K and under concentrated AMO-like spectral input. Optical configuration and junction temperature are parameters.

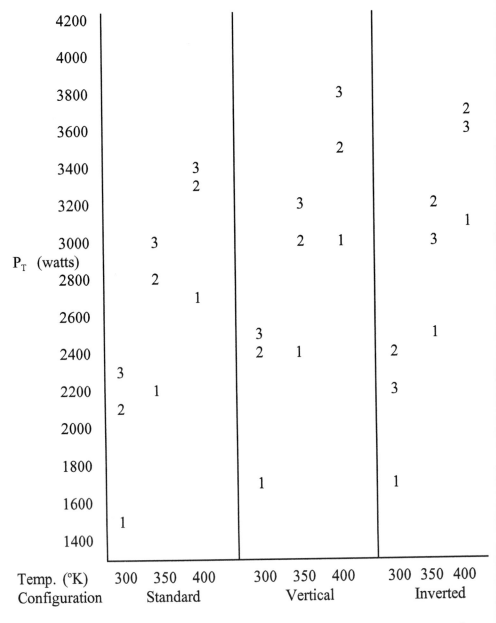

Figure VIII.5. The total power, P_T, supplied (in theory) to customers by a second stage solar cell system utilizing n-type substrate heterojunction solar cells of Si (1), GaAs (2) or CdTe (3) with a cold temperature heat engine reservoir at 300°K and under concentrated AM1-like spectral input. Optical configuration and junction temperature are parameters.

Study of Tables VIII.2 through VIII.4 and Figures VIII.2 through VIII.5 clearly indicates that the delivery of heat energy, where possible, enhances overall system performance significantly. Whether it is possible to make use of a system which delivers both thermal and electrical power to consumers depends on the economics of the situation and will be discussed in a later chapter. For the moment, note that the overall delivered power from these second order systems is much higher than for the simple first stage systems discussed in Chapters IV through VI.

Third Generation Solar Cell Systems

Like second generation solar cell systems, third generation solar cell systems deliver both electrical and thermal power to consumers, use concentrated light as inputs and are required to track the sun. Third generation systems differ from second generation solar cell systems because of a single important modification. Study of Figures IV.4 and IV.5 as well as the theoretical investigations of Chapters V through VII indicates that the solar spectrum (AMO, AM1 or any other minor modification of the spectral output of the sun--such as might be caused by the presence of smog) is far from ideal when it comes to interaction with the semiconductors that make up the solar cells. A major reason for this non-ideality is visible in Figure II.1. From this figure it is clear that the solar spectrum contains photons of a wide range in wavelength (energy). This means that many of the solar photons do not have sufficient energy to interact with the semiconductor and produce the hole-electron pairs. Many other photons do interact but have a large surplus of energy so that the conversion process of solar to electrical energy is poor.

What a third generation solar cell system does is to modify the spectral nature of the light which impacts the semiconductor solar cells. There are several ways in which this can be done. We will concentrate, in this section, on an approach known as thermophotovoltaics [5 and 6]. A typical photovoltaic system is presented in Figure VIII.6.

The thermophotovoltaic system of Figure VIII.6 consists of a heavily insulated container which encloses a vacuum. Sunlight enters the container through a lens system which heavily concentrates the light and focuses it on the component known as the converter. The converter, in turn, absorbs the concentrated photons and becomes quite hot--sufficiently hot to act as (approximately) a black body emitter of photons. These

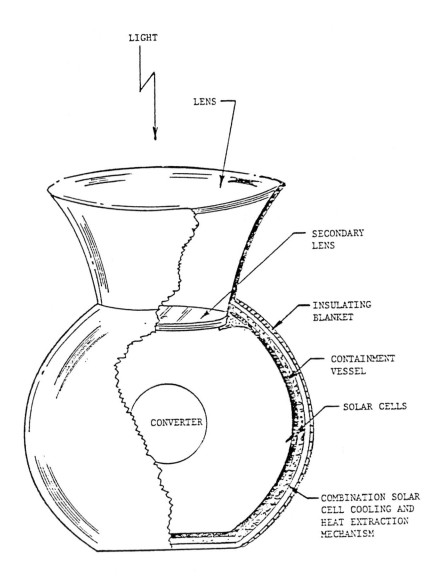

Figure VIII.6. A cut-away view of a thermophotovoltaic solar energy conversion system. See text for details.

secondary photons have a spectral output which is characteristic of the temperature of the converter [7]. Since the natural spectrum of the sun has too many high energy photons, it is clear that we wish a converter temperature which is less than that of the surface of the sun (~6000°C). The photons emitted from the converter, the secondary photons, are irradiated in all directions. A few escape from the thermophotovoltaic system by emerging through the input optical system. The majority of the secondary photons impact the internal walls of the thermophotovoltaic system. These walls are lined with a multi-layered structure. The innermost layer (and, therefore, the layer closest to the converter) is a bandpass optical filter. To perform satisfactorily, this filter must pass photons of an energy ranging from approximately the energy gap of the semiconductor making up the solar cells to a value approximately 10% greater than the energy gap. The secondary photons in this range of energies, when converted to hole-electron pairs in the solar cells, yield almost all of their energy in the form of electrical energy, with very little surplus thermal energy. Thus the photons in this energy range provide for the most efficient generation of electrical energy. The optical bandpass filter allows the secondary photons in this energy range to pass through and to reach the next layer. This second layer is composed of semiconductor solar cells and is exposed to a highly concentrated flux of photons; photons of an ideal energy for conversion to electrical energy. Finally, behind the solar cells, and next to the outermost wall of the thermophotovoltaic system, is a heat sink/heat pump layer. The function of this layer is to maintain the solar cells at their design operating temperature while, at the same time, delivering thermal energy to some external consumer.

Let us return, briefly, to the bandpass filter. The bulk of the photons emitted by the converter does not possess the energy required for passage through the bandpass filter. These photons (either too high or too low in energy) are reflected (for the most part) back towards the converter. They impact on and are absorbed by the converter, helping to maintain its temperature. If one considers Figure VIII.6 and what has been said here, the end result of this approach is that almost all of the energy in sunlight is eventually converted to photons of the desired energy.

No one, to date, has completely succeeded in fabricating a complete thermophotovoltaic system. There exist a number of interesting technical problems which require operational solutions before a working thermophotovoltaic system is established. First, the converter needs to be suspended approximately in the center of the thermophotovoltaic device by some mechanism which minimizes thermal losses from the converter. (One possibility is a thin, highly thermal resistive stalk supporting the spherical converter--highly thermal resistive because we need to minimize thermal convection losses. We have already evacuated the thermophotovoltaic system in order to reduce convection losses and the bandpass filter reduces radiation losses from the converter.) If this is done the principal energy loss from the converter will consist of those photons which pass through the bandpass filter and are absorbed within the solar cells.

A second technical problem is the design and fabrication of the bandpass filter. This filter must have extremely sharp "corners" so that only the desired photons reach the solar cells. The reflectivity of the bandpass filter must be high, both to keep it cool and to maintain the temperature of the converter. Furthermore, the bandpass filter must not adversely affect the surface properties of the solar cells.

A third problem lies in the design of the solar cells themselves. They are exposed to a high density of incoming photons and will generate a significant photocurrent. The twin requirements of high electrical and thermal energy delivery place severe limitation on the resistance (both electrical and thermal) of the solar cells. Taking into consideration the theoretical studies of Chapters V through VII, it appears that the ideal optical configuration for the solar cells in a thermophotovoltaic system would be inverted. This allows for maximum photocurrent with the junctions located close to the heat sink. The cooler junctions permit a lower saturation current density and higher optical-to-electrical energy conversion efficiency.

Finally, we must select a thermal energy extraction engine. This device is responsible for maintaining a proper solar cell junction temperature as-well-as delivering thermal energy to consumers.

Let us consider some predicted performance characteristics for potential thermophotovoltaic systems. To do so, in view of the, as-yet, unrealized nature of the device, let us take a theoretical thermophotovoltaic system with the following properties:

A] Cooling System--This system uses a Carnot heat engine with a high temperature heat energy reservoir of 300, 350 or 400°K and works into a low temperature heat energy reservoir (the consumer) at 300°K. Thus this heat engine exhibits the delivery efficiencies furnished in Table VIII.1.

B] Solar Cells--Based on our previous studies, let us analyze thermophotovoltaic systems utilizing solar cells made from silicon, gallium arsenide, or cadmium telluride. We will assume these solar cells to be of an inverted configuration with n-type substrates. These devices are like those discussed in Chapter VII with properties as in Table VIII.8.

Table VIII.8

Solar cell characteristic parameters for the inverted configuration, n-type substrate, heterojunction solar cells to be used in thermophotovoltaic systems

Semiconductor	Si	GaAs	CdTe
Resistance times area product (Ω-cm^2)	.0066	.0015	.0106
Saturation current density (A/cm^2)			
300°K	4.50×10^{-12}	7.20×10^{-19}	1.31×10^{-20}
350°K	4.95×10^{-09}	5.62×10^{-15}	1.22×10^{-16}
400°K	8.00×10^{-07}	5.40×10^{-12}	1.31×10^{-13}

To compute the performance of the thermophotovoltaic solar cells (using Equations VI.16 through VI.19, we need the photocurrent. To obtain the photocurrent we must examine the bandpass filter and converter.

C] Bandpass Filter--Assume a bandpass filter with the characteristics provided in Table VIII.9 [8], where λ is the photon wavelength.

Table VIII.9

Proposed characteristics of the bandpass filter in terms of the wavelength of photons with an energy equal to the semiconductor energy gap, λ_g

Wavelength	$\lambda \leq .9 \, _g$	$.9 \, _g \leq \leq \, _g$	$\geq \, _g$
		Filter Characteristics	
Absorptivity	0.02	0.02	0.02
Reflectance	0.98	0.01	0.98
Transmittance	0.00	0.97	0.97

D] The Converter--The converter is assumed to have the spectral output of a black body as given by Equations VIII.3 and VIII.4 [9].

$dP_p/d\lambda = \{3.7\times10^4/ {}^5\}\{1/(exp[1.44\times10^4/ T] - 1)\}$ watts/cm$^2\mu$m   (VIII.3)

and

$P_p = 5.67\times10^{-12}T^4$ watts/cm^2 ,   (VIII.4)

where P_p is the photon power density in watts/cm^2 and the photon wavelength, λ, is in microns. Equation VIII.3 provides the incrimental power per unit wavelength and Equation VIII.4 presents the total power density radiated per unit area of converter (the Stefan-Boltzmann law).

Using Equations VIII.3 and VIII.4 it is possible to determine the output power levels (the insolation) from the black body converter in the critical wavelengths for Si, GaAs and CdTe[*]. This information is presented in Table VIII.10. Recall that the total spectrum power output for AMO and AM1 conditions are .1353 and .105 watts/cm^2. Also note that the insolations are a function of converter temperature.

Table VIII.10

Bandpass filter window insolation (watts/cm^2), at the converter surface, as a function of the converter temperature

Semiconductor		Si	GaAs	CdTe
Converter Temperature (°K)	Total spectrum	Individual windows		
1000	05.67	0.00623	0.0003	.0001
1500	28.7	0.518	0.0925	.0491
2000	90.7	4.80	1.56	.986
2500	0221	018.4	008.53	05.96
3000	0459	045.3	026.6	19.8
4000	1450	142	112	89.7

---

[*] We can use the data in Appenidx B along with Equation II.3 to determine the exact wavelength values.

It is clear from Table VIII.10 that the insolation in general, and in the bandpass filter windows of interest, increases rapidly as the temperature of the converter increases. Note that the insolation furnished in Table VIII.10 is that at the surface of the converter. Should the inside surface of the solar cells be coincident with the outside surface of the converter we can compute the photocurrent density for the solar cells by accounting for the effects of the bandpass filter using the data of Table VIII.9. Assuming one electron-hole pair generated per absorbed photon and that each photon in a bandpass filter window has some average energy (1.05 eV for Si, 1.45 eV for GaAs and 1.55 eV for the CdTe windows) we arrive at the coincident photocurrents of Table VIII.11.

Table VIII.11

Photocurrents, based on the assumption that the emitting converter surface is coincident with the inner solar cell surface and correcting for the reflection, absorption and transmission effects of the bandpass filter

Semiconductor	Si	GaAs	CdTe
Converter Temperature (°K)	J_p (A/cm^2)		
1000	0.00575	0.00223	0.000078
1500	0.478	0.0619	0.0307
2000	4.433	1.043	0.617
2500	017.0	05.70	03.73
3000	041.8	17.8	12.4
4000	131	74.9	56.2

The photocurrent values in Table VIII.11 are not the photocurrent densities experienced by the solar cells. The photocurrent density for the solar cells depends on the ratio of the inner surface area of the solar cells to the radiating area of the converter. If, for example, this area ratio is two, then the photocurrent density at the solar cells is one half that indicated in Table VIII.11.

For the purposes of our analysis, let us consider three potential area ratios for solar cells to converter. Let us take a ratio of two (which we will designate as a tight ratio), a ratio of ten (which we will designate as a moderate ratio of areas) and a ratio of one hundred (which we will term as a loose area ratio). Taking these area ratios we can compute the photocurrent density experienced by the inverted, n-type substrate, heterojunction solar cells. Using Equations VI.16 through VI.19; we can determine the operating junction voltage, V_p; the loss factor, K'; the maximum delivered electric photopower density, P_{max}/A_D; and the maximum solar cell efficiency, η_s. Note that the incoming electrical energy is that in the bandpass filter window so that the efficiency of these solar cells is quite high.

Figures VIII.7 through VIII.9 present values of the maximum operating efficiency for conversion of solar to electrical energy for inverted configuration, n-type substrate, heterojunction solar cells in a thermophotovoltaic system--as a function of the converter temperature with solar cell junction temperature and the ratio of the solar cell area to the converter secondary photon emitting area as parameters.

Note that the optical-to-electrical energy conversion efficiency decreases with increasing junction temperature. Also note that the effects of converter temperature and solar cell-to-converter area ratio are more complex. For low converter temperatures, the smaller area ratios (producing, as they do, higher photocurrents) have higher efficiency levels. At the high converter temperatures, the photocurrents for smaller area ratios are high and internal resistance losses decrease the operating efficiency levels. The result is that a thermophotovoltaic system must be designed for semiconductor, junction temperature, converter temperature and area ratio. Furthermore, note that the tight area ratio systems place a severe operating temperature and photon flux load on the bandpass filter, while the loose area ratio systems are significantly bigger and, hence, use more materials than the smaller area ratio systems.

While the situation may not be optimum, it is a practical one and so we will examine silicon, gallium arsenide and cadmium telluride inverted configuration, heterojunction, n-type substrate solar cell theoretical performance in thermophotovoltaic systems with an area ratio of 10 (the moderate ratio systems) as a function of converter and junction temperatures. Table VIII.12 provides data (operating voltage, loss factor, delivered power density and efficiency) for selected instances.

Study of the potential performance for a thermophotovoltaic system, as indicated in Table VIII.12, and comparison with the results derived for second stage systems as shown in Chapter VII, indicate that

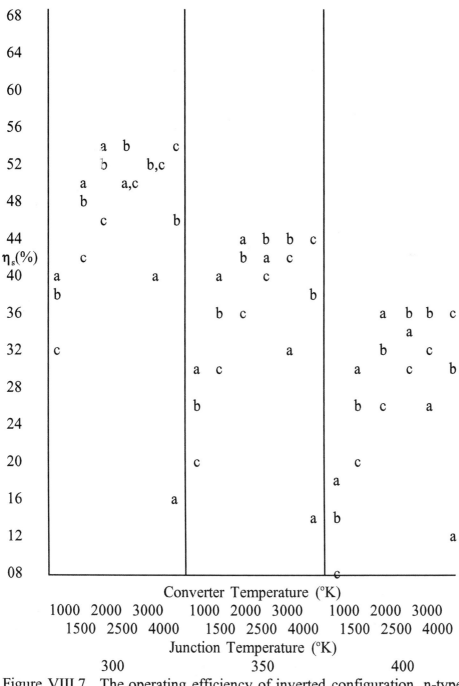

Figure VIII.7. The operating efficiency of inverted configuration, n-type substrate, heterojunction silicon solar cells operating in a thermophotovoltaic sytem. Area ratio legend: a-->2, b-->10 and c-->100.

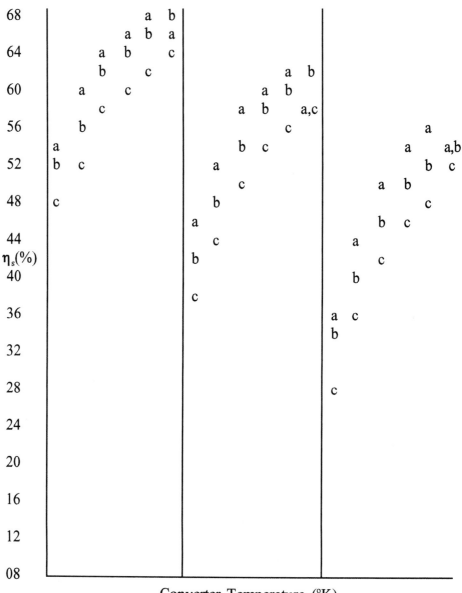

Figure VIII.8. The operating efficiency of inverted configuration, n-type substrate, heterojunction gallium arsenide solar cells operating in a thermophotovoltaic sytem. Area ratio legend: a-->2, b-->10 and c-->100.

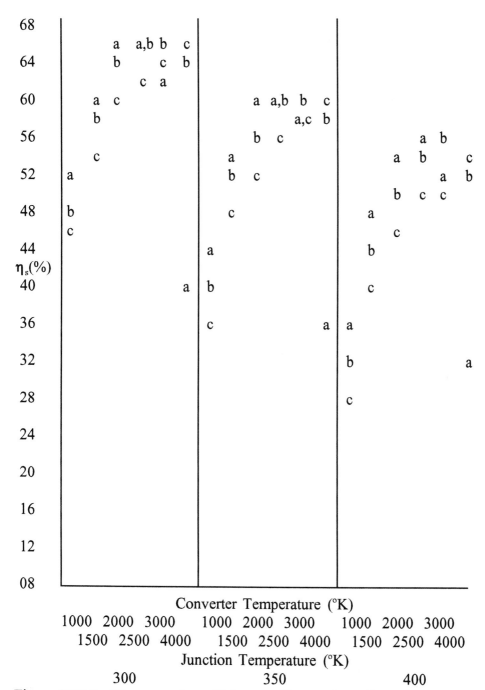

Figure VIII.9. The operating efficiency of inverted configuration, n-type substrate, heterojunction cadmium telluride solar cells operating in a thermophotovoltaic sytem. Area ratio legend: a-->2, b-->10 and c-->100.

Table VIII.12

Theoretical operating photovoltage, loss factor, maximum electrical power delivered and solar cell efficiency for inverted configuration, n-type substrate, heterojunction solar cells under selected operating conditions and assuming an area ratio (solar cell-to-converter) of 10

Converter temperature	Junction temperature	V_p volts	K' ---	P_{max}/A_D watts/cm^2	η_s %
			Silicon		
1500 °K	300 °K	0.535	0.00220	0.0475	47.1
	350 °K	0.423	0.00272	0.0368	36.5
	400 °K	0.322	0.00346	0.0270	26.8
2000 °K	300 °K	0.591	0.01884	0.4843	51.8
	350 °K	0.488	0.02250	0.3921	41.9
	400 °K	0.393	0.02734	0.3077	32.9
2500 °K	300 °K	0.625	0.07173	1.916	53.5
	350 °K	0.528	0.08458	1.581	44.1
	400 °K	0.438	0.10101	1.274	35.9
3000 °K	300 °K	0.650	0.18773	4.661	52.8
	350 °K	0.557	0.22143	3.871	43.9
	400 °K	0.472	0.26408	3.151	35.7
4000 °K	300 °K	0.689	0.84361	12.69	45.9
	350 °K	0.603	1.0300	10.44	37.8
	400 °K	0.526	1.2646	8.423	30.5
			Gallium Arsenide		
2000 °K	300 °K	0.946	0.00063	0.1872	61.6
	350 °K	0.841	0.00070	0.1651	54.3
	400 °K	0.733	0.00080	0.1424	46.8
2500 °K	300 °K	0.989	0.00330	1.070	64.4
	350 °K	0.891	0.00364	0.9563	57.5
	400 °K	0.789	0.00406	0.8394	50.5
3000 °K	300 °K	1.018	0.01005	3.417	66.0
	350 °K	0.924	0.01100	3.079	59.5
	400 °K	0.827	0.01219	2.731	52.8
4000 °K	300 °K	1.055	0.04223	14.72	67.4
	350 °K	0.967	0.04590	13.38	61.3
	400 °K	0.876	0.05045	12.02	54.9

Table VIII.12, continued

Converter temperature	Junction temperature	V_p volts	K' ---	P_{max}/A_D watts/cm^2	η_s %
			Cadmium Telluride		
2000 °K	300 °K	1.034	0.00241	0.1212	63.0
	350 °K	0.938	0.00264	0.1091	56.8
	400 °K	0.839	0.00293	0.0968	50.3
2500 °K	300 °K	1.080	0.01414	0.7613	65.5
	350 °K	0.991	0.01533	0.6937	59.7
	400 °K	0.899	0.01678	0.6243	53.7
3000 °K	300 °K	1.111	0.04726	2.570	66.3
	350 °K	1.027	0.05097	2.354	60.8
	400 °K	0.940	0.05548	2.135	55.2
4000 °K	300 °K	1.153	0.24379	11.09	63.3
	350 °K	1.077	0.26334	10.21	58.3
	400 °K	0.997	0.28714	9.296	53.1

the major feature for third stage solar cell systems is a vastly improved optical-to-electrical conversion efficiency.

Let us now consider the thermal conversion efficiency of a thermophotovoltaic system. Once again, let us assume that the thermal energy is available at the junction temperature of the solar cells, 300°K, 350°K or 400°K. This energy is delivered to consumers at a temperature of 300°K. Thus, the thermal conversion efficiency is that provided in Table VIII.1 Following the procedure used in investigating second stage systems earlier in this chapter, let us assume a solar power system (in our case a third stage or thermophotovoltaic system) with a solar power input of 10 kW. This energy is focused upon the converter, heats it, and causes the converter to radiate secondary photons. Some of the secondary photons leave the thermophotovoltaic system via the input optics. For the purposes of this analysis, let us assume that some 10% of the input energy is lost via this mechanism. The converter also loses some energy by thermal conduction down the stalk which supports it within the thermophotovoltaic system. The amount of this loss is dependent upon the temperature of the converter and the structure of the supporting stalk. Assuming that the energy loss down the stalk for a 1000°K converter is

200 joules per second, then, with a similar stalk/converter construction the losses for other converter temperatures are provided in Table VIII.13.

Table VIII.13

Potential thermal conduction losses out of the stalk supporting the converter, as a function of the converter temperature for a 300°K temperature at the base of the stalk

Converter temperature (°K)	Thermal losses (W)	Converter temperature (°K)	Thermal losses (W)
1000	200	1500	340
2000	490	2500	630
3000	770	4000	1050

The inclusion of these losses means that some 7950 to 8800 of the watts radiated from the converter fall upon the bandpass filter where some are passed onto the solar cells, some are absorbed by the filter and some are reflected back to the converter (then reradiated). Following our assumptions, the initial optical input provides energy to make up for the energy lost due to the secondary photons escaping via the output optics, the energy passed to the solar cells and the optical energy absorbed in the bandpass filter. Thus, under our assumptions, the entire 7950 to 8800 watts are available for thermal and electrical energy[*].

The electrical conversion efficiency provided in Figures VIII.7 through VIII.9 and in Table VIII.12 applies to the 7.95 to 8.8 kW value. Thus, we can compute the electrical energy output for our various thermo-photovoltaic systems. Given the electrical power delivered, we can determine the remaining energy (which is available as a thermal energy source) and, using Table VIII.1 compute the total energy available to potential consumers. The results are supplied in the following figures assuming a solar cell-to-converter area ratio of 10.

---

[*] Note that the thermal energy lost by conduction down the stalk also serves as a thermal energy source.

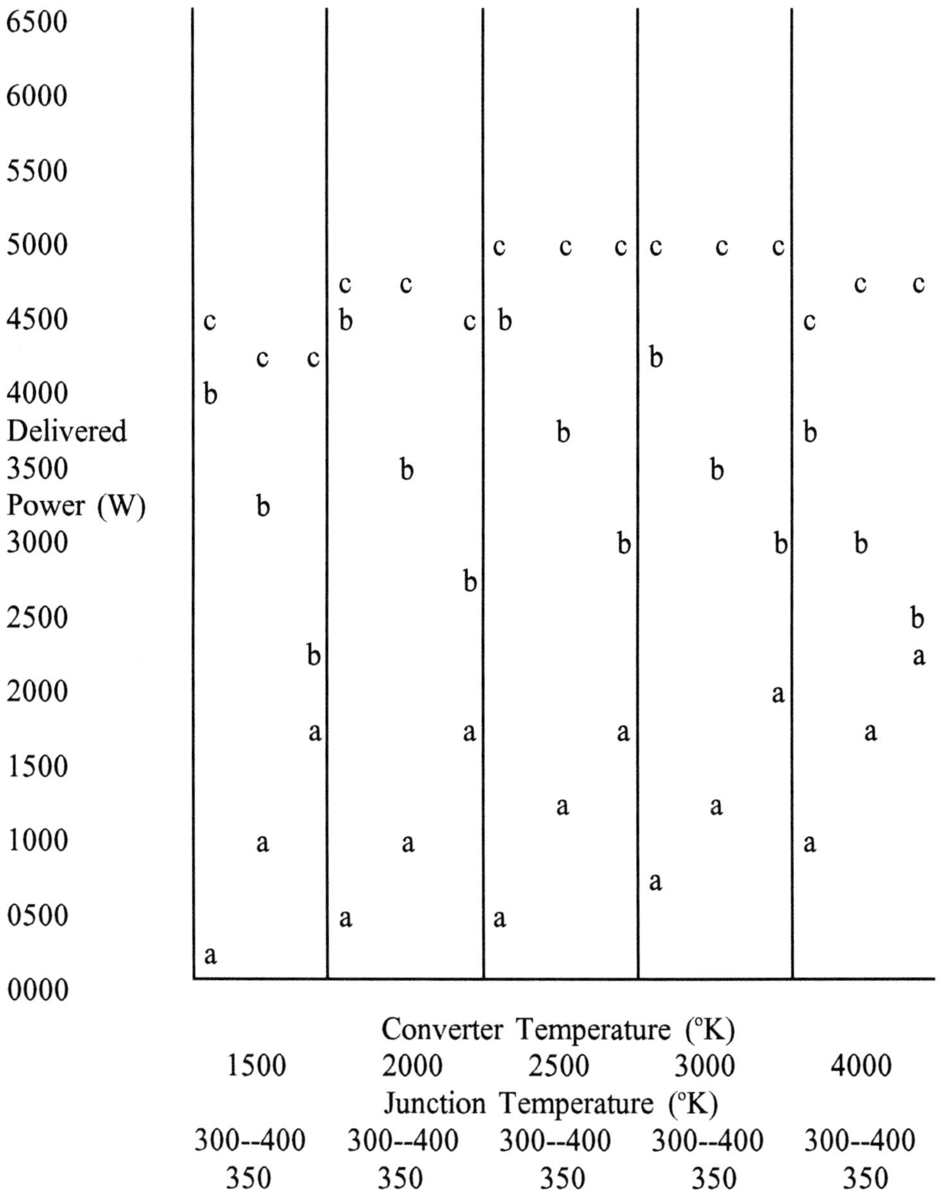

Delivered Power (W)

```
6500

6000

5500

5000                    c    c    c  c    c    c
                  c   c                              c    c
4500    c         b              c  b                     c
            c   c
4000    b                                  b
Delivered                         b                 b
3500             b                                  b
Power (W)   b
3000                               b         b    b    b
                        b
2500          b                                       b
                                                      a
2000          a                              a
                a         a         a              a
1500                         a         a
1000    a         a                          a
                                        a
0500          a         a

0000
```

Converter Temperature (°K)
1500   2000   2500   3000   4000
Junction Temperature (°K)
300--400   300--400   300--400   300--400   300--400
350   350   350   350   350

Figure VIII.10.   The potential delivered thermal, electrical and total powers from a thermophotovoltaic system employing silicon heterojunction inverted configuration solar cells with n-type substrates, an area ratio of 10 and a thermal consumer at 300°K as a function of converter and junction temperatures. The input power is 10 kW.
Legend: a-->P_{Th}, b-->P_E and c-->P_T.

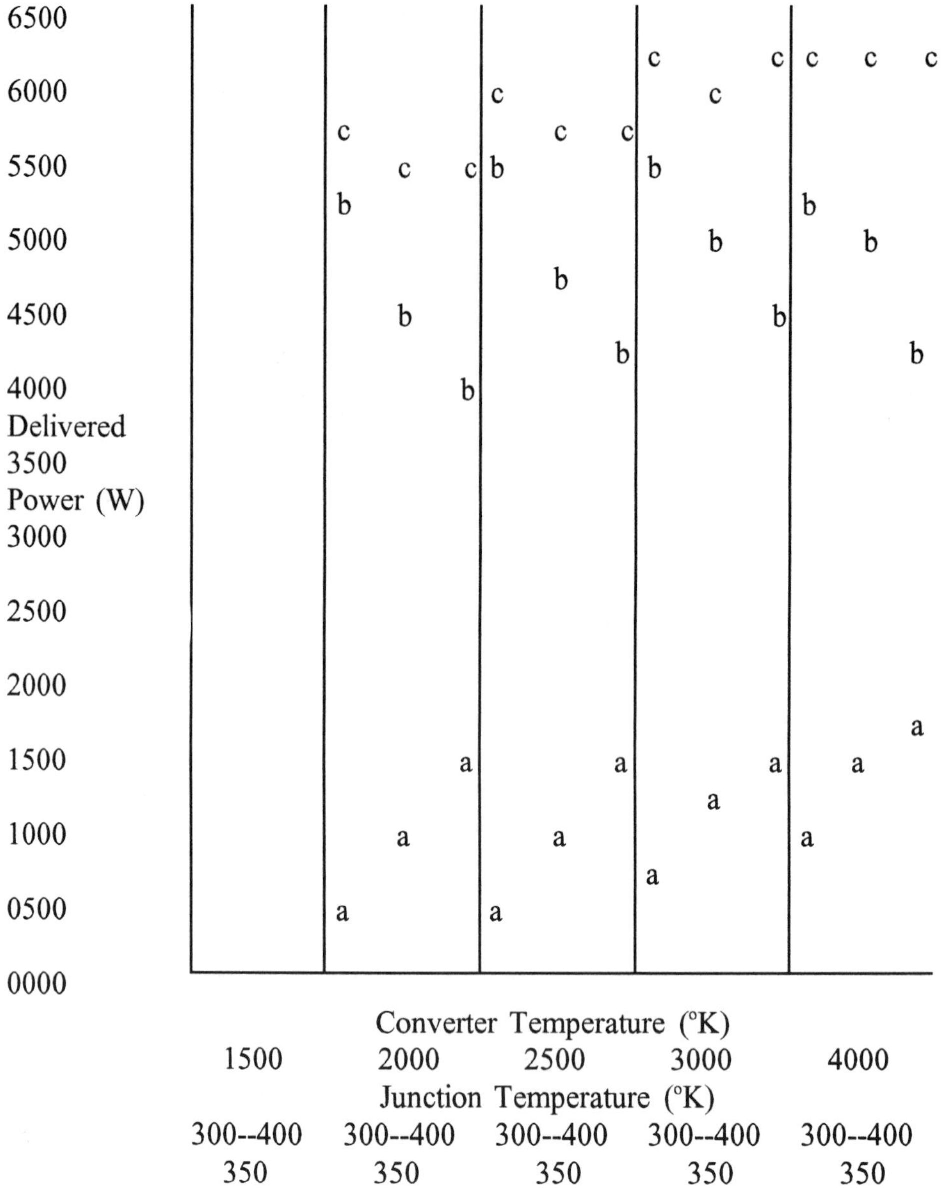

```
6500
                                                    c        c  c   c   c
6000                             c             c
                      c                  c   c           b
5500              c      c b                       b
                  b                                              b
5000                                          b                  b
                                     b
4500              b                                     b
                                          b                      b
4000                  b
Delivered
3500
Power (W)
3000

2500

2000
                                                                    a
1500                  a             a              a        a
                                              a
1000              a             a                          a
                                                   a
0500          a             a
```

```
0000 └─────────────────────────────────────────────────────
              Converter Temperature (°K)
        1500        2000        2500        3000        4000
              Junction Temperature (°K)
        300--400  300--400  300--400  300--400  300--400
          350        350        350        350        350
```

Figure VIII.11.   The potential delivered thermal, electrical and total powers from a thermophotovoltaic system employing gallium arsenide heterojunction inverted configuration solar cells with n-type substrates, an area ratio of 10 and a thermal consumer at 300°K as a function of converter and junction temperatures. The input power is 10 kW.
Legend: a-->P_{Th}, b-->P_E and c-->P_T.

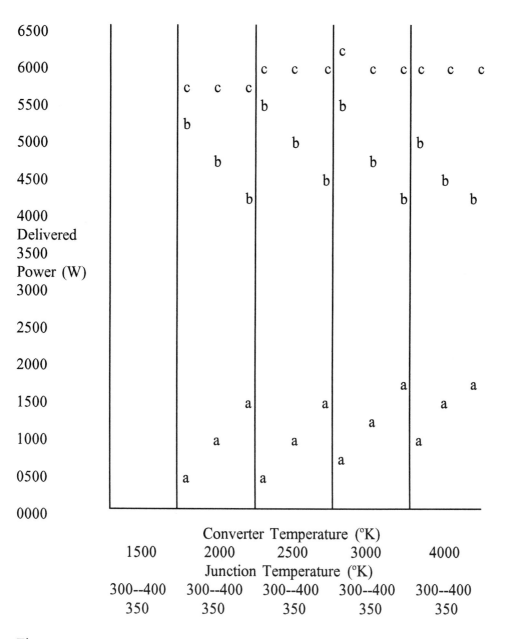

Figure VIII.12. The potential delivered thermal, electrical and total powers from a thermophotovoltaic system employing cadmium telluride heterojunction inverted configuration solar cells with n-type substrates, an area ratio of 10 and a thermal consumer at 300°K as a function of converter and junction temperatures. The input power is 10 kW. Legend: a-->P_{Th}, b-->P_E and c-->P_T.

Let us summarize the results of our analysis of thermophotovoltaic systems by looking at the maximum electrical, thermal and overall system efficiencies of such systems. This is done in Table VIII.13.

Table VIII.13

The thermophotovoltaic systems with highest thermal, electrical and overall system conversion efficiencies. The solar cells are heterojunction devices with inverted configurations and n-type substrates. The area ratio is assumed to be 10 and the consumer is taken to operate at 300°K. The system input is assumed to be 10 kW and efficiencies are computed relative to this value

Maximum thermal generation efficiency				
Semiconductor	Converter temperature (°K)	Junction temperature (°K)	Thermal output (Watts)	System efficiency (%)
Si	4000	400	2350	47.5
GaAs	4000	400	1870	62.3
CdTe	4000	400	1900	61.3
Maximum electrical generation efficiency				
Semiconductor	Converter temperature (°K)	Junction temperature (°K)	Electrical output (Watts)	System efficiency (%)
Si	2500	300	4480	50.3
GaAs	3000	300	5440	61.3
CdTe	2500	300	5480	60.3
Maximum system generation efficiency				
Semiconductor	Converter temperature (°K)	Junction temperature (°K)	Electrical output (Watts)	System efficiency (%)
Si	3000	300	4340	50.4
GaAs	4000	300	5360	63.3
CdTe	3000	300	5460	61.5

Study of Figures VIII.10 through VIII.12 and Table VIII.3 demonstrates that the efficiencies of thermophotovoltaic systems are high. Which of the several systems considered (treating converter temperature, junction temperature and semiconductor as variables) we should select depends on the consumers' requirements for electrical and thermal energy and upon the relative costs of the systems considered.

We will not consider thermophotovoltaics further at present. For the reader who is interested in additional information on this subject, there is the scientific literature [6, 10 and 11].

Miscellaneous Approaches

The mind of man is fertile and has made a considerable number of contributions aimed at improving the performance of solar cells and solar cell power systems. The literature is replete with various improvements [12]. By way of general illustration, let us consider a few ways in which improvements can be made.

Sunlight, even at sea level, has a wide range of wavelengths (see the discussion in Chapter II) and we have seen that, for highest efficiency in the conversion of optical to electrical energy, the semiconductors which make up solar cells exhibit their best performance for a strictly limited range of these wavelengths. One method for improving the performance of solar cells is to insure that only a limited range, the optimum wavelengths, actually reach the solar cells. To accomplish this spectrum splitting [13] we employ dichroic mirrors. A dichroic mirror [14] is designed to separate the sunlight photons of different wavelengths and direct these different photons in different directions. These photons impact a number of solar cells constructed of various semiconductors with selected optimum photon wavelength ranges. Thus each solar cell operates with a maximum conversion efficiency. Coupled with the use of lens or mirror systems to concentrate sunlight, this method holds promise.

Another improvement technique involves constructing a series of solar cells, one layered upon the other. This is known as a tandem cell [13] and the top solar cell has the widest energy gap. Only the highest energy photons are absorbed in this cell; the remainder being passed on to the subsequent solar cells. Theoretically, this process has been carried out with up to 11 solar cells in tandem [15]. For example, a combination of five cells with energy gaps of 0.6, 1.0, 1.4, 1.8 and 2.2 eV has a predicted efficiency of 56.3%. While an intriguing arrangement, this approach has a number of technological problems. In the first place the construction of one semiconductor solar cell upon a second semiconductor is not simple as the fabrication process has to adjust for varying atomic sizes and interatomic spacings. Ideally, the use of molecular beam epitaxy makes this process feasible, but only a few combinations have been successfully constructed to date. Even if one succeeds in placing a number of solar cells, constructed of different semiconductors, in a stack, there are electrical problems. Note, from Chapters V through VIII that the operating photovoltage and photocurrent for a solar cell depends, in part, on the semiconductor employed. This means that our stack of solar cells is composed of individual components which have different voltages and currents [16, 17 and 18]. We will consider this topic further when we address solar cells based on amorphous and polycrystalline materials in the following chapter.

A third approach involves improvements to the solar cells in order to increase the photocurrents[*], decrease the series resistance and enhance hole and electron lifetimes. As an advanced example of this process consider Reference [19]. In this reference Dr. Green investigates a number of potential improvements for silicon solar cells including differing growth techniques, exotic donor and acceptor impurities and texturing the surface using a chemical etch. This last procedure provides a surface which, under an electron microscope, resembles a field of small pyramids, some 10 microns on a side. Combined with a buried grid of illuminated-side contacts [20, 21 and 22] (formed by diffusion, etching and sputtering or, perhaps, molecular beam epitaxy) this process is capable of yielding much enhanced solar cell system performance. The problems with this process are largely matters of production and practicality--which equates to financial limitations. In any revised process it must be determined whether the work involved in making the changes is justified by the enhanced performance, and there is no doubt that these noted improvements do enhance the performance. For example, Dr. Green reports [21] operating efficiencies in excess of 23% for silicon solar cells with buried layer contacts and pyramidal, low reflectance, surfaces--efficiencies significantly higher than those predicted for solar cells in Chapters VI and VII.

---

[*] The reader should reconsider the discussions of Chapters V through VII. Note that the photocurrents used in this work to estimate solar cell performances are somewhat less than the ideal value.

So far, in this section, we have examined improvements in solar cell system performance brought about by breaking up sunlight, by stacking solar cells so that the photons are absorbed in a tandem set of cells, and by treating the cells themselves in order to reduce photon losses and improve lifetime. Let us close this section with a short examination of another stage three photovoltaic system.

Consider the disc shaped device shown in Figure VIII.13. The tiles are constructed of some suitable plastic material impregnated with a light-sensitive dye. The dye absorbs the incident photons in sunlight and

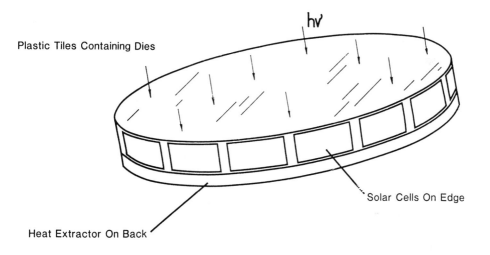

hν

Plastic Tiles Containing Dies

Solar Cells On Edge

Heat Extractor On Back

Figure VIII.13.   A third stage solar cell employing a luminescent concentrator.

then reradiates secondary photons over a narrow range of wavelengths. The solar cells, which are of inverted configuration, are mounted upon the lateral surfaces and heat extractors are mounted behind the solar cells and on the rear (non-illuminated) surface of the tiles. By careful selection of the dye and the semiconductor used in fabricating the solar cells it is possible for the secondary photons to optimally match with the semiconductor energy gap. The plastic of the tiles is selected for maximum input of solar photons and to allow for total internal reflection of the photons from the dye--total internal reflection except where the silicon solar cells are mounted on the tiles. Such a luminescent system does exhibit optical losses, some photons are reflected from the illuminated surface and some secondary photons escape from the tiles. There are also inefficiencies in the photon energy to electrical energy transformation. Such inefficiencies yield heat; heat which can be transferred to system "customers" by the heat extractors. As is the case for the thermophotovoltaic system, it is necessary to strike a balance between the efficiency of the solar cells and that of the heat energy delivery system.

We will not attempt a detailed analysis of a third stage solar system employing a luminescent concentrator. This has been done elsewhere [23 and 24] with theoretical overall system efficiencies approaching 33% (including both electrical and thermal output powers). These efficiencies are considerably less than those predicted for the thermophotovoltaic systems studied earlier. However, luminescent systems have some advantages, relative to the thermophotovoltaic variety of a third stage solar cell system. Luminescent systems do not require solar tracking; making for a far simpler mechanical portion of the system. Furthermore, the tiles themselves can be used to replace the roof upon which the luminescent system is mounted, further reducing the cost of such a system. Additionally, the energy output from these devices is primarily thermal, providing an energy mixture which closely matches the requirements of human dwellings [24].

References

1   A. B. Meinel and M. P. Meinel, Applied Solar Energy, Addison-Wesley, Reading MA, 1976, Chap. 5 and 6.
2   M. A. Green, Solar Cells, Prentice-Hall, Englewood Cliffs, NJ, 1982, p. 208.
3   Reference [1], Chap. 10 and 11.
4   S. W. Angrist, Direct Energy Conversion, 3rd Edition., Allyn and Bacon, Boston MA, 1976, Appendix D.
5   R. C. Neville, Annual meeting of International Solar Energy Society, Brighton, UK, Aug. 1981.
6   R. C. Neville, 8th Miami International Conference on Alternative Energy Sources, Miami Beach FL, 1987.
7   J. J. Loferski, in Applied Physics, Vol. 27, 1956, p. 777.

8 R. C. Neville, "Thermophotovoltaics - An Alternative to Simple Solar Cells", a report written for the state of Arizona, 1980.

9 R. L. Sproull, Modern Physics, John Wiley and Sons, New York, 1956, p. 104.

10 R. C. Neville, "Silicon Thermophotovoltaic Systems and Converter Temperature", National Solar Energy Conference, Denver, CO., 1989.

11 R. C. Neville, "Gallium Arsenide Thermophotovoltaic Systems and Converter Temperature", 9th Miami International Conference on Energy and the Environment, Miami Beach, FL, 1989.

12 Rather than make specific citations, the author recommends perusal of the proceedings of the annual meetings of the IEEE Photovoltaic Specialists Conference, the International PVSEC and other technical socities.

13 M. A. Green, Solar Cells, Prentice-Hall, Englewood Cliffs, NJ, 1982, p. 213.

14 S. M. Sze, Physics of Semiconductor Devices, 2nd Edition, Wiley Interscience, New York, 1981, p. 833.

15 A. Besnet and L. C. Olson, in Proceedings of 13th Photovoltaic Specalists Conference, Washington D. C., 1978, p. 868.

16 J. S. Wanless, et. al., in session 3P of the 23rd IEEE Photovoltaic Specialists Conference, Lousville, KY, 1993.

17 J. Ermer and D. Tarrant, in session 2P of the 23rd Photovoltaic Specialists Conference, Louisville, KY, 1993.

18 A. Freundlich, et. al., in session 3A of the 23rd Photovoltaic Specialists Conference, Louisville, KY, 1993.

19 Reference [13], Chap. 7.

20 A. Cuevas, in IEEE Electron Device Letters, Vol. 11, Jan. 1990, p. 6.

21 M. A, Green, S. R. Wenham and J. Zhao, in opening session of the 23rd Photovoltaic Specialists Conference, Lousville, KY, 1993.

22 S. Wenham, in Progress in Photovoltaics, Research and Applications, Jan. 1993, p. 3.

23 J. S. Batchelder, A. H. Zewail and T. Cole, in Applied Optics, Vol. 20, 1981, p. 3733.

24 Reference [6].

CHAPTER IX: POLYCRYSTALLINE AND AMORPHOUS SOLAR CELLS

Introduction

In the preceding chapters of this text we have considered a variety of solar cell designs and made predictions concerning their performance. All of these solar cell varieties were constructed on single crystal semiconductor materials. Major reasons for following this course of action are the higher efficiencies exhibited by solar cells constructed with single crystal semiconductors and the fact that a detailed theory of operational performance exists for single crystal materials.

However, it is possible to construct working solar cells from semiconductors that are not present as single crystals. This chapter is devoted to an examination of solar cells constructed from polycrystalline and amorphous materials[*]. As stated in previous chapters, single crystal materials exhibit an ordered structure for atomic location, while polycrystalline materials are composed of a number of single crystals oriented at random and separated by "grain boundaries". Amorphous materials are noted for their complete lack of order when atomic locations beyond the next nearest neighbors are considered. Over and above the lower efficiency exhibited by polycrystalline (pc) semiconductor solar cells and solar cells constructed of amorphous materials (a), the operating characteristics of polycrystalline based solar cells are dictated, in large part, by the characteristics of the grain boundaries. In the case of

---

[*] Following common practice we will delineate single crystal materials by c-X, where X is the normal chemical symbol for the semiconductor, or simply by X. For polycrystalline materials we will write pc-X and for amorphous materials we will write a-X. For example, Si or c-Si would denote single crystal silicon, pc-Si represents polycrystalline silicon and the use of a-Si denotes amorphous silicon.

amorphous semiconductor based solar cells the performance is not thoroughly understood, depending as it does on the existence of trace elements, deep lying energy levels within the energy gap and the lack of bonding symmetry.

In this chapter we will examine a number of well-known polycrystalline based solar cells and consider a variety of solar cells made from amorphous semiconductors. Because of the complexity involved in physically describing these materials and their behavior, we will do this, more by considering specific examples than from any deep theoretical basis. Solar cells fabricated from these types of semiconducting materials hold considerable promise as inexpensive sources of energy, but also require considerable "engineering" if the resultant devices are to be reliably constructed and employed for dependable energy conversion.

Polycrystalline Solar Cells

The underlying rationale for investigating and using solar cells constructed of polycrystalline semiconductors is their much reduced cost as compared with single crystal solar cells. This reduction in cost results, primarily, from the reduction in energy required to fabricate the devices; a reduction occasioned by the elimination of the requirement for single crystal material. We will keep this fact in mind and return to this topic later in this section and in the next chapter. For now, let us discuss the technical aspects of polycrystalline solar cells. In doing so we will rely, to a great extent, on the research reported in recent literature. The reader desirous of more information on the performance, design and construction of polycrystalline solar cells will do well to pursue this literature.

Figure IX.1 presents a typical polycrystalline based solar cell. This solar cell was constructed by coating a glass "slide" or substrate with a transparent metallic conductor (for example, tin oxide, indium tin oxide, etc.). Next, the process continues by spray "painting" a wide gap semiconductor on this "front" contact side; and applying, via evaporation, chemical ion-exchange, ion implantation, molecular beam epitaxy, or some other process, another polycrystalline semiconductor. One of the semiconductors in Figure IX.1 is n-type and the other, the substrate in this case, is p-type. Note that the processing has resulted in some preferential diffusion along the grain boundaries. Also note that the individual crystals in this polycrystalline structure are more vertical than horizontal, i.e., the structure is such that the individual crystals are in columns as seen from the illuminated surface. This structure has been shown to lead to a longer effective carrier lifetime and, hence, to improved photocurrent and enhanced solar cell performance [1 and 2].

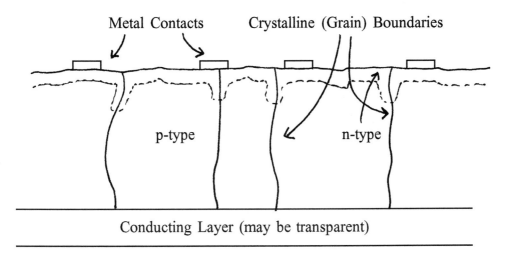

Figure IX.1. A typical polycrystalline material solar cell employing two semiconductors in a heterojunction configuration.

Light can enter such a polycrystalline solar cell from either side, or both sides [3, 4, 5]. If it enters from the top side the metal contacts partially block the incoming photons and the glass substrate may be replaced with a non-transparent, conducting substrate. Also, the input light can be natural sunlight, concentrated sunlight, the light within a stacked (or tandem) solar cell or the light due to a dichroic mirror. The surfaces of such a polycrystalline solar cell can be smooth, coated with an antireflecting substance, faceted to enhance photon entrapment within the semiconductors or any combination of the above [6].

The photon derived current in these cells flows both through the individual crystallites and along the grain boundaries. In polycrystalline semiconductors the majority of the grain boundaries grow randomly and are usually of low symmetry and large angle. The electrical properties of such grain boundaries are not well understood. It appears that the point defect density[*] at a typical grain boundary is so high that defect-defect interaction is common [7]. Electrical charges are trapped on these defect sites with the grain boundaries, affecting carrier lifetime and transport properties. Carrier motion at right angles to a grain boundary is particularly difficult in many compound polycrystalline materials. Such boundaries can be represented by a double Schottky barrier with the precise performance depending on details concerning the surrounding crystallites, the illumination level (photons may spill charges from defect levels as well as fill them with photon-generated charge carriers) and the nature of the semiconductor. This is why solar cells made from columnar polycrystalline semiconductors exhibit superior performance.

One effect of the grain boundaries is that the overall resistivity of polycrystalline materials does not, in general, decline linearly with increasing donor and acceptor concentrations, but tends to show little or no change over a wide range of impurity concentrations with sudden decreases evidenced at critical values of impurity concentration [8].

There are a wide variety of semiconductor materials which have been examined in polycrystalline based solar cells. Besides the elemental polycrystalline silicon (pc-Si) there are compounds such as pc-GaAs and many polycrystalline II-VI semiconductors such as pc-CdSe, pc-CdTe, pc-ZnTe, pc-CdS, pc-HgTe, etc.. There are also a number of interesting compounds with a chalcopyrite configuration such as $AgInS_2$ and $CuInSe_2$. Let us now consider a few examples of polycrystalline semiconductor solar cells and the problems and successes encountered with these devices, thereby outlining the technology of polycrystalline solar cells.

<u>Cadmium Sulfide/Copper Sulfide</u>

Solar cells made from these two II-VI semiconductors (CdS and Cu_2S) are clearly heterojunction diodes with the pc-CdS having an energy gap of 2.3 eV and the pc-Cu_2S having an energy gap of 1.2 eV. Considering the nature of sunlight, it is clear that the Cu_2S layer is

---

[*] Defects arise due to interstitial atoms, vacancies and antisite location of atoms in compound semiconductors. For example, defects due to the deposition of SiO_2 with excess oxygen present can result in donors in compound semiconductors, and SiC has been found in ribbon silicon.

responsible for the bulk of photocurrent generation. As with heterojunctions in general, these solar cells have a tendency towards multiple recombination centers located in the vicinity of the pn-junction. This leads to a situation wherein the photocurrent increases in a non-linear manner as the light intensity is increased. This is a result of the fact that with higher light intensities there is an increased concentration of photo-generated charge carriers and they succeed in saturating the recombination centers.

These thin film polycrystalline solar cells were first fabricated in 1954 [9]. Then and now CdS/Cu$_2$S solar cells are most often constructed by the Clevite process [10]. Here, approximately 20 μm of pc-CdS is vacuum deposited (evaporated) onto a metal, a metal coated plastic or a metal covered glass sheet. This layer, with its ˜5 μm sized crystallites, is then dipped into a cuprous chloride solution (˜20 seconds at 90° C). A chemical process exchanges copper for cadmium in a thin surface region some 2000 Å thick forming the pc-Cu$_2$S layer and a heterojunction. Efficiencies in excess of 9% have been reported for these solar cells [11].

The current voltage characteristics of these solar cells differ from single crystal devices in a number of ways. For example, as shown in Figure IX.2, the non-illuminated (dark) and illuminated (lighted) forward biased current-voltage characteristics of pc-CdS/Cu$_2$S cross each other

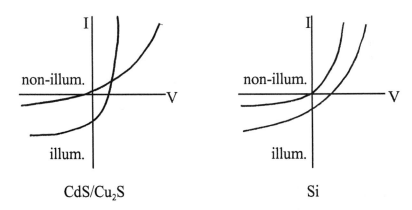

Figure IX.2. The current-voltage characteristic for illuminated and non-illuminated pc-CdS/Cu$_2$S solar cells and illuminated and non-illuminated c-Si solar cells.

while those of most solar cells (for example, c-Si) do not. In addition, the photovoltage depends on the spectral characteristics of the incident light and the solar cell capacitance depends, in a major way, upon the illumination level*. All of these variations from single crystal material device behavior are dependent on the fabrication processes employed.

The principal advantage of these polycrystalline solar cells is one of cost. Being polycrystalline and thin, they can be fabricated on a wide variety of substrates by mass production techniques. Note that the effective optical absorptivity of most polycrystalline materials is considerably higher than that for the single crystal form of the same material. This enables us to construct much thinner solar cells and still capture the incoming photons.

The major disadvantages of these solar cells are their relatively low efficiency [12], coupled with an inherent instability [13]. A number of degradation or destabilization modes have been observed in cadmium sulfide/copper sulfide solar cells: (1) air, particularly moist air, can oxidize the pc-Cu_2S; (2) high temperatures (which may be as low as 60°C) can alter the stoichiometry of the pc-Cu_2S to pc-Cu_xS where x is less than two; and (3) the performance decreases whenever the load voltage exceeds 0.33 volts [14].

In summary, cadmium sulfide/copper sulfide thin film polycrystalline solar cells suffer from poor efficiency and instability. However, if conversion efficiencies in excess of 10% can be reliably sustained, then such cells, owing to their low fabrication cost, offer a strong possibility of being economically feasible. For solar cells of efficiencies less than 10% the costs of the supporting structure, the energy conditioning equipment and other sub-systems are sufficient to make single crystal semiconductor based systems preferable.

Copper Indium Selenide (CIS)

Polycrystalline $CuInSe_2$, widely known as CIS, is another semiconductor, used to construct thin film solar cells. Solar-to electrical energy conversion efficiencies close to 16% have been reported [15].

Substrates of glass, metalized glass and molybdenum foil have been used in the construction of heterojunction solar cells employing

---

* The rapid increase in capacitance with illumination level is consistent with a decrease in space charge width caused by the trapping of photon generated holes by defects in the pc-Cu_2S layer.

this material. The layers of this polycrystalline solar cell material are prepared by selenization with H_2Se, reactive co-sputtering from planar metal targets or multiple-source in-line evaporation; all with some success [16].

Major factors limiting minority carrier lifetime, hence limiting the photocurrent and solar cell efficiency, appear to be point and chemical defects, the grain boundaries and various impurities. Thus, solar cells based on pc-CuInSe$_2$ exhibit performance strongly dependent on fabrication techniques and operational history. Trapping transitions involving selenium appear to be important for both recombination and carrier transport. Heat treatment of CIS in selenium appears to be efficacious in removing deep trapping levels [17]. Removal of these defects also seems to heavily reduce the illumination intensity dependence of CuInSe$_2$ solar cells [18].

Modifications to basic CIS are often found to be useful in solar cells. Fuji Electric Corporation has reported 10% to 12% efficiencies for selenized pc-CuIn(Ga)Se$_2$, an efficiency level which may be considered to be a more "practical" operating efficiency than the "star" device value of 16% reported above [15].

Many other variations on the basic CIS solar cell have been performed. For example, copper indium selenide is often used in conjunction with pc-CdS in heterojunction structures. By using pc-Cu(In,Ga)(Se,S,Te)$_2$ and varying the stoichiometry, the energy gap of the modified CIS-like absorbing layer can be varied from one to 1.7 electron volts and solar conversion efficiencies of up to 13% have been obtained. Efficiencies of 10% have been observed with conventional CIS sputtered on c-GaAs [19]. International Solar Electric Technology reports efficiencies in excess of 12% for flexible solar cells using a ZnO/CdS/CIS/Mo structure and having surface areas on the order of a square centimeter. In this case, the zinc oxide layer is acting as a "window" to allow entrance of the sunlight. Large area (3900 cm^2) modules of cadmium indium selenite have exhibited operating efficiencies in excess of 10% [20]. Polycrystalline zinc selenide has also been used with CIS, but as a window material, with observed device efficiencies less than 10 percent [21].

It is clear from the preceding discussion that much work is being carried on with cadmium indium selenide and its derivatives. However, no single direction or process has emerged as a clear winner in terms of device reliability and efficiency.

Polycrystalline Silicon

Polycrystalline silicon solar cells have been reported to exhibit efficiencies in the neighborhood of 15% [15]; a value close to the conversion efficiencies demonstrated with standard single crystal silicon solar cells. Combined with amorphous silicon in a tandem solar cell, polycrystalline silicon solar cells have exhibited conversion efficiencies in excess of 20% [22]. Study of the absorption coefficient of pc-Si indicates that a thin film of pc-Si some 100 microns in thickness would have the same sunlight absorbing capability as a single crystal film of silicon some three times as thick.

Polycrystalline silicon layers are constructed in many ways--a widely used technique being the deposition of pc-Si on metallurgical polycrystalline substrates (thus allowing the use of the substrate as a low resistance conductor) from a Cu/Si solution. Plasma-chemical vapor deposition (known as PVCD) has also been used, as well as various straight casting processes [23]. This latter type of process has been known to lead to non-linear surface photovoltages. This phenomenon has been attributed to variations in the absorption coefficient [24]. Other methods of device formation include dendritic webs, edge defined film fed growth, ramp assisted foil techniques and ribbon growth processes [25]. All of these techniques have shown some promise. However, considerable development work is required before any single technique is singled out as preferable.

Other methods of improving performance which have been demonstrated are gettering to delete impurities such as the transition metals [26], hydrogen passivation and the formation of back surface electric fields [27] (see also Chapters V and VI since many of these improvement techniques are also valuable for single crystal material solar cells). Additional improvements such as the concentrator cell using polycrystalline silicon with glass fibers (suggested by Cole of George Mason University) and the use of multijunction stacked solar cells also hold considerable promise.

Despite a considerable amount of experimentation, and the obtaining of occasional efficiencies in excess of 17% [28] the use of polycrystalline silicon for solar cells has not been as satisfactory as many other polycrystalline materials nor have the overall results (including the costs of fabrication) for manufactured devices been as successful as those for single crystal silicon solar cells, or for amorphous silicon based solar cells.

Thin Film Cadmium Telluride

Despite the relative scarcity of cadmium and tellurium, polycrystalline CdTe is a promising material for thin film solar cells. Conversion efficiencies in excess of 15% have been reported for pc-CdTe/pc-CdS solar cells constructed on transparent tin oxide coated glass [29]. Large area (approximately 0.7 square meters) panels have demonstrated conversion efficiencies of 7.8% at output levels in excess of 53 watts [30].

Construction of pc-CdTe films by close spaced sublimation, radio-frequency planar magnetron sputtering and laser physical vapor deposition has been reported [29, 31]. The area of the substrate, in many cases, is reported as approaching a square meter. Frequently, the solar cells are heterojunction structures employing pc-CdTe and pc-CdS. Properties of the overall solar cells in these cases are often subject to the influence of the CdS layer. For example, tellurium depletion from the CdTe side of a CdS/CdTe heterojunction solar cell is theorized to be the cause of observed long term solar cell degradation in NREL* operational life studies [32].

In order to improve overall performance, substrates of titanium oxide and silicon dioxide are often used. Treatment of the CdS layer with $CdCl_2$ has been shown to reduce the observed increase, with time, in saturation current density of these devices. This has been demonstrated to improve long term efficiency [33].

Other Possibilities for Polycrystalline Solar Cells

In addition to the specific examples briefly cited above, a considerable number of polycrystalline semiconductor combinations have been or are under investigation. We mention a few below.

Thin film heterojunctions involving the polycrystalline semiconductors, pc-CdSe/pc-CdTe and pc-ZnSe/pc-CdTe [34], pc-GaAs thin film heterojunctions, and solar cells based on pc-Zn_3P_2, pc-$CuGaSe_2$ and pc-$GaAs_xSb_{1-x}$ [35] have also shown considerable promise. Recent work on thin film polycrystalline heterojunctions composed of p-type pc-$CuGa_{0.25}In_{0.75}Se_2$ on n-type pc-$Zn_{0.35}Cd_{0.65}O$ has yielded initial solar cell efficiencies of 9% [36].

---

* NREL is the National Renewable Energy Laboratory near Denver, Colorado.

In many other combinations of polycrystalline semiconductors, lattice mismatch between the semiconductors, even in the polycrystalline state, generates sufficient trapping sites to drastically reduce the photocurrent and conversion efficiency, making the combination impractical.

Final Comments on Polycrystalline Solar Cells

We have stated that the underlying reason for investigation of polycrystalline solar cells is their potential economic value. In theory, thin film photovoltaic solar cells are extremely inexpensive. The problem, to this date, is that thin film polycrystalline solar cells are still expensive (in part, because the cost of a solar cell system includes much more than the cost of the solar cells alone--see the next chapter). To achieve the wide commercial acceptance expected of them, the polycrystalline solar cells need to be inexpensive to manufacture, economical to operate and stable in long term performance. Currently this is not true of polycrystalline based thin film solar cells. For thin film pc-CdTe, the only major current market is approximately one megawatt a year of solar cells fabricated for calculators, and this volume of production cannot, in all fairness, be considered to be mass production. If mass production is to be achieved progress must be made in substrate fabrication, absorber (the narrow gap side of the heterojunction, thin film, polycrystalline solar cell) deposition, and window (the wide gap portion of the heterojunction solar cell, the side through which light enters the solar cell) construction, as well as improvements in metalization, heat removal and reducing the device-environment interaction.

We will resume our discussion of thin film polycrystalline solar cells in the next, and final chapter. For now, let us consider the remaining class of atomic organization for solar cells; the amorphous material based solar cells.

Amorphous Material Based Solar Cells

Amorphous materials are characterized by lack of long range order (long range, in this context, may be considered to be any distance in excess of one lattice constant) when the locations of the constituent atoms are considered. The resulting devices, in general, exhibit wider energy gaps than the crystalline varieties of the same material, higher (often dramatically higher) absorption coefficients and, owing to a large number of non-symmetrical interatomic bonds, a significant number of energy states within the forbidden gap.

The wider energy gap for amorphous silicon facilitates a better match between the light absorbing semiconductor and the solar spectrum. The larger absorption coefficient permits the solar cell manufacturer to construct thinner films than with crystalline or polycrystalline solar cells.

The large number of energy states within the forbidden gap result in poor overall material charge carrier transport properties, excessive recombination and a much reduced optical-to-electrical energy conversion efficiency. Also, on the debit side of the situation, most amorphous material based solar cells exhibit a degradation in efficiency with prolonged exposure to light [37]. This decrease in performance can exceed 30% under conditions commonly encountered in the field. Considerable effort is being made to understand and combat this problem (see comments under amorphous silicon).

A variety of amorphous materials have undergone investigation as solar cell candidates. The preponderance of such materials has involved silicon. Among materials showing some promise are a-Si, a-SiGe and a-SiC. Because the majority of effort on amorphous materials has been devoted to understanding amorphous silicon, we will concentrate our efforts, here, on amorphous silicon.

Amorphous Silicon

In many ways, it is easier to prepare amorphous silicon (a-Si) samples than to fabricate polycrystalline samples [38]*. Interestingly, elemental amorphous silicon does not appear to possess any commercially valuable photovoltaic properties. Solar cells constructed of elemental amorphous silicon exhibit conversion efficiencies of less than a percent and have electrical characteristics which vary with time. The reason for this is the existence of large numbers of vacancies, and other imperfections within the semiconductor. When coupled with the nonperiodic

---

* Three commonly employed methods for fabricating a-Si are electron beam evaporation, rf sputtering and deposition from silane (SiH_4) in a radio frequency discharge. Electrical and optical properties of the resultant films depend on the method of deposition. We will discuss this subject, at greater length, later.

arrangements of the silicon atoms, this creates vast numbers of allowed energy states. These states, more-or-less, span the energy gap and impede the manufacturer's ability to adequately "dope" (introduce impurities leading to controlled amounts of extra holes and electrons) the amorphous semiconductor. These states also drastically reduce the carrier lifetime.

The density of states observed within the energy gap of intrinsic amorphous silicon is outlined in Figure IX.3. Note that the density of the induced states is extremely high, being approximately one state for

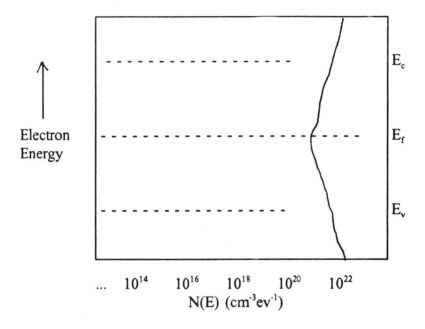

Figure IX.3. The volume density of states in the Energy Gap for intrinsic amorphous silicon.

every ten silicon atoms, and that this density of states is roughly constant across the forbidden gap. As noted earlier, this high density of states makes the use of added impurity atoms (doping) to change the type of semiconductor effectively impossible.

In 1975 [39] it was reported that amorphous silicon films produced by the glow discharge decomposition of silane (SiH_4) could be doped to produce pn junctions. Such films clearly contain hydrogen, a byproduct of the decomposition of silane, at a five to ten percent atomic proportion level. It is theorized that these hydrogen atoms saturate the dangling bonds that are a feature of the internal vacancies and atomic structure.

This saturation reduces the state density within the energy gap [40]. By 1976 amorphous silicon (or more precisely amorphous silicon containing hydrogen or a-Si:H) based solar cells with an efficiency of 5.5% were produced [41, 42]. It was predicted, in 1976, that efficiencies as high as 15% were feasible [42]. Efficiencies in the neighborhood of 12% were reported at the 1993 Photovoltaic Specialists Conference [43].

 Let us consider in more detail some of the properties of hydrogen containing amorphous silicon. Many of these properties depend on the density of states within the forbidden gap. This density of states is shown for an undoped a-Si:H sample in Figure IX.4.

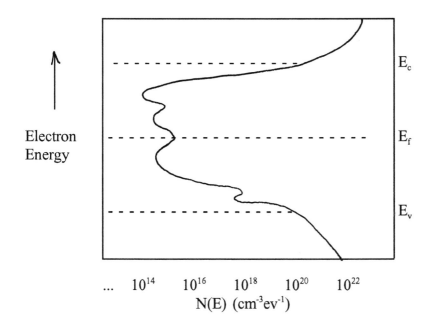

Figure IX.4. The volume density of states in the Energy Gap for intrinsic amorphous silicon containing hydrogen (a-Si:H).

 The density of states exhibited in Figure IX.4 for intrinsic a-Si:H is approximately similar to that given for doped a-Si in Figure IX.3 at the conduction and valence band edges, but is far less within the energy gap. Note that a-Si:H films typically exhibit an energy band gap (E_c - E_v) on the order of 1.5 to 1.7 eV when optical phenomena are considered [44, 45]. This effective energy gap is considerably wider than the 1.1 eV exhibited by single crystal silicon and represents a much better match for the solar spectrum. As a result, the observed absorption coefficients for a-Si:H are significantly higher than for c-Si when photon energies in excess of the a-Si:H optical energy gap are considered [44]*.

The increased absorption coefficient for a-Si:H means that solar cells can be constructed of films of this material with less than a micron in thickness of absorbing material. Such films can be deposited on many different substrates using a variety of processes. Glow discharge of silane, sometimes known as PECVD or plasma enhanced chemical vapor deposition, appears to be favored by many experimenters [47]. Other techniques for the production of a-Si:H include sputtering [48], pyrolysis [49], chemical vapor deposition [50] and photodecompositon combined with chemical vapor deposition [49, 51].

The conductivity of intrinsic a-Si:H is less than 10^{-10} mhos/cm and the observed mobility is under 0.1 cm^2/volt-sec [52]. In fact, the high density of states within the forbidden gap makes for a "mobility gap" wherein the mobility of electrons and holes having energies between E_v and E_c is, to all effective purposes, zero [53].

The addition of donor and acceptor impurities to a-Si:H to improve the conductivity is not easily accomplished. The conductivities experienced in boron doped a-Si:H vary from approximately .01 mho/cm for impurity concentrations in excess of 10^{21}/cm^3 to almost a flat 10^{-10} mho/cm for boron concentration levels less than 10^{16} cm^{-3}. A similar variation is observed for phosphorous concentration changes [54].

However, it is possible to form a pn junction by employing trace amounts of appropriately doped gases during the a-Si:H film deposition. The minority carrier diffusion lengths in the resulting devices tend to be very short, much less than a micron. This presents significant difficulties in extracting the photocurrent from a-Si:H solar cells [55]. As a result, the typical a-Si:H solar cell has a p-type region and an n-type region separated by an intrinsic region. The electric field generated by the pn junction fills the intrinsic region and serves as the mechanism whereby the photon generated charge carriers are collected [56].

---

* At a photon energy of 1.5 eV, the absorption coefficient for c-Si is 10^3 cm^{-1} while that for a-Si:H is 10^2 cm^{-1}. At 1.7 eV photon energies both materials exhibit absorption coefficients of 2 x 10^3 cm^{-1}. However, at a photon energy of 2 eV the absorption coefficient for c-Si is 4 x 10^3 cm^{-1} while that for a-Si:H is an order of magnitude higher [46].

At this juncture, let us consider the typical a-Si:H solar cell depicted in Figure IX.5. By considering the reasons for the existence of each region of this solar cell and the effects of the several regions on overall performance we can best convey the present state of affairs.

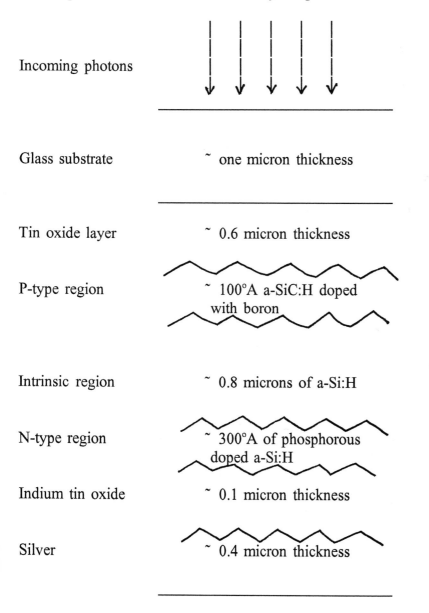

Incoming photons

Glass substrate ~ one micron thickness

Tin oxide layer ~ 0.6 micron thickness

P-type region ~ 100°A a-SiC:H doped with boron

Intrinsic region ~ 0.8 microns of a-Si:H

N-type region ~ 300°A of phosphorous doped a-Si:H

Indium tin oxide ~ 0.1 micron thickness

Silver ~ 0.4 micron thickness

Figure IX.5. The construction of a state-of-the-art amorphous silicon solar cell. Not to scale and with approximate thicknesses given.

The solar cell is constructed, in an up-side-down fashion, by depositing various semiconductor and contact layers on the glass substrate. This substrate is fabricated of a material which is transparent to the solar spectrum. Reflection of photons from the surface of this layer is responsible for the majority of the solar cell losses[*]. A multi-layered antireflection coating on the glass is capable of reducing the reflection losses to almost zero. However, the cost of such a layer is often found to be prohibitive.

The first layer deposited upon the glass is a sheet of transparent conducting tin oxide. It is responsible for some 20% of the photon losses as a result of absorption. This layer forms the electrical contact to the p-type side of the solar cell. Note that the lower side of this region is shown as being irregular (see Figure IX.5). To some extent this irregularity is deliberate as the surface is roughened or "textured" [58] to decrease photon reflection from the p-type semiconductor. Note that other transparent window/conducting contact materials such as ZnO and indium tin oxide have been employed as substitutes for tin oxide and with considerable success.

The p-type region selected in Figure IX.5 is made of amorphous silicon carbide containing hydrogen (a-SiC:H). This means that the device we are considering is a heterojunction device. It is possible to fabricate the p-type from a-Si:H, but this generally results in higher charge carrier losses due to recombination of charge carriers in the p-type region adjacent to the intrinsic region. The use of a-SiC:H gives rise to a heterojunction, and the resulting electric field keeps the photon generated electrons within the intrinsic layer, reducing recombination and improving device efficiency. This effect can be improved upon by altering the energy band gap of the a-SiC:H during fabrication [59] creating a graded energy gap within the p-type layer. Note that this layer has a relatively low charge carrier lifetime and is kept thin to minimize absorption of the photons.

The relatively thick ($\tilde{}0.8$ μm) intrinsic layer of a-Si:H is deposited directly upon the thin p-type layer. Note that it is in this layer that the bulk of the solar photon absorption is carried out. The built-in electric field (generated by the heterojunction) is responsible for extracting the

---

[*] Estimates for losses from this glass layer range as high as 30% [57].

photogenerated charge carriers, for the lifetime, and hence the diffusion length, in the intrinsic region is very short. Note that hydrogen concentrations in this region, of approximately 15%, appear to provide optimum performance [60]. Considerable care needs to be exercised in the production of this layer to assure that the silicon-to-hydrogen bonds are Si-H rather than Si-H$_2$ or Si-H$_3$. The presence of these alternate bonds induces excess forbidden gap states which result in reduced photocurrent and carrier mobility [61].

The next layer in Figure IX.5 is the n-type a-Si:H region. This layer does experience some photon absorption (primarily of long wavelength photons), but contributes little to the overall converted energy. Experimental work is being conducted on this layer, considering the use of some other semiconductor as the n-type region. The reasons for such work are similar to those put forward for using a-SiC:H for the p-type layer.

On "top" of the n-type region is a layer of indium tin oxide. This is a semitransparent contact to the n-type layer. It also serves as a buffer between the solar cell and the next, and final, layer; a buffer made necessary by the difference in mechanical expansion with temperature variations.

The final layer in Figure IX.5 is one of silver. This layer exists primarily to reflect photons that have passed entirely through the solar cell, reflect them back through the solar cell and so increase overall conversion efficiency. Of course, it also acts as a low resistance electrical contact.

Note that the amorphous silicon solar cell in Figure IX.5 has the light entering the intrinsic layer through the substrate glass layer and the p-type region. It is also possible to design a configuration with the light entering through a transparent layer and the n-type region [62]. Devices constructed in this manner have not been as efficient as the type outlined here. It is also possible, clearly, to design a device in which light enters the solar cell from both sides. This configuration can be employed with optical concentration, and the higher photon flux densities results in enhanced conversion efficiency.

The relatively high resistivity of amorphous silicon means that the interconnection of many such solar cells in electrically stacked modules is a difficult undertaking [63]. A number of approaches have been suggested which add little to the complexity of the manufacturing process while overcoming such series resistance effects [63, 64]. When small modules (panels) of amorphous silicon based solar cells appeared in the early 1980s, the majority were used to power watches and pocket calculators and they exhibited effective efficiencies of a few percent.

Currently, other approaches to fabrication of amorphous solar cell modules include situations wherein: (1) opaque substrates such as plastic or stainless steel (often flexible) are employed with the p-type layer and its transparent contact uppermost or (2) both the upper and lower sides of the solar cell are contacted by transparent conductors and the incoming photons enter the device from both sides. As mentioned earlier this is a promising approach.

In 1978 [65] the use of fluorine, from the decomposition of SiF_4, in conjunction with hydrogen in a-Si was tested. Whether the observed improvement in performance was due to improved doping efficiency [66, 67] or to the formation of microcrystallites is not yet clear.

There exists a significant problem with a-Si:H solar cells. The conversion efficiency of these cells exhibits degradation in performance with time when exposed to sunlight. This phenomenon is known as the Staebler-Wronski effect and was first observed in the late 1970s [68]. Decreases in conversion efficiency and obtained photocurrent production in excess of 10% are frequently observed in these photovoltaic devices [69-71]. The observed loss in solar cell performance can be reversed by exposure (in the dark) to temperatures of approximately 150°C for periods of time amounting to an hour [72]. Experiments aimed at eliminating this problem involve stabilization by initial soaking in sunlight and annealing at temperatures in the 60°C to 70°C range [73]. Another approach to reducing this problem, that has met with some success, is the use of energy gap grading to produce an electric field which assists in hole collection [74]. In general, reducing structural defects has been shown to be advantageous. The literature is replete with studies investigating the energy states and defect densities within the energy gap in a-Si:H based solar cells [75].

Multijunction cells using a-Si:H, a-SiC:H and a-SiGe:H have been reported with efficiencies in excess of 13% [76-78]. The reported light sensitive areas of these devices range as high as four square feet. There is some evidence that such multijunction amorphous solar cells may be more stable under illumination (the Staebler-Wronski effect) than single junction cells [79].

A typical triple junction, stacked amorphous solar cell is displayed in Figure IX.6. Note that the substrate is glass and that, as in evidenced

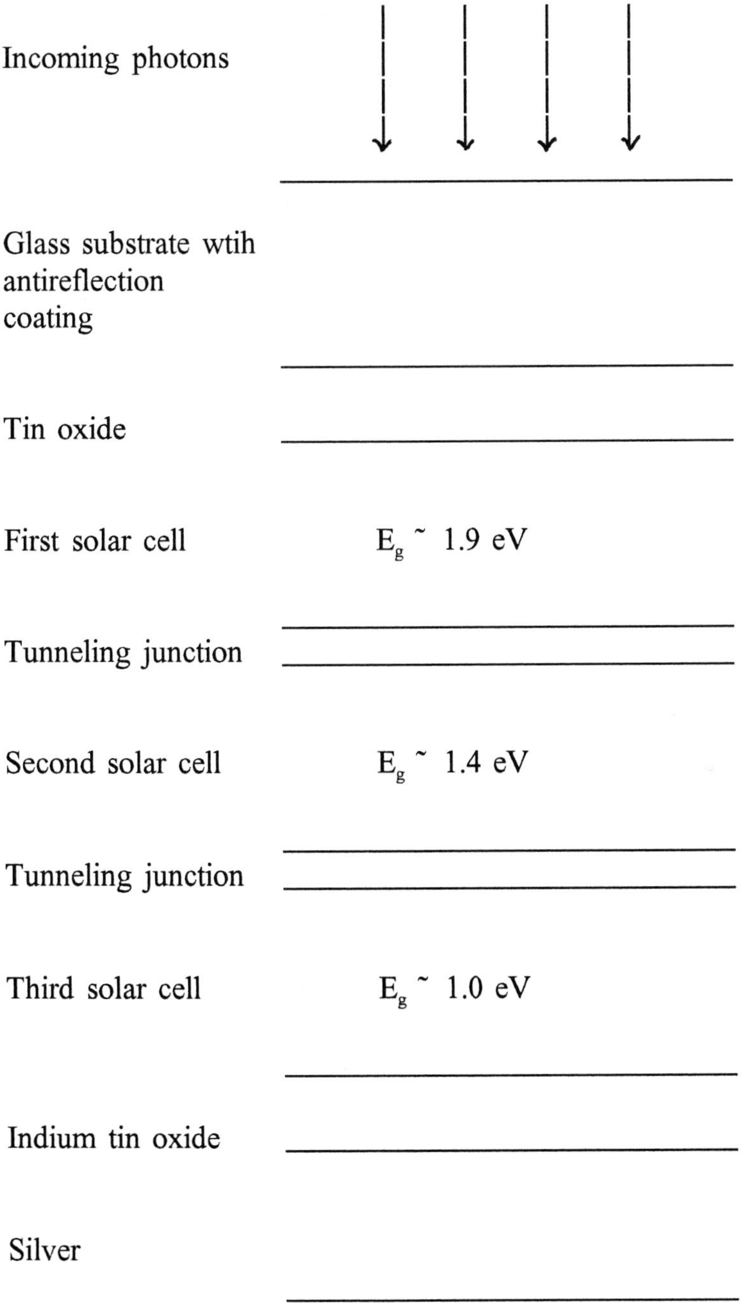

Incoming photons

Glass substrate wtih
antireflection
coating

Tin oxide

First solar cell $E_g \sim 1.9$ eV

Tunneling junction

Second solar cell $E_g \sim 1.4$ eV

Tunneling junction

Third solar cell $E_g \sim 1.0$ eV

Indium tin oxide

Silver

Figure IX.6. A typical triple junction, stacked, and amorphous material solar cell. Not to scale.

in Figure IX.6, over 20% of the optical losses of such a device are due to reflection from the front surface of this device and absorption of the solar photons within the glass layer.

The tin oxide layer is a transparent front contact which is also electrically conducting. Unlike the solar cell shown in Figure IX.5, the individual layers of this triple junction solar cell are depicted as being smooth. In point of fact, a rough surface for the tin oxide and subsequent layers is far more common than the smooth surfaces shown.

Note that the three solar cells in Figure IX.6 are deposited on the glass substrate with the wide energy gap amorphous solar cell being deposited first and the narrow gap cell (which absorbs the longest wavelength photons) being deposited last. In between the photon absorbing solar cells (which, in actuality, consist of an intrinsic region surrounded by n-type and p-type regions) we have tunneling junctions, using very heavily doped amorphous semiconductors. This procedure allows us to electrically stack the three devices. It is important to equalize the photocurrent delivered by the three solar cells. This can be done by adjusting the energy gaps of the three solar cells. Since it is possible to adjust the energy gap of a-SiC:H by adjusting the deposition process [78], two of the three amorphous semiconductor solar cells may be constructed from a-SiC:H while the third layer could easily be made from a-Si:H.

The indium tin oxide region at the bottom acts as an electrical contact to the third solar cell and a transparent window, while isolating the solar cells from the silver layer at the bottom. The underlying silver layer acts both to reflect the photons that have succeeded in transiting the solar cells back through the solar cell, so that the absorption is more complete, and as the bottom layer electrical contact and support.

Concluding Remarks

From the descriptions of the various polycrystalline and amorphous based solar cells, and the multitude of techniques for their fabrication, it is clear that we face complex problems. The discussions in this chapter are but a preliminary view into a complicated and fascinating field.

For the purposes of this work in reviewing the status of solar energy and solar cells, let us now proceed to a concluding chapter and consider a number of additional aspects concerning photovoltaics.

References

1 D. L. Pulfrey, Photovoltaic Power Generation, Van Nostrand Reinhold, New York, 1978, p. 122.
2 C. Lanza and H. J. Hovel, in the Proceedings of the IEEE Photovoltaic Specialists Conference, Baton Rouge, LA, 1976, p. 96.
3 M. A. Green, Solar Cells, Prentice-Hall, Englewood Cliffs, NJ, 1982, p. 188.
4 Reference [1], p. 121.
5 R. A. Sinton, in the Conference Record of the 23rd IEEE Photovoltaic Specialists Conference, Louisville, KY, 1993, p.157.
6 The reader is referred to a large number of articles in the Conference Records of the IEEE Photovoltaic Specialists Conferences, The Photovoltaic Solar Energy Conferences, The Photovoltaic Science and Engineering Conferences and the Annual Program Reviews of the National Renewable Energy Labs (now NREL, formerly SERI).
7 H. J. Möller, Semiconductors for Solar Cells, Artech House, Boston, MA, 1993, p. 90.
8 H. J. Möller and R. Podbielski, in the Proceedings of the 13th Conference on Defects in Semiconductors, Metallurgical Society of AIME, 1985, p. 435.
9 F. A. Shirland, Advanced Energy Conversion, Vol. 6, 1966, p. 201.
10 Reference [3], p. 196.
11 J. A. Braganolo et. al., in IEEE Transactions on Electron Devices, Vol. 27, 1980, p. 645.
12 Despite the relatively high efficiency for Cu_2S/CdS thin film solar cells listed in Reference [10], the efficiency for standard production devices is closer to 5%.
13 J. Bessen, et. al., in the Proceedings of the 11th IEEE Photovoltaic Specialists Conference, Scottsdale, AZ, 1975, p. 468.
14 M. A. Green, Reference [3], p. 199.
15 M. A. Contreras, et. al., at the 23rd IEEE Photovoltaic Specialists Conference, Louisville, KY, 1993, reported an efficiency of approximately 14%. A Gabor at the NREL (see Reference [6]) Conference in 1993 reported 16% for $Cu(In,Ga)Se_2$ based pc solar cells.
16 Efficiencies close to 13% have been reported for three source in-line evaporator systems, Reference [15].
17 H. R. Moutinho, et. al., in the Conference Record of the 23rd Photovoltaic Specialists Conference, Louisville KY, 1993, p. 572.

18 D. Willet and S. Kuriyagawa, in the Conference Record of the 23rd Photovoltaic Specialists Conference, Louisville KY, 1993, p. 495.

19 L. C. Yang, L. J. Chour and A. Rockett, at the 23rd IEEE Photovoltaic Specialists Conference, Louisville, KY, 1993.

20 B. M. Basol, et. al., in the Conference Record of the 23rd IEEE Photovoltaic Specialists Conference, Louisville, KY, 1993, p. 426.

21 M. Fujinnka, in Sun World, Vol. 17, Aug. 1993, p. 14.

22 W. Ma of the faculty of Engineering Science, Osaka Univeristy, Osaka, Japan has described a complex structure consisting of both a-Si and pc-Si in conjunction with indium tin oxide and SiC.

23 P. Schatzle, et. al., at the 23rd IEEE Photovoltaic Specialists Conference, Louisville KY, 1993.

24 I. G. Hwang, D. K. Schroder and J. H. Wohlgemuth, in The Conference Record of the 23rd IEEE Photovoltaic Specialists Conference, Louisville, KY, 1993, p. 178.

25 Reference [7], Chap. 6.

26 J. I. Pankove and N. M. Johnson, volume editors, Hydrogen in Semiconductors, Semiconductors and Semimetals, Vol. 34, Academic Press, New York, 1991, Chaps. 4 and 5.

27 M. Y. Ghannam, et. al., IMEC Kapeldreef 75, 3001 Leuven, Belguim.

28 P. Sana, J. Salami and A. Rohatgi, in IEEE Transactions on Electron Devices, Vol. 40, 1993, p. 1461.

29 C. Ferekides, J. Britt and Y. Ma, in Session 2A of the 23rd IEEE Photovoltaic Specialists Conference, Lousiville KY, 1993.

30 D. Sandwich, at the 1993 NREL (see Reference [6]) meeting.

31 A. D. Compaan, in Session 2A of the 23rd IEEE Photovoltaic Specialists Conference, Louisville, KY, 1993.

32 A Cazaderna, Session III, in the NREL meeting, Oct. 1993.

33 F. A. Abou-Elfotouh, in session 2A of the 23rd IEEE Photovoltaic Specialists Conference, Louisville, KY, 1993.

34 A. L. Fahrenbruch, in the Proceedings of the IEEE Photovoltaic Specialists Conference, Baton Rouge, LA, 1978, p. 529.

35 Articles by D. Warschauer, F. C. Treble and M. Rodot in the Proceedings of the IEEE Photovoltaic Specialists Conference, Baton Rouge, LA, 1978.

36 V. Aparna, in the Conference Record of the 23rd IEEE Photovoltaic Specialists Conference, Louisville KY, 1993, session 2D.

37 R. Ross, et. al., in session 4D of the 23rd IEEE Photovoltaic Specialists Conference, Louisville KY, 1993.

38 R. W. Griffith, in the Proceedings of the Joint Conference of the American Solar Energy Society and the Solar Energy Society of Canada, Winnipeg, Canada, 1976, p. 205.

39 W. E. Spear and P. G. LeComber, in Solid State Communications, Vol. 17, 1975, p. 1193.

40 Reference [26], Chap. 12 and Electronic Materials, edited by L. S. Miller and J. B. Mullin, Plenum Press, London, 1991, Chap. 11.

41 D. E. Carlson, et. al., in the Conference Record of the 12th IEEE Photovoltaic Specialists Conference, Baton Rouge, LA, 1976, p. 893.

42 D. E. Carlson and C. R. Wronski, in Applied Physics Letters, Vol. 28, 1976, p. 671.

43 Ni, T. Kase and P. Sichanugrist, in The Conference Record of the 23rd IEEE Photovoltaic Specialists Conference, Louisville, KY, 1993, p. 941.

44 Reference [1], p. 137.

45 A. Catalano, Amorphous and Microcrystalline Semiconductor Devices and Optoelectronic Devices, J. Kanicki, editor Artech House, Boston, MA, 1991, p. 9.

46 Reference [7], p. 15.

47 F. J. Gupta, Semiconductors and Semimetals, Vol. 21, A. J. Pankove, editor, Academic Press, New York, 1984, p. 168.

48 W. J. Möller, in Solar Energy Materials, Vol. 10, 1984, p. 171.

49 K. Takahashi and M. Konagai, Amorphous Silicon Solar Cells, North Oxford Academic Publishers, Ltd., London, 1986, Chap. 4.

50 S. Hegedus, in the Journal of Applied Physics, Vol. 60, 1986, p. 1046.

51 Reference [45], p. 17.

52 Reference [7], Chap. 9.

53 Reference [49], Fig. 4.3.

54 S. Kalbitzer, et. al., in the Philosophical Magazine, Vol. B41, 1980, p. 439.

55 Reference [3], p. 191.

56 Reference [45], Fig. 2.5.

57 S. Wiedeman, J. Morris and L. Yang, in Proceedings of 21st IEEE Photovoltaic Specialits Conference, New York, 1990.

58 It is suggested that the reader further pursue the topic of a-Si:H solar cell construction by perusing such other References as [6, 45 and 49].

59 R. R. Arya, A. Catalano and R. S. Oswald, in Applied Physics Letters, Vol. 49, 1986, p. 1089.

60 Reference [49], p. 105.

61 T. D. Moustakes, <u>Semiconductors and Semimetals</u>, Vol. 21, A. J. Pankove, editor, Academic Press, New York, 1984, p. 47.

62 Reference [49], Chap. 5.

63 Reference [45], p. 60.

64 A. E. Delahoy, in Solar Cells, Vol. 27, 1989, p. 39.

65 S. R. Ovshinsky, in Nature, Vol. 276, 1978, p. 482.

66 A. Madan, S. R. Ovshinsky and W. Czubatyj, in the Journal of Electronic Material, Vol. 9, 1989, p. 385.

67 A. Matsuda, in Japanese Journal of Applied Physics, Vol. 19, 1980, p. 305.

68 D. L. Staebler and C. D. Wronski, in Applied Physics Letters, Vol. 31, 1977, p. 292.

69 W. B. Luft, et. al., in the Conference Record of the 22nd IEEE Photovoltaic Specialists Conference, Vol. II, 1991, p. 1393.

70 W. Herbst, et. al., in the Conference Record of the 23rd IEEE Photovoltaic Specialists Conference, 1993, p. 913.

71 NREL meeting, session III, Reference [6], Denver, CO, Oct. 1993.

72 Reference [7], p. 292.

73 W. Luft, et. al., in the Conference Record of the 23rd IEEE Photovoltaic Specialists Conference, 1993, p. 850.

74 V. Dalal, in session 4A of the 23rd IEEE Photovoltaic Specialists Conference, 1993.

75 See the Conference Record of the 23rd IEEE Specialists Conference, 1993 and the 1993 NREL meeting in Denver, CO.

76 A triple stacked solar cell, using a-Si:H and a-SiGe:H with an observed efficiency of 11% was reported by E. Maruyama in the Conference Record of the 23rd IEEE Photovoltaic Specialists Conference, p. 827. The indicated stability of this device upon exposure to sunlight is considerably better than the average solar cell.

77 M. Izu, et. al., reported, in session 4D of the 23rd IEEE Photovoltaic Specialists Conference, 1993, on a-Si:H devices employing spectrum splitting, and an area of four square feet. This device initially exhibited efficiencies of 11.1%, but degraded to 9.5% efficiencies with time and under solar illumination.

78 R. Arya, in session II of the 1993, NREL review meeting in Denver, CO. Reference [6].

79 A. Catalano, in the Proceedings of 21st IEEE Photovoltaic Specialists Conference, New York. 1990.

CHAPTER X: CONCLUDING THOUGHTS

Introduction

The initial chapter of this work is devoted to a discussion of the various factors involved in the "energy crisis"; surveying various sources of energy, considering the ecological consequences of their usage; and indicating the potential of a solar cell driven solar-electric energy system. Chapter II provides insight into the amount of energy available in sunlight as a function of the site location on the earth's surface, the weather and the time of year. This second chapter also includes preliminary information on the ways and means of manipulating and concentrating sunlight. The third chapter is devoted to a brief review of solid state physics, semiconductor materials and those phenomena of importance when considering solar cells. The interaction of light and semiconductors, including reflection, transmission and photon absorption is examined in Chapter IV. A general model for estimating the energy that can, potentially, be converted from optical form (sunlight) to electrical nature is developed in this chapter.

In the fifth chapter, methods for separating and collecting the hole-electron pairs generated by the incoming photons are discussed. An expression for the electrical power delivered to a matched external resistive load by a solar cell, in terms of the parameters of that cell, is derived. Chapter VI is concerned with specifying the properties of a semiconductor solar cell in terms of the methods used in its construction and the optical orientation of the solar cell. The seventh chapter investigates the effects of temperature and optical concentration on the performance of a solar cell. Chapter VIII studies additional advanced techniques including techniques for altering the input photon spectrum such as dichroic mirrors and thermophotovoltaic systems.

To a great extent, the material covered in Chapters I and II is introductory in nature, concerning energy as a general field as well as solar energy in particular. The third through eighth chapters are devoted to solar cells constructed of single crystal semiconductors--both in theory and practice. The ninth chapter, expands our viewpoint by examining solar cells constructed with polycrystalline and amorphous semiconductors. Because the theory of operation for such solar cells is not, as yet, thoroughly grounded, Chapter IX is primarily one of example, discussing the observed performance of recent amorphous and polycrystalline solar cells.

As intimated in the preceding chapters, there remains considerable research and engineering development to be done before we can consider the solar cell to be as a routine a device to construct and use as a rectifier diode. It is not even certain whether single crystal, polycrystalline, or amorphous material orientations would be preferable, or even which semiconductor would be most desirable. Rather than spend additional time on other solar cell optical orientations, other materials, special heat sinking schemes (all of which would be variations on the various technologies and designs considered to date) let us spend what space remains in this work on some of the system aspects of solar energy.

We cannot simply "plunk" down a solar cell, attach a load and expect a completely satisfactory energy production scenario. We need to consider the economics of fabrication (of the solar cells and of the other portions of the solar cell modules, including such items as a solar tracking systems) and of assembly (taking individual solar cells and placing them in electrical circuits, installing the circuits in modules or collections of solar cells, and placing the modules in physical structures that protect the cells from ambient conditions, allowing for electrical and thermal conduction and enabling the solar cells to follow the motion of the sun). We also must consider the cost of solar cell system operation. (Do the solar cell systems require energy to enable them to move to face the sun, is cleaning of the solar cells required, and is some of the obtained solar energy required for other operating purposes?)

Furthermore, sunlight is not available on a twenty-four hour basis (we are considering solar cell systems installed on the earth's surface, as opposed to those employed in various artificial satellite operations) so that we need to give thought and commit resources to storing this solar-derived energy against periods of darkness and inclement weather. We should also give some consideration to potential future developments in photovoltaic energy production and storage. We begin this final survey of solar cells and solar cell systems by considering the economics of solar cell power production.

Economics

As is true when we consider every other phase of this complex subject, it is not possible to conduct a discussion of solar cell system energy costs in isolation. Whether used as a "specialty" energy source to energize a wrist watch or power a pocket calculator, or as an energy source for an orbiting communications satellite, or to provide power to a remote desert site in North America, or as commercial of energy for general consumers, solar cells must compete with already existing energy sources and with other potential energy supplies. If solar-electric energy from solar cells is to be widely used it must be cost competitive. The reader is well aware of the extensive use of solar cells in earth orbiting spacecraft (orbiting communications relay devices, geophysical survey and weather satellites) and in various remote locations on the earth's surface [1]. In all of these situations the cost of delivering the electrical energy produced using conventional systems (burning oil, natural gas and coal; hydroelectric; wind energy; etc....) makes the value of conventional energy prohibitive. However, if we are to use photovoltaic energy in urban settings, the energy must be produced and delivered at a price which is competitive with the "general" power grid on its own ground where delivery is not prohibitively expensive.

In considering the cost of a conventional electric power system, be it fossil fuel, nuclear or hydropower, it is commonplace to speak of the capital cost per kilowatt of generating capacity. Sometimes this implies merely the cost of the central generating station and sometimes this includes the electrical grid which distributes energy to the various consumers. The cost of the electric power to the end consumer must include provision for recovering the initial capital expenditure on the generating station and distribution grid, replacing the equipment at the conclusion of its useful life, the fuel used, taxes, salaries and other employee expenses, and a profit on the capital investment.

Conventional power generating systems have a further complication in that these primary electrical generating systems are of three basic types. The most fundamental type of electric plant is the so-called base load power plant. These generators are designed to operate almost constantly, for close to twenty-four hours a day for months at a time and at fully rated output levels. Plants of this kind are frequently fossil fuel powered, hydroelectric or nuclear (fission) plants that are designed for durability and reliability. Next, we have intermediate load power plants. Such plants operate approximately 40% of the time, running in the early evening hours and other times when the power demands of civilization exceed the power delivered by the base load plants. These conventional sources of electricity are designed to be turned on and off albeit with some difficulty. Such plants are, for the most part, older base load facilities whose efficiency has diminished. Fossil fuel fired plants are often used in this category. Finally, we have the peak load electrical power generators. These facilities operate for brief periods of time (10% or less of each day) and are only utilized during those periods such as to be found during a hot and humid summer afternoon when energy demands are exceptionally high. Many of these generators are adaptations of large jet aircraft engines. Such plants are not very efficient, but can be turned on and off quickly.

Clearly, fuel, operating and capital costs vary as functions of the type of power plant and its location. It is not the purpose of this chapter to perform a detailed economic analysis of the energy situation. Our purpose is to point out trends and potential problem areas. However, we do need some gauge against which the potential of solar cell produced electric power can be measured. Table X.1 lists capital costs and consumer prices for hypothetical average conventional power plants in central North America.

Table X.1

Average capital cost and energy pricing for consumers supplied by conventional electrical power plants [2, 3] (see accompanying text)

Type of power plant	Capital cost $/kW	Consumer pricing $/kWh
Base load plant		
Nuclear	1,300	0.08
Fossil fuel	1,100	0.07
Hydropower	1,200	0.06
Intermediate plant		
Fossil fuel	700	0.10
Peak load plant		
Fossil fuel	300	0.18

Note that the expenses provided in Table X.1 are only approximate averages. The costs for a specific power plant and a particular consumer may vary by anything up to a factor of four. This is particularly true of the price for energy to the small consumer. In underdeveloped countries the cost of energy to a consumer may run as high as $15 per kilowatt-hour when battery power is utilized. Even in an industrialized country such as the United States delivered energy costs vary from numbers as low as three to four cents per kilowatt-hour to in excess of $0.30/kWh in remote locations in the southwestern deserts and Alaska. The data presented in Table X.1 are meant for bench mark comparisons, not to be used as definitive values.

When we consider solar cell energy production we must recognize one essential fact--the sun does not shine for 24 hours each day. Referring back to Chapter II, consider Figure II.5 and Table II.4. Clearly, the amount of sunlight being received depends on the time of day, the season of the year and upon the geographic location. Even if we utilize a tracking solar energy collection system, we cannot guarantee uniform solar energy input. Does this imply that we can use solar cells as an energy source only during the day and as a peak or intermediate load type of source? Of course not! It does mean that we require some form of energy storage to provide energy during sunless periods[*]. This fact also has led to a rating system for solar cell arrays[#]. Each solar cell array is rated to produce so many kilowatts of electrical power under some specific set of illumination conditions (AMO for spacecraft, and AM1 or AM1.5 most commonly employed for earth based systems).

---

[*] We will discuss this topic further, later in this chapter.

[#] From the earlier portions of this book, in particular Chapters V through IX, it is clear that a single solar cell (of an area, perhaps as much as 200 cm^2) does not provide very much electrical power. It is, therefore, necessary to assemble many individual solar cells into arrays (panels or modules). These arrays currently range from several hundred milliwatts in size to tens of thousands of kilowatts [4]. Typical designs for photovoltaic power plants for the future range from 5 kW (suitable for a small house) to 100 MW peak power levels. In the latter case, the power plant would be constructed of several hundred (or thousand) individual panels or arrays.

As an illustration of the effect of variability in solar conditions as a function of location on the North American continent and local weather conditions, consider Table X.2.

Table X.2

Site, latitude and maximum energy received each year considering an average 12 hours of unimpeded AM1 sunlight each day; the maximum insolation per year including non-AM1 conditions (such as sunrise and sunset); and the actual insolation experienced including non-AM1 and weather effects

Location	Latitude	Energy received for 12 hours of AM1 sunlight	Maximum insolation with non-AM1 sunlight	Actual insolation with weather
Honolulu, HI	21.3	4,690	3,610	2,340
Santa Barbara, CA	34.4	4,690	3,510	2,530
Flagstaff, AZ	35.2	4,690	3,490	2,510
Boston, MA	42.2	4,690	3,420	1,980

Note the significant decrease in insolation values when weather and prevailing non-AM1 conditions are taken into consideration. Table X.2 leads to the effective power ratings for a one kW peak photovoltaic power system in the indicated locations provided in Table X.3.

Study of Table X.3 indicates that we can approximately rate a one kW peak power photovoltaic system at 0.25 kW average power over a 24 four hour period. Keeping this effective photovoltaic derating value in mind, let us now consider the cost of photovoltaic power systems.

In the early 1960s, photovoltaic systems intended for use in earth orbiting spacecraft cost in excess of $200,000 per peak kW. By 1974 a one kW peak photovoltaic power system could be purchased for less than $30,000. Later in the same decade the price per peak kilowatt of photovoltaic power plant was ~$15,000, and by 1993 it was possible to purchase a kilowatt of peak photovoltaic power system for less than $3,000 [5-9]. Even this price is considerably in excess of the prices listed in Table X.1 for the capital costs of conventional plants. This is more strikingly so when we consider the fact that a peak kilowatt photovoltaic plant produces something closer to 0.25 kW when night periods, seasonal variations and weather are taken into consideration.

Table X.3

The effective average power delivered (in kW) of a one peak kW rated photovoltaic power system at several locations and under the conditions of Table X.2

Location	Lati-tude	Energy received for 12 hours of AM1 sunlight	Maximum insola-tion with non-AM1 sunlight	Actual insolation with weather
Honolulu, HI	21.3	0.5	0.38	0.25
Santa Bar-bara, CA	34.4	0.5	0.37	0.27
Flagstaff, AZ	35.2	0.5	0.37	0.27
Boston, MA	42.2	0.5	0.36	0.21

The decrease in costs per peak kilowatt for photovoltaic systems given in Table X.3 depends partially on the fact that solar cell efficiencies have increased from something on the order of eight percent in 1960 to values better than 15% for commercial devices (better than 23% for the "best" single crystal silicon devices [10] and over 30% for some of the optimum performance, III-V based semiconductor solar cells [11]). It is also partially dependent on the increasing use and efficiency of amorphous silicon solar cells. Despite the Wronski-Staebler effect [12], photovoltaic cells from this material now account for close to 30% of the total peak solar power generated. A major reason for this is that, while the efficiencies of amorphous silicon solar cells are approximately eight to nine percent at present [13], the construction techniques employed require far less energy to fabricate the solar cells than the conventional processes employed in fabricating single crystal based solar cells [14].

Further capital cost reductions for solar cells have been realized as a result of the increasing production of photovoltaic devices. The total shipments of photovoltaic modules have risen from approximately 0.25 peak megawatts in 1976, to 3.33 MW peak in 1980, 18.5 peak megawatts in 1984 and an estimated volume of over 23 peak megawatts in 1992. When we consider that a 100 MW fossil fuel plant produces so much carbon dioxide that an area of ˜2700 sq km (about the size of Rhode Island in the USA) would have to be planted in trees to remove the carbon dioxide from the atmosphere, we see that photovoltaics are beginning to have a beneficial environmental effect.

It will require some time, and a considerably increased market for photovoltaic cells before the solar cell industry will be in a position to close the cost gap between photovltaic produced power and conventionally produced power. However, consider the fact that the source of energy in conventional power plants is obtained at a constantly increasing, and by-no-means inconsiderable expense. Further consider that the maintenance and operation of a conventional power plant requires considerably more in the way of personnel than the photovoltaic plants constructed to date [4]. The growing movement to include environmental costs as a tax on conventional electric plant fuels will further increase the cost of conventional electric power. On the negative side, we are not yet sure of the long term reliability of photovoltaic power systems. The National Renewable Energy Laboratories in Denver Colorado, USA and a number of other facilities around the world are conducting extensive reliability studies on both individual solar cell designs (as functions of cell design and fabrication techniques) and on photovoltaic systems [15]. As of this time, it is unwise to attempt too precise a prediction of the moment when the capital and operating costs of photovoltaic derived power will match the expenses involved in more conventional systems. A preliminary estimate of some time period during the second quarter of the 21st century would probably be as accurate a prediction as possible at this time. Despite these uncertainties, let us attempt to consider some additional facets of this topic.

Solar cell production levels continue to increase[*]. The May 1993 issue of the Photovoltaic Insiders Report mentions the construction, by United Solar Corporation, of a 10 MW (The peak power potentially produced by a year's worth of solar cell output) plant in the state of Virginia. Even larger production capability plants were discussed at the 1993 NREL meeting [16]. It is thought that the cost versus volume output of the solar cell industry will resemble that of its parent semiconductor industry [17] yielding a sharp decrease in device cost with increasing production volume.

The basic material for most solar cell production is currently silicon. The cost of pure polycrystalline silicon in 1972 was approximately $60 per kilogram [18]. Since then, the cost has decreased by approximately 50 percent. For single crystal solar cells this polycrystalline silicon has to be converted to single crystal form and the crystals sliced into wafers, losing 1/2 to 3/4 of the silicon in the process. The single crystal wafer must then be fabricated into solar cells (including metalization and testing)--processes consuming much time and involving considerable cost. For several years research and development has been ongoing into the production of ribbon silicon, dendritic web silicon, surface web silicon, hexagonal tube silicon and other material fabrication technologies [8-26]. These technologies have been primarily aimed at producing thin film crystalline and polycrystalline silicon (as well as other semiconductors in these crystalline forms). It is felt that by extruding silicon in a continuous ribbon or sheet, the single crystal or polycrystalline semiconductor will be immediately available for solar cell fabrication with little or no intermediate, and costly, processing required. Cost reductions of up to a factor of three over those experienced with the standard Czochralski process for single crystal silicon production have been predicted for these techniques [21, 27][#].

---

[*] Note that this statement is not quite the same as saying that the market for solar cells continues to increase. If the photovoltaic industry is to be solvent, any increase in production capacity must be matched by increases in market demand.

[#] While single crystal silicon is by far the most inexpensive single crystal semiconductor, the polycrystalline and amorphous forms of silicon and other semiconductors have shown signs of excellent solar energy performance, as demonstrated in Chapter IX.

Amorphous silicon and other amorphous semiconductors have been reported as exhibiting photovoltaic energy conversion efficiencies as high as 10.5% for modules with areas of 1,200 cm^2 [28]. Estimated theoretical costs for tandem (stacked) amorphous silicon solar cells are as low as $50 (U.S.) per peak watt. As demonstrated in Chapter IX, amorphous semiconductors are, as yet, imperfectly understood [29]. Despite the low cost of the amorphous form of semiconductor we still have the costs of ancillary components, such as module interconnects, solar tracking systems, etc. As a result the cost of amorphous silicon solar cells is not, yet, substantially different from that of single crystal solar cell systems.

Polycrystalline solar cells are somewhat more expensive than amorphous solar cells and, as stated in the previous paragraph, the cost of the required non-semiconductor components can be the deciding factor in overall system value. Even with potential operating efficiencies in excess of 16% [30] the cost per peak kW still remains in excess of three times the early-on reported prediction of $140 per peak kW [31]. The energy payback time for both polycrystalline and amorphous material based solar cells is generally on the order of months, rather than years as for single crystal semiconductors [32]*.

Either (often because it is the only way in which a working solar cell can be fabricated) as a pn junction, as a heterojunction, as Schottky barriers; or in vertical, standard or inverted configuration; or using single crystal, polycrystalline or amorphous semiconductors; it is clear from Chapters V through IX that a number of compound semiconductors are potentially better at photovoltaic conversion of the energy in sunlight than is silicon. The examples used in this book, GaAs, CdTe, InP, AlSb and CdSe are by no means the only non-silicon semiconductor possibilities and, as seen in these previous chapters, these materials may or may not

---

* The concept of energy-payback has evolved from the requirement that energy must be used in the fabrication of any solar cell and attendant supporting structures. This energy must be returned before the photovoltaic cell can become a net contributor of energy. Estimates of the time required to pay back the energy used in construction, installation and maintenance of a solar cell range from 12 to 18 months for amorphous and polycrystalline solar cells to five to seven years for earth based single crystal solar cells, and up to 30 years for the orbiting solar cells used on communications satellites in synchronous orbits.

be superior to silicon, depending on type and level of illumination. The potential number of semiconductor materials for solar cells is very large today [33]. Such single crystal based solar cells as stacked (tandem) $GaInP_2/GaAs$ have shown considerable promise [34] with efficiencies in excess of 25% under AMO conditions. Double and triple junction stacked amorphous solar cells with efficiencies in excess of 10% and surface areas of over a square foot now appear to be realistic [35].

The basic question remains one of cost[*]. For example, recent increases in module costs have driven up prices of polycrystalline solar cells and have resulted in a loss of market share [37]. Technology is the prime stumbling block behind cost, particularly for compound semiconductors. It is more difficult, for compound semiconductors, to grow single crystals, fabricate polycrystalline structures or provide amorphous semiconductors than it is to do these for a single material such as silicon. In turn, this makes it more difficult to attain maximum theoretical performance. The fact that silicon is 10,000 times more plentiful on the earth than gallium, a 100,000 times more plentiful on earth than arsenic and a million times more available than cadmium, means that even obtaining the raw materials for some compound solar cells can be very expensive. Other potential situations need considering. For example, suppose we have a solar cell system erected on some building which then burns down. What are the environmental effects? Silicon oxidizes (burns) to form sand (quartz), gallium arsenide oxidizes to form gallium and arsenic oxides, both of which are toxic. Other service conditions and semiconductors pose other environmental effects.

This brings us to another method of reducing the cost of solar energy. As discussed in Chapter VII, there is no a priori reason for the entire area of a solar energy collector/converter to be constructed entirely of solar cells. We can reduce the amount of expensive semiconductor used in a solar energy conversion system by concentrating the incoming sunlight using less expensive mirrors or lenses (see Chapter II). The resulting system may even be more efficient!

---

[*] The best solar cell material is not necessarily that with the highest efficiency solar cells. It is not even that which provides the largest number of kilowatts per dollar. We must also include environmental effects and energy-payback analysis in our considerations [36].

As an example, consider and compare solar energy conversion systems involving: (1) a compound semiconductor with a square meter of solar cell costing approximately $1,500 (US) and exhibiting an operating efficiency of approximately 20%, (2) a square meter of single crystal silicon solar cell costing $300 (US) and with a modular efficiency of 15% and (3) an amorphous solar cell of one square meter area with a fabrication cost of $100 (US) and an operating modular efficiency of 10%. What would be the capital cost for an optical concentrator system employing these solar cells where the capital cost for the concentrator portions of the system varies between $50/m^2 and $500/m^2 (US)[*]? The actual cost value, clearly, depends on many technical, geographical, political and economic factors. However, we can make some, rough, preliminary estimates.

Figures X.1 through X.3 present capital cost estimates for assembled photovoltaic concentrator systems assuming a spectral input similar to AM1 conditions of 1.07 kW/m^2 but reduced, on average, to 25% of this value as the result of weather and the earth's rotation. Each system consists of enough solar cells and appropriate amounts of concentrators, based on the desired optical concentration, to provide 10 kW of averaged delivered solar power. The assumed costs of optical concentration system are the somewhat arbitrary values of the previous paragraph. Study of these figures provides additional insight into the advantages and disadvantages of concentrator systems. Should the reader desire to perform a detailed cost analysis for a specific photovoltaic system, he/she must acquire not only solar cell and other component cost figures (note that as the concentration ratio increases, the cost of the tracking system attached to the solar cells also increases), but data on taxes, real estate costs, environmental restrictions, and government permit procedures and incentives. Such an analysis is far beyond us here, and the reader is referred to such sources as the nearest college of business, discussions in the IEEE Photovoltaic Specialists Conference Proceedings and various government publications.

---

[*] These figures are preliminary estimates based on both old cost studies [38] and many conversations with attendees at the various meetings of the IEEE Photovoltaic Society and the annual National Renewable Energy Laboratories photovoltaic review meetings.

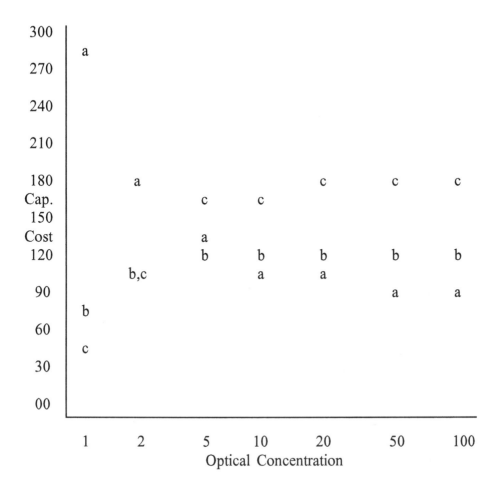

Figure X.1. The total capital cost (in thousands of dollars) for a solar cell system capable of delivering an averaged power of 10 kW, including the effects of weather and the earth's rotation. Three cost and performance levels of solar cell are considered: (a) $1,500/m^2$ for 20% efficiency (possibly a compound semiconductor), (b) $300/m^2$ for 15% efficiency (possibly a single crystal silicon solar cell) and (c) $100/m^2$ for 10% efficiency (possibly an amorphous silicon solar cell). The system cost is provided as a function of the concentration level with the light concentrating system having a cost of $500/m^2$.

Consider the information depicted in Figure X.1. The assumed cost of the semiconductor solar cells varies considerably depending on the crystal structure. It could vary even more if we consider the variability of the fabrication process. Table X.4 provides a sampling of the overall

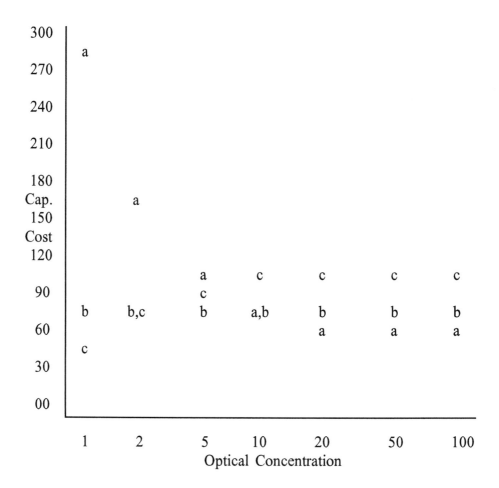

Figure X.2. The total capital cost (in thousands of dollars) for a solar cell system capable of delivering an averaged power of 10 kW, including the effects of weather and the earth's rotation. Three cost and performance levels of solar cell are considered: (a) $1,500/m^2 for 20% efficiency (possibly a compound semiconductor, (b) $300/m^2 for 15% efficiency (possibly a single crystal silicon solar cell) and (c) $100/m^2 for 10% efficiency (possibly an amorphous silicon solar cell). The system cost is provided as a function of the concentration level with the light concentrating system having a cost of $300/m^2.

system costs for various solar cell expense levels for this degree of concentrating expense.

Consider the results of Figure X.2. Once again the assumed investment in solar cells could vary from the assumed values. A sam-

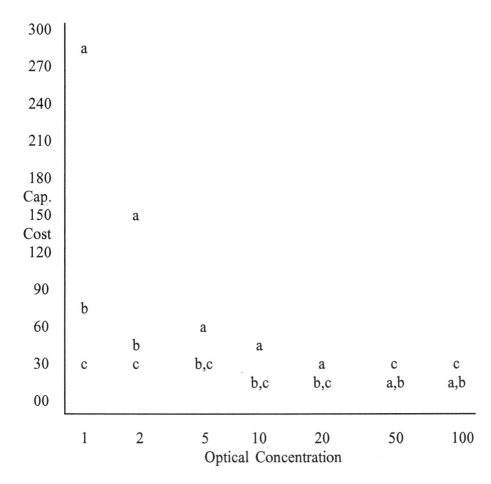

Figure X.3. The total capital cost (in thousands of dollars) for a solar cell system capable of delivering an averaged power of 10 kW, including the effects of weather and the earth's rotation. Three cost and performance levels of solar cell are considered: (a) $1,500/m^2$ for 20% efficiency (possibly a compound semiconductor, (b) $300/m^2$ for 15% efficiency (possibly a single crystal silicon solar cell) and (c) $100/m^2$ for 10% efficiency (possibly an amorphous silicon solar cell). The system cost is provided as a function of the concentration level with the light concentrating system having a cost of $50/m^2$.

pling of overall system costs for variations in solar cell cost in Figure X.2 is presented in Table X.5.

Study of Figures X.1 through X.3 leads to the conclusion that the result of having a very inexpensive solar cell is to make the optical concentration portion of the overall system more economically important. Note that, at high optical concentration levels, the compound semiconductor with its assumed higher conversion efficiency, appears to have an economic advantage. However, recall that, in practical situations, the per unit area cost of the concentrator portions of this system will increase as the optical concentration level is increased and the demand for accurate solar tracking rises. In constructing Figures X.1 through X.3 it was assumed that the per unit area cost of the concentrator was a constant. This change in assumptions may negate the perceived economic advantage of the compound semiconductors.

Table X.4

Estimated overall photovoltaic concentrator system costs for selected solar cell values, efficiencies and optical concentration levels. The optical concentration subsystem is estimated to have a per unit area price tag of $500/m^2$. The system, accounting for weather and the earth's rotation, is producing an averaged power level of 10 kW

Solar cell cost ($/m^2$)	Opt. conc.	System capital cost ($)	Opt. conc.	System capital cost ($)	Opt. conc.	System capital cost ($)
			Solar cell efficiency 20%			
2,000	2	234,000	10	121,000	50	99,000
1,500	2	187,000	10	112,000	50	97,000
1,000	2	140,000	10	103,000	50	95,000
			Solar cell efficiency 15%			
400	2	112,000	10	122,000	50	124,000
300	2	100,000	10	120,000	50	123,500
200	2	87,000	10	117,000	50	123,000
			Solar cell efficiency 10%			
150	2	121,000	10	174,000	50	184,500
100	2	112,000	10	172,000	50	184,000
70	2	107,000	10	171,000	50	183,500

Table X.5

Estimated overall photovoltaic concentrator system costs for selected solar cell values, efficiencies and optical concentration levels. The optical concentration subsystem is estimated to have a per unit area price tag of $300/m^2. The system, accounting for weather and the earth's rotation, is producing an averaged power level of 10 kW

Solar cell cost ($/m^2)	Opt. conc.	System capital cost ($)	Opt. conc.	System capital cost ($)	Opt. conc.	System capital cost ($)
			Solar cell efficiency 20%			
2,000	2	215,000	10	88,000	50	62,000
1,500	2	168,000	10	78,500	50	61,000
1,000	2	121,500	10	69,000	50	59,000
			Solar cell efficiency 15%			
400	2	87,000	10	87,000	50	87,000
300	2	75,000	10	75,000	50	75,000
200	2	62,000	10	62,000	50	62,000
			Solar cell efficiency 10%			
150	2	84,000	10	106,500	50	111,000
100	2	75,000	10	105,000	50	110,700
70	2	69,000	10	104,000	50	110,400

Consider Figure X.3. As in the situations covered in Figures X.1 and X.2 the assumed cost of the semiconductors could vary considerably depending on solar cell processing variables. Table X.6 presents a sampling of overall system costs for various capital costs of the semiconductor solar cells.

The reader is encouraged to study Tables X.4 through X.6 as well as Figures X.1 through X.3. Such perusal leads one to an appreciation of the variation in capital costs engendered by changes in optical concentration, alterations in mirror, lens and tracking systems and by changes in the solar cells and the semiconductors from which they are made.

Table X.6

Estimated overall photovoltaic concentrator system costs for selected solar cell values, efficiencies and optical concentration levels. The optical concentration subsystem is estimated to have a per unit area price tag of $50/m^2$. The system, accounting for weather and the earth's rotation, is producing an averaged power level of 10 kW

Solar cell cost ($/m^2$)	Opt. conc.	System capital cost ($)	Opt. conc.	System capital cost ($)	Opt. conc.	System capital cost ($)
			Solar cell efficiency 20%			
2,000	2	192,000	10	46,000	50	17,000
1,500	2	145,000	10	36,000	50	15,000
1,000	2	98,000	10	27,000	50	13,000
			Solar cell efficiency 15%			
400	2	56,000	10	21,000	50	14,000
300	2	44,000	10	19,000	50	13,700
200	2	31,000	10	16,000	50	13,200
			Solar cell efficiency 10%			
150	2	37,000	10	22,000	50	19,400
100	2	28,000	10	21,000	50	19,100
70	2	22,000	10	19,000	50	18,800

Over and above the trends suggested by study of these tables and figures, note that the optical concentrating systems are not 100% efficient. This decrease in efficiency, which depends on the condition of the mirrors or lenses and the accuracy of the solar tracking apparatus, results in the need for a larger, and more expensive solar energy collecting/converting system. However, to balance this detriment, consider that, with the optically concentrated systems discussed here it is possible to use the waste heat for space and water heating, increasing the overall performance of these solar systems and reducing the capital cost per joule of delivered energy [39-42].

Assuming a 20 year system lifetime* and solely being concerned with the electrical energy delivered we have an estimated capital cost per kWh of delivered energy of ~21.3 cents (US) for a photovoltaic system employing compound semiconductors costing $2,000/m² and operating at 20% efficiency with no optical concentration. An "intermediate cost" system, employing single crystal silicon costing $300/m² and operating at 15% with an optical concentration level of 10 and utilizing a tracking lens or mirror system costing $300/m² could supply electricity at an approximate cost of 4.3 cents (US) per kWh. Potentially low cost photovoltaic systems such as one employing $70/m² amorphous solar cells operating at 10% and optically concentrated to a factor of 50 with a tracking lens or mirror system costing $50/m² offer the possibility of electrical power at a cost of ~1.1 cents/kWh (US). Compare these values with those of a conventional electric power plant. From Table X.1, assuming a 20 year plant life, operating for 90% of the time, a conventional electric power plant will have an initial capital cost in the neighborhood of 0.5 cents/kWh (US).

Note that with the use of optical concentration and assuming solar cells and optical concentration levels and equipment of minimal cost, this section demonstrates that solar energy is approaching conventional power plants in-so-far as capital costs are concerned. In the situation of optical concentration levels approaching 50, further note that (see Chapters VII and VIII), as the concentration level is increased thermal and other effects act to reduce the solar cell operating efficiency. However, the thermal energy available for delivery to customers in these situations increases overall system efficiency and reduces the capital cost per joule of delivered energy. It is this two-faceted delivery of energy which makes solar energy, in sum, so attractive.

In discussing the actual cost of the electrical (and the thermal) energy delivered to the customer we have noted that many other factors need to be considered. For example, with conventional plants the energy needs to be transported over considerable distances, while with on-site

---

* Reliability studies are underway at NREL and many other locations, but it will be another decade before we can confidently assign system lifetimes for the multiplicity of photovoltaic energy conversion systems and the vast array of environmental conditions under which they must operate.

solar cell panels no energy transport is required. However, a practical approach to solar cell systems involves electrical interties between the individual converting systems and hence the transport of electrical energy is required. Note that it is much more difficult to transport thermal energy over long distances, so-much-so that such an undertaking is not economical and any thermal energy generated must be used locally.

There is one subsystem in these solar energy systems which we have not yet addressed--energy storage. With conventional power systems energy is stored in chemical (coal, oil); nuclear (U^{235}) or potential (water behind a dam) forms. The conversion of these forms of energy to electrical energy is done at the power plant, more-or-less independently of the time of day. The energy is then shipped directly to the consumer. However, in a photovoltaic system the input energy is in the form of sunlight and this is only available, on-the-average, for 12 hours each day. Thus extra electrical energy must be generated during each sunny day, for use in nocturnal and poor weather periods. This electrical energy must be stored. If we are using a hybrid system, generating both electrical and thermal energy, there must also be provision for the storage of thermal energy.

Since this work is primarily devoted to solar cells and the generation of electrical energy we will concentrate, in the next section, on the storage, by various means of electrical energy. For the reader who is interested in the storage of thermal energy, the following references are suggested [43-45].

Electrical Energy Storage

Classically, the storage of electrical energy has been achieved in two ways. First, utility companies have long used "pumped storage". In this technique excess electrical generating capacity has been utilized to supply energy to pump water into some reservoir behind a dam. During periods of high electrical demand, the stored water is released through conventional hydro-electric generating turbines converting the potential energy to electrical energy. The second conventional storage method stores electrical energy as chemical energy within a storage battery. The secondary, or rechargeable, battery has long been used in moving vehicles to supply starting power, and in the case of electric vehicles, motive power. During the last ~30 years the fuel cell has also been popular as an energy storage and production device. Its use has ranged from the Apollo moon exploration program through local, small area power plants [46-48].

The secondary battery most commonly encountered is the lead-sulfuric acid battery found in use in automobiles. This combination of electrodes and electrolyte has a nominal open circuit voltage of 2.1 volts and an energy storage density of approximately 20 watt-hours per kilogram of weight. Recent advances in battery technology have led to the sealed lead-acid battery which also has a 2.1 open circuit voltage, but a much higher energy storage density, close to 30 watt-hours per kilogram. The energy efficiency, or the amount of stored energy which can be recovered, usefully, from a lead acid battery depends on the charging and discharging conditions and is known to be approaching 75% [49].

The are a multitude of potential secondary batteries including the well known nickel-cadmium battery with an individual cell voltage of 1.3 and a storage capacity of 50 watt-hours per kilogram. Several secondary battery characteristics play roles in the selection of storage elements in a photovoltaic system. Energy density is important in those applications that require mobility (such as in electrical cars, both those which are under development by automobile manufacturers and those featured in the "Sunrace" competitions of solar power electric cars in Australia. In photovoltaic systems we also need a storage battery that can undergo many charging and discharging cycles. If our storage battery is to supply nighttime energy to a home over 20 years, we can expect a discharge/charge cycle which is repeated in excess of 7,000 times. This ability to undergo numerous charge/discharge cycles should be coupled with the ability to return a preponderance of its stored energy to the consumer. The 75% efficiency values listed above for lead-acid batteries is quite good, but the number of deep discharge cycles available with these batteries is only a few hundred. However, secondary batteries such as nickel-zinc and lithium-metallic sulfide come close to this efficiency value and, additionally, appear to be capable of up to 800 to 1,000 deep discharge cycles. Note that, while a considerable improvement, this number of cycles is much lower than the target value of 7,000.

The cost of secondary batteries is, clearly, another important selection parameter. The batteries mentioned in the preceding paragraphs have a cost of approximately $50 (US) per kWh of storage capacity [50]. If the consumer needed to store 20 kWh of energy during the day for use during the ensuing night, the consumer would require, at a minimum,

$1,000 worth of batteries. Given the current potential number of deep discharge cycles for secondary batteries, the consumer would need to replace the batteries every two to three years. This would result in a capital cost of close to seven cents (US) per kWh--a cost incurred over and above the expenses required to purchase solar cells, lenses (or mirrors) and a tracking system.

Another method of energy storage is in the form of hydrogen. One can use the direct current from a solar cell system to electrolyze water. The output from such an operation is hydrogen and oxygen. The oxygen can be released into the atmosphere where it aids in breathing, and the hydrogen is stored. This electrolytic process can either be performed at a remote location [51-53] or directly in contact with the solar cells [54-56]. The hydrogen is stored, either under pressure or adsorbed in some material. To recover the stored energy we could burn the hydrogen in air, yielding heat energy and drinkable water along with a small amount of nitrous oxides [57]. If the hydrogen were burned (oxidized) in pure oxygen, water would be the only by-product of the production of electrical energy. The water is potable, if care is taken to assure the cleanliness of the equipment. In any case, hydrogen, while it must be handled with care, is fundamentally no more dangerous than natural gas.

To convert hydrogen directly to electricity we can either burn it as above and use the thermal energy to generate electricity or we can utilize a fuel cell. In principle, a fuel cell unites a fuel (hydrogen, in this instance) and an oxidizer (most commonly air) to form a by-product (pure water in our case) and electrical energy. The potential efficiency of these devices is high (in excess of 90%) and investigative and developmental work on these devices is an ongoing process [58]. An excellent introduction to fuel cell technology and the problems involved in their operation may be found in Angrist [59].

The System

As discussed in Chapter I, modern civilization consumes thousands of megawatts of electrical power. If solar cells are to supply a significant portion of this energy requirement, we must give some thought to overall system organization. Fundamentally, we require a source of heat energy and a source of electric energy to power our homes, factories and businesses. A single solar cell supplies direct current at approximately one volt. If this electrical energy is to be used efficiently we must electrically "stack" several solar cells in order to reach the 12 to 200 volts line voltages in common usage. In addition, many kinds of electrical equipment require considerably more than a single ampere of current, so that several series "stacks" must be wired in parallel to provide for needed voltage and current values.

When we consider the dc aspect of solar cell system output, most modern electrical equipment can be converted to direct current operation, though the use of inverters to "condition" dc solar cell system power outputs and convert them to ac is widespread. If the solar-electric power is generated close to the point of use, then low voltage and dc power transmission is feasible. There are, however, many locations in the world where weather and geography discourage year around photovoltaic energy generation. As a result, we must consider long distance transmission of solar cell derived energy. The existing electrical power transmission network is available, but requires the use of inverters on solar cell systems in order to produce ac power. Once ac power is available, then transformers can be employed to produce the required high voltages for long distance transmission[*].

Rather than transport the electrical energy, an alternative is to use the solar cell generated electrical energy to produce hydrogen. In turn, the hydrogen can be locally stored or transmitted by tanker or pipeline to the point of use, where it can be burned or used in a fuel cell. Presumably this hydrogen could also be used as an automotive fuel (compressed, liquified or adsorbed in some material) and as feed stock in a number of chemical processes, resulting in materials ranging from cooking oils to nylon stockings.

In practice, all of these options will most likely be employed depending on location (of both solar cell system and of the consumer) and circumstances. For example, consider the possible applications of a solar energy conversion/storage system to an on site usage. Figure X.4 portrays such a system using both thermal and photovoltaic conversion systems.

---

[*] To reduce long distance power transmission losses, electrical energy is transmitted at very high voltages. It would be impractical to "stack" enough solar cells to reach a million volts, so transformers are used to "boost" voltages to this level and this "boosting" requires ac signals.

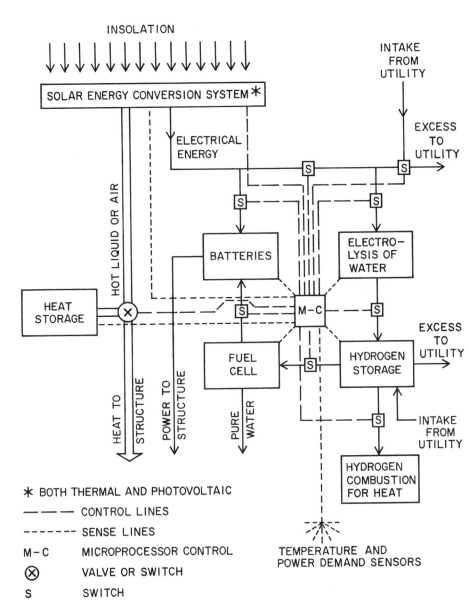

Figure X.4.   An integrated domestic solar power system. Copyright Earthlab, Santa Barbara, CA, USA . Reproduced by permission.

Note that heat is stored and used for low quality heating of water and building space. Electrical energy is directly used (lighting, cooking, cooling, entertainment, communication, etc.) and is also stored in batteries (for high current, short term applications) as well as being converted to hydrogen by electrolysis of gray water. The hydrogen is then stored on site until required for burning to produce heat energy or being reconverted to electricity via fuel cell. Provision is also made for interconnection of both produced hydrogen and electricity to public utilities. Note that the entire system is computer controlled. Whether such a system applies to a single family residence* a shopping center, a village, an industrial or commercial plant or an entire geographic region, the solar energy supplying system must have the same assortment of components. Efficient operation is made possible by the dedicated computer which is responsible for minute-to-minute control of the energy demands and supplies. The computing power required for such an operation is well within the range of the modern personal computer. Clearly, systems such as those described in Figure X.4 are not inexpensive and the cost will, in general, be beyond the resources of the average individual [61, 62]. This leads to a potential future utility company; a company composed of many small, widely scattered generating systems, some of which are leased to individuals or groups of consumers. Such a utility company will act more as an "energy broker" assisting in the transfer of energy from one part of its service area to another, than it will as a centralized producer of electrical energy. A considerable number of studies have been conducted examining the aspects of merging conventional electric utilities with small photovoltaic energy producers. To provide the reader with a start on this topic, the following references are given [63-65].

Final Words

This work is primarily about solar cells and only secondarily concerned with the milieu in which the solar cells exist. Solar cells have been shown to be useful devices which convert sunlight directly to electricity. We have considered how they operate, and have constructed

---

* Estimates of the energy required by a single family residence have been made by a number of sources [60]. On the average, it appears that the typical U.S. household requires ~1.8 kW of electrical power and 4.6 kW of thermal energy input. Note that these values may triple or half depending on environmental and time constraints.

a realistic model which allows us to make predictions of both their efficiency and the power delivered by such devices. Such predictions are available for solar cells constructed from single crystal semiconductors. Chapter IX considers solar cells fabricated on polycrystalline semiconductors and with amorphous semiconductors. In both of the situations, detailed theories are not available since the electrical properties of the devices depend on variables such as the nature of polycrystalline grain boundaries and the methods by which hydrogen is introduced into the semiconductor. With detailed theories unavailable, the examination of these devices is conducted primarily by citing various examples of solar cells and their exhibited performance.

In terms of solar cells fabricated on and with single crystal semiconductors, it is clear that the design, both optical orientation and physical layout, of a photovoltaic cell must be varied as a function of the intensity of the incoming illumination and the configuration of the optical input. It is also clear that much research and development remains to be done if the promise of photovoltaic energy conversion is to be fully met. Meanwhile, annual photovoltaic module shipments continue to increase-- from 26.1 MW in 1986 to 60.2 MW in 1992 [66]. In dollar amounts, it is estimated that the photovoltaic market will reach five billion by the year 2010 [67].

At this time, it is not obvious whether solar cells should be constructed with polycrystalline, amorphous, or single crystal semiconductors. Single crystal photovoltaic cells have, in general, the advantages of greater efficiency, less sensitivity to environmental stress and greater susceptibility to theoretical analysis. On the other hand, the polycrystalline and amorphous semiconductor forms of solar cells are potentially less expensive to fabricate and require significantly less energy to manufacture. In all cases, we need to consider the various methods of fabrication of the starting material. For single crystal materials we have the Czochralski, float zone and Bridgeman techniques as well as liquid and gas phase epitaxy. Polycrystalline and amorphous materials often employ plasma enhanced chemical vapor deposition (PCVED), dendritic web, ribbon, edge defined film processes and other fabrication techniques [68 and 69]. The employment of molecular beam epitaxy, though not yet thoroughly understood nor completely reliable, appears to be promising.

Nor is it obvious which junction type, which semiconductor or which type of optical system (configuration and concentration) should be utilized. The heterojunction has the advantages of potentially greater efficiency as a result of reduced surface recombination and lower internal series resistance. However, the added complexity of device fabrication, owing to the presence of two or more semiconductors, is a serious handicap. Of course, as seen in Chapter IX, a number of solar cell configurations depend on heterojunctions and there is no option. The Schottky barrier, either in metal on semiconductor or in metal on insulator on semiconductor forms, is potentially a simple, inexpensive to fabricate, class of solar cell. However, in general, the built-in potential barrier for Schottky diodes is smaller than desirable and the existence of the non-semiconductor layer reduces the experienced photocurrent. These phenomena produce a lower operating efficiency than is generally desirable. The pn junction, utilizing but a single semiconductor, has been the standard technology of the silicon based semiconductor industry. The class of junction is energy intensive to fabricate and in some semiconductors is not even possible.

As a semiconductor, silicon is inexpensive, widely available and environmentally benign. This semiconductor, however, is not optimum from an efficiency viewpoint, having too small an energy gap. Other single crystal semiconductors considered in this work, including gallium arsenide and cadmium telluride, have wider energy gaps and potentially exhibit greater conversion efficiencies. In general, these materials are more difficult to fabricate. Also it is harder to locate the raw ores needed to construct the solar cells and use of the semiconductors is often accompanied by some environmental penalty. Chapter IX is filled with promising polycrystalline and amorphous semiconductors, often appearing as heterojunctions.

We have demonstrated that the orientation of the solar cell with respect to the incoming sunlight is critical. The standard orientation solar cell is simple to assemble, but photon losses, charge carrier recombination and device series resistance result in performance degradation. Additionally, reflection in Schottky barriers causes added losses. The vertical configuration potentially reduces the recombination and resistance losses, but presents assembly and construction difficulties. The inverted configuration type of solar cell reduces the recombination and series resistance engendered performance degradation and, given sufficient carrier lifetime, provides enhanced performance. It is also a useful configuration when the electrical "stacking" of solar cells is considered. This "stacking" is required if solar cell systems are to produce electrical energy at practical working voltages.

The nature of the light which illuminates the solar cell is also of importance. If the solar cell uses concentrated sunlight, the type of solar cell employed needs to be carefully considered. Moreover, the means of concentrating the sunlight involves more than a choice between lenses and mirrors. Consideration of the optics involved makes clear the requirement that any concentrating optical system be kept "pointed" at the sun. This can be accomplished with an active system involving motors and some sort of computer control, or via some type of passive system such as a bimetallic strip that expands and drives the concentrator upon exposure to sunlight. since the concentrator system is exposed to ambient conditions it must be capable of operating under various conditions of wind, rain, snow, sand storms and varying temperatures. Ideally, such a tracking system also makes provision for removing the excess heat, which is a useful facet in such optical concentrator systems. Chapter VIII studies other options concerning solar cells and the nature of the light which illuminates them. Such techniques as optically "stacking" solar cells so that the initial semiconductor absorbs, efficiently, the high energy short wavelength photons; the next semiconductor absorbs intermediate energy, medium wavelength photons with good efficiency; and a third solar cell based on another semiconductor absorbs the long wavelength, low energy solar photons are often employed. Dichroic mirrors are also used to insure that each solar cell material is illuminated by photons which are most efficiently absorbed by that semiconductor. Finally, Chapter VIII discusses thermophotovoltaics. This technique, and others like it, alters the fundamental nature of the spectrum of the photons which illuminate the solar cells. If the alteration is cleverly done, the resulting photons interact much more efficiently with the solar cells. Whether one of the spectrum-modifying techniques becomes the method of choice depends, very strongly, on overall system economics.

Once energy is converted to heat and electrical forms, this energy must be transported to points of use and or storage. The types of energy usage and storage impose constraints on the organization and types of energy converter. As indicated earlier in this chapter, it does appear that any solar energy conversion system capable of supplying civilization with a significant portion of its energy demands must be composed of a number of different converters and that energy storage must also be accomplished in various ways (without storage, renewable energy can, at most, account for 42.5% of civilization's requirements [70]).

The entire energy generation energy use system imposes constraints on the solar cells. This, clearly, includes the politics of energy [71]. The existence of all these influences encourages continuing research into the performance of solar cells--research ranging from fundamental materials studies, through operating considerations involving temperature, ambient conditions and device reliability. Furthermore, much additional work needs to be done on systems aspects. Such problems as the assembly of individual solar cells into modules and the storage and dissemination of solar derived energy must be reliably solved if solar cell power is to become economically important [72].

It would be the "height of folly" to state that all possible photovoltaic energy conversion schemes have been outlined in this work; or that all possible solar cell constructions, configurations and junction designs have been studied in detail. The commercial sale of solar cells continues to expand in absolute size (MW) and in types of solar cells. For example, in 1989 56% of all solar cell shipments were single crystal silicon devices and 28% were cast and ribbon silicon devices. Also in 1989 4% of solar cell shipments were concentrator solar cells of silicon [73]. In terms of photovoltaic shipments, there are a wide variety of sources with 1992 photovoltaic shipments of 18.4 MW from the USA, 18.8 MW from Japan, some 17 MW from various European sources and 6 MW from various other locations. It will be interesting to follow the progress of this fast moving field into the twenty-first century.

References

1   It is suggested that the perusal of the Annual Conference Records of the IEEE Photovoltaic Specialists Conferences, the meetings of the Photovoltaic Energy Conference, the Photovoltaic Science and Engineering Conference and the publications of the International Solar Energy Society will provide the reader with much applications information on solar cells.

2   The data in this table is derived in part from documents published by the Electric Power Research Institute and by the United States Department of Energy.

3   Thanks are expressed to Dr. F. Goodman, Jr. of the Los Angeles, CA Light and Power Corporation for comments and information concerning Table X.1.

4 The reader is referred to Solar Today, a journal published by the International Solar Energy Society, for a continuing series of
   articles on solar photovoltaic systems. Such journals as the American Scientist (the voice of Sigma Xi) and Scientific American frequently have articles on solar cell systems.

5 R. M. Winegarner, in the Proceedings of the 1977 Annual Meeting of the American Section of the International Solar Energy Society, June 1977, p. 8.

6 Machine Design, 4 Oct. 1976, p. 8.

7 L. Curran, in Electronics, 11 Nov. 1976, p. 91.

8 W. D. Johnston, Jr., in American Scientist, Vol. 65, 1977, p. 1976.

9 R. Taylor, in session VI, of the 1993 National Renewable Energy Laboratory Review Meeting, Denver, CO, Oct. 1993.

10 M. A. Green, S. R. Wenham and J. Zhao, in The Conference Record of the 23rd IEEE Photovoltaic Specialists Conference, Louisville, KY, May 1993, p. 8.

11 M. Umeno, et. al., in The Conference Record of the 23rd IEEE Photovoltaic Specialists Conference, Louisville, KY, May 1993, p. 741.

12 D. L. Stabler and C. D. Wronski, in Applied Physics Letters, Vol. 31, 1977, p. 292.

13 K. Rabago, in the keynote address of the National Renewable Energy Laboratories Review Meeting, Denver, CO, Oct. 1993.

14 See the discussion in Chapter IX.

15 National Renewable Energy Laboratories Review Meeting, session III, Denver, CO, Oct. 1993, had a number of discussions/presentations on this subject.

16 National Renewable Energy Meeting, the opening session and session VI, Denver, CO, Oct. 1993, contained considerable discussion on this subject.

17 F. M. Smits, in IEEE Transactions on Electron Devices, Vol. 23, 1976, p. 640.

18 C. G. Currin, et. al., in The Record of the 9th IEEE Photovoltaic Specialists Conference, May 1972, p. 363.

19 A. L. Hammond, in Science, Vol. 197, 1987, p. 445.

20 P. Rappaport, in the Proceedings of the Workshop on Photovoltaic Energy for Terrestial Applications, Oct. 1973, p. 8.

21 T. Surek, in the Journal of Applied Physics, Vol. 47, 1976, p. 4384.

22 A. L. Hammond, in Science, Vol. 184, 1974, p. 1359.

23 A. I. Mlavsky, in the Proceedings of Annual Meeting of the American Section of the International Solar Energy Society, 1977.

24 C. T. Ho, R. O. Bell and F. V. Wald, in Applied Physics Letters, Vol. 31, 1977, p. 463.

25 H. J. Möller, Semiconductors for Solar Cells, Artech House, Boston, MA, 1993, Chap. 6.

26 J. Kalejs, in session IV of the National Renewable Energy Laboratory Photovoltaic Review Meeting, Denver, CO, Oct. 1993.

27 National Renewable Energy Photovoltaic Review Meeting, session IV, Denver, CO, Oct. 1993.

28 Y. Ichikawa, et. al., in the opening session of the 23rd IEEE Photovoltaic Specialists Meeting, Louisville, KY, May 1993.

29 The reader is referred to Reference [1] for very recent data. The following texts may also be of interest. (1) H. J. Möller, Semiconductors for Solar Cells, Artech House, Boston, MA, 1993; (2) K. Takahashi and M. Konugai, Amorphous Silicon Solar Cells, North Oxford Academic Publishers, Ltd., London, 1986; and (3) J. I. Pankove and N. M. Johnson, volume editors, Hydrogen in Semiconductors, Semiconductors and Semimetals, Volume 34, Academic Press, New York, 1991.

30 K. Knapp, reporting on CuInSe$_2$, in session II of the National Renewable Energy Laboratories Photovoltaic Review Meeting, Denver, CO, Oct. 1993.

31 E. L. Ralph, in the Conference Record of the 8th IEEE Photovoltaic Specialits Conference, Aug. 1970, p. 326.

32 This estimate was first made, to the author's knowledge during session F.2. of the Annual Meeting of the American Section of the International Solar Energy Society, Jun. 1977, by K. W. Böer. Present day estimates are esentially the same.

33 An early look at the incredible number of potential solar cell material and fabrication technique combinations was made by J. J. Loeferski, in the Proceedings of the Workshop on Photovoltaic Energy Conversion of Solar Energy for Terrestial Applications, Oct. 1973, p. 27. A more modern survey of materials and processes may be found by scanning the table of contents of the Conference Record of the 23rd IEEE Photovoltaic Specialists Conference, Louisville, KY, May 1993.

34 J. M. Olsen, in session II of the NREL Photovoltaic Review Meeting, Denver, CO, Oct. 1993.

35 K. Rabago, keynote address at the NREL Photovoltaic Review Meeting, Denver, CO, Oct. 1993.

36 For an initial survey of the environmental effects of solar cells see J. G. Holmes in the Proceedings of the Annual Meeting of the American Section of the International Solar Energy Society, June 1977, p. 28. For a more recent study consider <u>Alternate Energy Sources VIII</u>, T. N. Veziroglu, editor Hemisphere Publishing Co., New York, 1989. The 23rd IEEE Photovoltaic Specialists Conference, Louisville, KY, May 1993, featured special tutorials on the hazards of solar cell production and use.

37 Photovoltaic Insiders' Report, May 1993, front page.

38 B. Cole, et. al., in the Proceedings of the 1977 Annual Meeting of the American Section of the International Solar Energy Society, June 1977, p. 23.

39 See sessions C.2, H.2, L.2 and L.3 of the Proceedings of the Annual Meeting of the American Section of the International Solar Energy Society, June 1977, for useful, if early, economic studies in this area.

40 M. A. Duguay, in American Scientist, Vol. 65, 1977, p. 422.

41 R. Taylor, in session V of the National Renewable Energy Laboratories, Photovoltaic Program Review Meeting, Denver C0., Oct. 1993.

42 K. Zweibel, in American Scientist, Vol. 81, 1993, p. 362.

43 "Sharing the Sun", the Proceedings of the Joint Conference of the American Section of the International Solar Energy Society, and the Solar Energy Society of Canada, Pergamon Press, Oxford, 1976, Vol. 5.

44 A. B. Meinel and M. P. Meinel, <u>Applied Solar Energy</u>, Addison-Wesley, Reading, MA, 1976, part 3.

45 There are papers on thermal energy storage in effectively every issue of Solar Energy, a journal published for the International Solar Energy Society by Pergamon Press.

46 E. P. Barry, R. L. A. Fernandes and W. A. Messner, in IEEE Spectrum, Nov. 1978, p. 47.

47 R. B. Aronson, in Machine Design, Vol. 49, 1985, p. 7.

48 C. Starr, M. F. Searl and S. Alpert, in Science, Vol. 256, 1992, p. 981.

49 J. F. Manwell and J. G. McGowan, in Solar Energy, Vol. 50, 1993, p. 399.

50 The overall cost depends, in part, on for how long a period of time the energy must be stored. See C. S. Thomas, in Solar Today, Sep./ Oct. 1993, p. 11.

51 C. E. Bamberger and J. Braunstein, in American Scientist, Vol. 63, 1975, p. 438.

52 S. Sunevason and F. J. Salzaro, in International Journal of Hydrogen Energy, Vol. 2, 1977, p. 53.

53 B. D. McNicol and D. A. J. Rand, editors, Power Sources for Electric Vehicles, Elsevier, Amsterdam, 1984, Chaps. 1, 2 and 4.

54 J. O'M Bockris and K. Uosaki, in International Journal of Hydrogen Energy, Vol. 2, 1977, p. 123.

55 A. K. Gosh and H. P. Maraska, in Journal of the Electrochemical Society, Vol. 124, 1977, p. 1516.

56 M. Tomkievicz and J. W. Woodhall, in Journal of Electrochemical Society, Vol. 124, 1977, p. 1436.

57 W. Hausz, General Electric-Tempo Report P-681, Santa Barbara, CA, 1975.

58 The reader is referred to the Proceedings of the 9th Miami International Congress on Energy and the Environment, edited by T. N. Veziroglu, Miami Beach, FL , Dec. 1989, for a number of articles on various aspects of hydrogen energy.

59 S. W. Angrist, Direct Energy Conversion, Allyn and Bacon, Boston, MA, 1976, Chap. 8.

60 R. C. Neville, in Solar Energy, Vol. 19, 1977, p. 539.

61 W. Stahl, K. Voss and A. Goetzberger, in Solar Energy, Vol. 52, 1994, p. 111.

62 L. Rawlings and M. Kapner, In Solar Today, Jan./Feb. 1994, p. 26.

63 K. Khouzam, et. al., Proceedings of the Annual Conference of the American Solar Energy Society, Denver, CO, 1989, p. 262.

64 K. Khouzam, P. Groumpos and F. E. Villaseca, Proceedings of the Annual Conference of The American Solar Energy Society, Denver, CO, 1989, p. 275.

65 W. Wallace, J. Serfass and A. Vesey in presentations during session VI, of the 1993 National Renewable Energy Laboratories, Program Review Meeting, Denver, CO, Oct. 1993.

66 Photovoltaic Insiders' Report, fall 1993.

67 Solar Today, Jan./Feb. 1994, p. 20.

68 L. S. Miller and J. B. Mullin, Electronic Materials - From Silicon to Organics, Plenum Press, London, 1991, Chaps. 9-11.

69 K. Takahashi and M. Konuyai, <u>Amorphous Silicon Solar Cells</u>, North Oxford Academic Publishers Ltd., London, 1986, Chap. 2.

70 C. E. Thomas, Solar Today, Sep./Oct. 1993, p. 11.

71 C. Starr, M. F. Searl and S. Alpert, in Science, Vol. 256, 1992, p. 981.

72 C. Flavin and N. Lenssen, Sun World, May/June 1991, p. 10.

73 Michael Nicklas in a guest editorial in Solar Energy, Vol. 50, 1993, p. 287.

APPENDIX A: CONVERSION FACTORS

The energy-environmental field in general and the photovoltaic field in particular operate in many locations and utilize a very wide group of descriptive parameters. This appendix provides a few of the more commonly encountered quantities.

Quantity (One):	Equals:
A	
Acre	43,560 square feet
Acre	4,047 square meters
Angstrom	3.937×10^{-9} inches
Angstrom	1.0×10^{-10} meters
Are	100 square meters
Are	119.6 square yards
B	
Barrel (U.S. dry)	7,056 cubic inches
Barrel (U.S. liquid)	31.5 gallons
Barrel (oil)	42 gallons of oil
British thermal unit (Btu)	1.055×10^{10} ergs
Btu	778.3 foot-lbs
Btu	252 gram-calories
Btu	3.931×10^{-4} horsepower-hours
Btu	1,054.8 joules
Btu	0.2520 kilogram-calories
Btu	107.5 kilogram-meters
Btu	2.926×10^{-4} kilowatt-hours
Btu/hour	0.0700 gram-calories/second
Btu/hour	3.929 horsepower
Btu/hour	0.2931 watts
Btu/square foot/minute	0.1221 watts/square inch

C

Candle/square centimeter	3.142 lamberts
Centigrade	$1.8 \times °C + 32$ in degrees Fahrenheit
Centimeter	3.281×10^{-2} feet
Centimeter	0.3937 inches
Centimeter	0.01 meters
Centimeter	393.7 mils
Centimeter/second	1.1969 feet/minute
Centimeter/second	0.02237 miles/hour
Cubic centimeter	0.0612 cubic inches
Cubic centimeter	1×10^{-6} cubic meters
Cubic centimeter	0.001 liters
Cubic foot	0.8036 bushels (dry)
Cubic foot	1,728 cubic inches
Cubic foot	0.02832 cubic meters
Cubic foot	28.32 liters
Cubic inch	1.639×10^{-5} cubic meters
Cubic inch	4.329×10^{-3} gallons
Cubic inch	0.01639 liters
Cubic meter	35.31 cubic feet
Cubic meter	264.2 gallons (U.S. liquid)

D

Day	24 hours
Day	1,440 minutes
Day	86,400 seconds
Degree (angle)	0.01745 radians

E

Erg	9.48×10^{-11} Btu
Erg	7.367×10^{-8} foot-pounds
Erg	2.389×10^{-8} gram-calories
Erg	3.725×10^{-14} horsepower-hours
Erg	1×10^{-7} joules
Erg	2.389×10^{-11} kilogram-calories
Erg	2.778×10^{-14} kilowatt-hours
Erg/second	1.341×10^{-10} horsepower
Erg/second	1×10^{-10} kilowatts

F

Foot	30.38 centimeters
Foot	1.645×10^{-4} miles (nautical)
Foot	1.894×10^{-4} miles (statute)
Foot	1.2×10^{4} mils
Foot-candle	10.764 lumens/square meter
Foot-pound	1.286×10^{-3} Btu
Foot-pound	1.356×10^{7} ergs
Foot-pound	0.3238 gram-calories
Foot-pound	5.050×10^{-7} horsepower-hours
Foot-pound	1.356 joules
Foot-pound	3.24×10^{-4} kilogram-calories
Foot-pound	3.766×10^{-7} kilowatt-hours
Foot-pound/minute	2.260×10^{-5} kilowatts
Foot-pound/second	1.818×10^{-3} horsepower

G

Gallon	3,785 cubic centimeters
Gallon	231 cubic inches
Gallon	3.785 liters
Gallon of water	8.3453 pounds of water
Gallon/minute	2.228×10^{-3} cubic feet/second
Gallon/minute	0.06308 liters/second
Gill	142.07 cubic centimeters
Gill	0.1183 liters
Gill	0.25 pints (liquid)
Gram-calorie	3.9863×10^{-3} Btu
Gram-calorie	4.1868×10^{7} ergs
Gram-calorie	1.5596×10^{-6} horsepower-hours
Gram-calorie	1.1630×10^{-6} kilowatt-hours

H

Hectare	2.471 acres
Hectare	1.076×10^{5} square feet
Horsepower	42.44 Btu/minute
Horsepower	550 foot-pounds/second
Horsepower	10.68 kilogram-calories/minute
Horsepower	0.7457 kilowatts
Horsepower-hour	2,547 Btu

Horsepower-hour	2.6845×10^{13} ergs
Horsepower-hour	1.98×10^{6} foot-pounds
Horsepower-hour	641,190 gram-calories
Horsepower-hour	2.6845×10^{6} joules
Horsepower-hour	0.7457 kilowatt-hours
Horsepower-hour	2.547×10^{-15} Q

I

Inch	2.54 centimeters
Inch	2.54×10^{-2} meters
Inch	1.578×10^{-5} miles
Inch	1,000 mils

J

Joule	9.48×10^{-4} Btu
Joule	1×10^{7} ergs
Joule	0.7376 foot-pounds
Joule	2.389×10^{-4} kilogram-calories
Joule	2.778×10^{-7} kilowatt-hours
Joule	9.480×10^{-22} Q

K

Kilogram	2.205 pounds
Kilogram-calorie	3.968 Btu
Kilogram-calorie	3,088 foot-pounds
Kilogram-calorie	1.560×10^{-3} horsepower-hours
Kilogram-calorie	4,186 joules
Kilogram-calorie	1.163×10^{-3} kilowatt-hours
Kilometer	3,281 feet
Kilometer	0.6214 miles
Kilowatt	56.92 Btu/minute
Kilowatt	1.341 horsepower
Kilowatt-hour	3,413 Btu
Kilowatt-hour	3.6×10^{13} ergs
Kilowatt-hour	2.655×10^{6} foot-pounds
Kilowatt-hour	1.341 horsepower-hours
Kilowatt-hour	3.6×10^{6} joules
Kilowatt-hour	860.5 kilogram-calories
Kilowatt-hour	3.413×10^{-15} Q

Knot	6,080 feet/hour
Knot	1.8532 kilometers/hour

L

League	3.0 miles (approximately)
Light year	9.4609×10^{12} kilometers
Light year	5.9×10^{12} miles
Lumen	0.001496 watts
Lumen/square foot	1.0 foot-candles
Lux	0.0929 foot-candles

M

Meter	3.281 feet
Meter	6.214×10^{-4} miles (statute)
Micron	1×10^{-6} meters
Mile (nautical)	1.853 kilometers
Mile (nautical)	1.1516 miles (statute)
Mile (statute)	1.609 kilometers
Mile (statute)	1,760 yards
Millimeter	3.281×10^{-3} inches
Millimeter	0.001 meters
Mil	2.54×10^{-3} centimeters
Mil	0.001 inches
Minute (angle)	0.01667 degrees (angle)
Minute (angle)	60 seconds (angle)

P

Parsec	3.084×10^{13} kilometers
Parsec	1.9×10^{13} miles
Peck (British)	554.6 cubic inches
Peck (British)	9.0919 liters
Peck (U.S.)	537.605 cubic inches
Peck (U.S.)	8.8096 liters
Pint (dry)	33.60 cubic inches
Pint (liquid)	28.87 cubic inches
Pint (liquid)	0.4732 liters
Pound	0.4536 kilograms
Pound	16 ounces
Pound	1.2153 pounds (troy)

Q
Q 1×10^{18} Btu
Q 3.9262×10^{14} horsepower-hours
Q 1.0548×10^{21} joules
Q 2.93×10^{14} kilowatt-hours
Quart (dry)   67.20 cubic inches
Quart (liquid)   946.4 cubic centimeters
Quart (liquid)   57.75 cubic inches
Quart (liquid)   0.26 gallons
Quart (liquid)   0.9463 liters

R
Radian   57.30 degrees (angle)
Revolution   360 degrees (angle)
Revolution   6.283 radians

S
Second 2.778×10^{-4} degrees (angle)
Second 4.848×10^{-6} radians
Square centimeter 1.973×10^{5} circular mils
Square centimeter 1.076×10^{-3} square feet
Square foot 2.296×10^{-5} acres
Square foot   0.09290 square meters
Square inch 1.273×10^{6} circular mils
Square inch   6.452 square centimeters
Square kilometer   247.1 acres
Square kilometer 10.76×10^{6} square feet
Square meter 2.471×10^{-4} acres
Square meter 3.861×10^{-7} square miles
Square mile   640 acres
Square mile   2.590 square kilometers

T
Temperature (°Kelvin)   °C + 273.
Temperature (°Kelvin)   5/9 x °F + 241

W
Watt   3.4129 Btu/hour
Watt   107 ergs/second

Watt-hour	3.4129 Btu
Watt-hour	3.6×10^{10} ergs
Watt-hour	1.341×10^{-3} horsepower-hours

Y

Yard	3.0 feet
Yard	0.9144 meters
Yard	5.682×10^{-4} miles (statute)

APPENDIX B: SELECTED PROPERTIES OF SOME SEMICONDUCTORS WITH PHOTOVOLTAIC CELL POTENTIAL

Semiconductor

Ge	Si	InP	GaAs	CdTe	AlSb	CdSe	GaP
			Molecular Weight				
72.59	28.09	145.79	144.64	240.00	148.73	191.36	100.69
			(Based on Carbon 12)				
			Density (g/cm³) [1]				
5.32	2.33	4.79	5.32	5.81	4.26	4.82	4.13
			Lattice Constant (Å) [1]				
5.66	5.43	5.87	5.65	4.30	6.14	4.14	5.45
			Melting Point (°C) [1]				
936	1415	1070	1238	1258	1080	1475	1470
			Energy Gap (eV)				
at 0°K (estimated)							
0.74	1.17	1.42	1.52	1.60	1.68	1.85	2.34
at 300°K							
0.67	1.10	1.35	1.42	1.50	1.58	1.70	2.26
[1,2]	[1,2]	[2]	[2]	[2,3]	[2]	[2,4]	[1,2]

Semiconductor							
Ge	Si	InP	GaAs	CdTe	AlSb	CdSe	GaP

Direction and Type of Conduction Band Minimum
Direction [4]

[111]	[100]	[000]	[000]	[000]	[100]	[000]	[100]

Type [4]

ind	ind	dir	dir	dir	ind	dir	ind

Energy Gap Shift (ev/°K x 10^4)
(at 300°K) [4]

-3.7	-2.3	-4.6	-5.0	-4.1	-4.0	-4.6	-5.4

Effective Densities of States Mass
(relative to the free electron mass) [3,5,6]
Conduction Band

0.55	1.08	0.07	0.072	0.11	0.11	0.13	0.13

Valence Band

0.31	0.56	0.40	0.51	0.35	0.38	0.41	0.81

Effective Densities of States (#/cm³ x 10^{-19})
(at 300°K)
Conduction Band, N_c

1.03	2.82	0.465	0.047	0.092	0.0717	0.118	0.118

Valence Band, N_v

0.43	1.04	0.636	0.700	0.520	0.612	0.636	1.805

Intrinsic Carrier Concentration (#/cm³) [7]
(at 300°K)

2.40 x10^{13}	1.45 x10^{10}	2.55 x10^7	2.20 x10^6	5.65 x10^5	1.15 x10^5	1.48 x10^4	0.764 x10^0

(at 500°K)

2.04 x10^{16}	2.74 x10^{14}	3.38 x10^{12}	6.94 x10^{11}	2.29 x10^{11}	1.84 x10^{11}	6.39 x10^{10}	1.88 x10^8

Relative Static Dielectric Constant
(Permittivity) [4,8]

16	12	14	12.5	10.9	11	10.6	10

Semiconductor							
Ge	Si	InP	GaAs	CdTe	AlSb	CdSe	GaP

Optical Refractive Index
[4, 8, 9]

Ge	Si	InP	GaAs	CdTe	AlSb	CdSe	GaP
4.0	3.42	3.10	3.30	2.67	3.18	2.75	3.32

Electron Mobility (cm^2/volt-sec.)
(Impurity concentration near to the intrinsic levels)
[4, 8, 9-16]
(at 300°K)

Ge	Si	InP	GaAs	CdTe	AlSb	CdSe	GaP
3900	1700	4600	8500	1050	200	800	110

(at 400°K)

| 1200 | 0920 | 2400 | 3800 | 0750 | 100 | 500 | 100 |

(at 500°K)

| 0900 | 0440 | 2000 | 3200 | 0650 | 050 | 350 | 080 |

(at 300°K)
(Impurity concentration of 10^{14} cm^{-3})

| 3600 | 1700 | 4000 | 7000 | 1000 | 180 | 620 | 100 |

(Impurity concentration of 10^{16} cm^{-3})

| 3300 | 1200 | 2200 | 6000 | 0900 | 080 | 580 | 090 |

(Impurity concentration of 10^{18} cm^{-3})

| 1800 | 0320 | 1800 | 3000 | 0800 | 070 | 500 | 060 |

Hole Mobility (cm^2/volt-sec.)
(Impurity concentration near to the intrinsic levels)
[4, 8, 10-17]
(at 300°K)

Ge	Si	InP	GaAs	CdTe	AlSb	CdSe	GaP
1900	600	150	400	100	420	---	75

(at 400°K)

| 0550 | 300 | 060 | 150 | 070 | 180 | --- | 50 |

(at 500°K)

| 0400 | 200 | 040 | 070 | 040 | 150 | --- | 35 |

(at 300°K)
(Impurity concentration of 10^{14} cm^{-3})

| 1900 | 600 | 140 | 380 | 100 | 420 | --- | 70 |

Semiconductor							
Ge	Si	InP	GaAs	CdTe	AlSb	CdSe	GaP
(Impurity concentration of 10^{16} cm^{-3})							
1300	460	050	320	080	410	---	50
(Impurity concentration of 10^{18} cm^{-3})							
0360	200	020	160	060	240	---	35
Electron Minority Carrier Lifetime (sec.) (At 300°K and near intrinsic impurity levels) [10, 15, 18, 19]							
1×10^{-3}	3×10^{-3}	7×10^{-7}	1.2×10^{-7}	5×10^{-6}	3×10^{-7}	---	1×10^{-7}
Hole Minority Carrier Lifetime (sec.) (At 300°K and near intrinsic impurity levels) [19, 12, 15, 18, 19]							
1×10^{-3}	3×10^{-3}	5×10^{-7}	1×10^{-7}	1×10^{-6}	3×10^{-7}	3×10^{-9}	1×10^{-7}

References

1 B. G. Streetman, Solid State Electronic Devices, Prentice-Hall, Englewood Cliffs, NJ, 1980, p. 443.

2 S. M. Sze, Physics of Semiconductor Devices, Wiley-Interscience, New York, 1981, p. 849.

3 J. L. Loferski, in Journal of Applied Physics, Vol. 27, 1956, p. 777.

4 J. I. Pankove, Optical Processes in Semiconductors, Prentice-Hall, Englewood Cliffs, NJ, 1971, p. 412.

5 Using the perpendicular mass for CdSe, see Reference [4].

6 Reference [2], pp. 17 and 849.

7 The values given here are commonly accepted (see, for example, A. S. Grove, Physics and Technology of Semiconductor Devices, Wiley, New York, 1967, p. 102 or S. M. Sze, Reference [2]). However, application of Equation III.18 yields slightly lower values if the effective densities of states listed in this appendix are used. It is probable that the difference in values arises from the uncertainty in the energy gap for the indirect semiconductors and from uncertainties in effective mass for direct gap semiconductors.

8   J. L. Moll, <u>Physics of Semiconductors</u>, McGraw-Hill, NY, 1964, p. 70.

9   This data is a mixture of averaged values from various sources, unpublished data and such reference works as R. K. Wilardson and A. C. Beer, <u>Semiconductors and Semimetals, Vol. 10</u>, Academic Press, New York, 1975, p. 84.

10 Reference [2], p. 27.

11 S. M. Sze and J. C. Irwin, in Solid State Electronics, Vol. 11, 1968, p. 599.

12 M. B. Prince, in Physics Review, Vol. 92, 1953, p. 681.

13 The literature contains significant variations in values for electron and hole mobility in indium phosphide. For example, μ_n = 4,600 cm^2/volt-sec. [8] and 3,500 cm^2/volt-sec. [14].

14 R. K. Wilardson and A. C. Beer, <u>Semiconductors and Semimetals, Vol. 10</u>, Academic Press, New York, 1975, Chap. 1.

15 M. Neuberger, <u>II-VI Semiconducting Compounds, Data Tables</u>, AD 698341, U. S. Department of Commerce, 1969.

16 M. Aven and J. S. Presener, editors, <u>Physics and Chemistry of II-VI Compounds</u>, Wiley, New York, 1967, Chap. 11.

17 The hole mobility furnished for gallium phosphide is an average of the values furnished in references [4] and [8].

18 Estimates based on diffusion length data in E. H. Stupp and A. Milch, writing in the Journal of Applied Physics, Vol. 48, 1977, p. 282.

19 Unpublished data by author.

APPENDIX C: THE SATURATION CURRENT IN PN JUNCTION SOLAR CELLS

In Chapters V, VI and VII, the saturation current density utilized for pn junction solar cells is given by Equation V.2:

$$J_S = (qD_{pn}p_{no})/L_{pn} + (qD_{np}n_{po})/L_{np} , \qquad (C.1)$$

where q is the absolute value of the electric charge, D_{pn} and D_{np} are the minority carrier diffusion constants in the n- and p-type regions, respectively. L_{pn} and L_{np} are the minority carrier diffusion lengths in the n- and p-type regions and n_{po} and p_{no} are the equilibrium minority carrier concentrations in the p- and n-type regions. This equation can be written:

$$J_S = (qD_{pn}n_i^2)/L_{pn}N_D + (qD_{np}n_i^2)/L_{np}N_A , \tag{C.2}$$

where n_i is the intrinsic carrier concentration of the semiconductor, N_D is the net donor concentration on the n-type side of the semiconductor junction and N_A is the net acceptor concentration on the p-type side of the semiconductor.

The above expression was originally derived by Shockley [1] and is founded on the basic assumptions: (1) the surfaces of the semiconductor are greater than a diffusion length in distance from the junction, and (2) the impurity levels, N_D and N_A, are constant with distance in the solar cell.

In a pn junction solar cell the discussions in Chapter VI lead us to the conclusion that the junction is not necessarily much greater than a diffusion length from the semiconductor surface. Under these conditions, Equation C.2 must be written [2]:

$$J_S = ((qD_{pn}n_i^2)/L_{pn}N_D)\Gamma + ((qD_{np}n_i^2)/L_{np}N_A) \Lambda, \tag{C.3}$$

where the factors Γ and Λ depend on the minority carrier surface recombination velocities, s_{pn} and s_{np}, of the n- and p-type regions. These factors are given by:

$$\Gamma = \frac{s_{pn} \cosh(W_n/L_{pn}) + (D_{pn}/L_{pn}) \sinh(W_n/L_{pn})}{(D_{pn}/L_{pn}) \cosh(W_n/L_{pn}) + s_{pn} \sinh(W_n/L_{pn})}, \tag{C.4}$$

and,

$$\Lambda = \frac{s_{np} \cosh(W_p/L_{np}) + (D_{np}/L_{np}) \sinh(W_p/L_{np})}{(D_{np}/L_{np}) \cosh(W_p/L_{np}) + s_{np} \sinh(W_p/L_{np})}, \tag{C.5}$$

where W_n and W_p are the physical thicknesses of the n- and p-type regions.

In practice, neither the p- nor the n-type region impurity concentration is constant. In the heavily doped portion of a solar cell (recall, from our earlier discussions, that junction solar cells are either p+n or n+p in structure, indicating that one region of the solar cell is heavily doped relative to the other) the form of the impurity concentration versus distance relationship depends on the method of introduction of the impurity. It is possible to achieve impurity concentrations that are effectively constant on the heavily doped side of the junction [3]. However, in Chapter VI we deliberately used a variable impurity concentration in the substrate, the more lightly doped side. This was done as a means of enhancing charge carrier collection efficiency. The substrate impurity concentration was graded to induce an electric field, which, in turn, serves to sweep the minority charge carriers generated by photon absorption towards the junction. Such a "reflecting" contact was first treated by Cammerow [4] and has the additional advantage that it leads to a much reduced minority carrier surface recombination velocity for the substrate.

The fact that the substrate impurity concentration increases with increasing distance from the junction forces further modifications on the saturation current formulas (Equations C.1 - C.5). In Chapter III, the general expression for the minority carrier concentration density in a semiconductor was obtained (Equation III.60). In the absence of optical illumination and for no surface recombination, this expression reduces to:

$$\partial p_n / \partial t = -(p_n - p_{no})/\tau_{po} + D_{pn} \partial^2 p_n / \partial x^2 - \partial(\mu_{pn} \mathscr{E} p_n)/\partial x , \qquad (C.6)$$

and

$$\partial n_p / \partial t = -(n_p - n_{po})/\tau_{no} + D_{np} \partial^2 n_p / \partial x^2 - \partial(\mu_{np} \mathscr{E} n_p)/\partial x . \qquad (C.7)$$

Equations C.6 and C.7 are written for a simplified one dimensional current flow, τ_{po} and τ_{no} are the small signal minority carrier lifetimes in the n- and p-type regions, μ_{pn} and μ_{np} are the minority carrier mobilities and $\mathscr{E}$ is the electric field.

Consider an n-type semiconductor. The hole concentration, p_n, is a function of the location as is p_{no}. The minority carrier lifetime and mobility are functions of the impurity concentration (see Chapter III). The discussions in Chapter VI lead to the following expression for the impurity concentration of the substrate, $N_D(x)$, if a constant substrate electric field, $\mathscr{E}_s$, is desired:

$$N_D(x) = N_D(o) \exp\{q \mathscr{E}_s x/kT) , \qquad (C.8)$$

where T is the absolute temperature, x is the distance from the junction in the substrate region and $N_D(o)$ is the substrate impurity concentration at the junction. In Chapter VI we assumed that $N_D(o) = 10^{15}/cm^{-3}$ and the substrate electric field, $\mathscr{E}_S$, was 15 volts/cm, leading, at 300°K, to:

$$N_D(x) = 10^{15} \exp(614.4x) .$$

(C.9)

Under steady-state conditions we may now rewrite Equation C.6:

$$\partial^2 p_n/\partial x^2 = (p_n(x) - n_i^2/N_D(x))/D_{pn}(x)\tau_{po}(x) + (1/D_{pn}(x)(\partial\mu_{pn}(x)\,\mathscr{E}_S p_n(x))/\partial x.$$

(C.10)

Clearly, a similar expression also applies where the substrate is p-type and $n_p(x)$ is the variable. If $\mathscr{E}_S$ is zero and the impurity concentration, minority carrier mobility, lifetime and diffusion constant are all fixed, expression C.10 leads to one of the terms shown in Equation C.2. However, if $\mathscr{E}_S$ is not zero and the minority carrier mobility and lifetime are dependent on the displacement, x, the solution is extremely complex-- even more so because the variability of the minority carrier diffusion constant, mobility and lifetime are not represented by simple analytic expressions [5].

Approximate numerical solutions using the material parameters for the semiconductors that served as examples in Chapters IV, VI and VII [6] lead to the conclusion that, for graded substrates, the saturation current density can be conservatively approximated by the use of Equation C.2. Furthermore, the saturation current density component that arises from the substrate (assuming that the donor impurity concentration is constant and on the order of $10^{15}/cm^3$) is one to three orders of magnitude larger than that originating in the heavily doped "front layer".

Therefore, the saturation current density expression used in calculations in Chapter VI and subsequently is that due to the substrate component of Equation C.2. These values are conservative and, therefore, "practical" in an engineering sense. On the other hand, the results of Chapter V are based on minimum leakage current estimates, optimizing and idealizing Equation C.2, and, therefore, represent a maximum performance approach.

References:
1   W. Shockley, in Bell System Technical Journal, Vol. 28, 1949, p. 435.
2   J. P. McKelvery, <u>Solid State and Semiconductor Physics</u>, Harper and Row, New York, 1966.

3   The saturation current component from the heavily doped region of a solar cell can be shown to be less than five percent of the total saturation current, making this situation of little importance. For a discussion of junction formation by diffusion, from which one we obtain an estimation of the accuracy of this assumption, the reader is referred to F. M. Smits, in the Proceedings of Institute of Radio Engineers, Vol. 46, 1958, p. 1049.

4   R. L. Cammerow, in Physics Review, Vol. 95, 1954, p. 95.

5   See the discussion of mobility and lifetime for minority charge carriers in Chapter III and recall that: $D = kT\mu/q$.

6   See the discussion in Chapter VI for the physical parameters used.

APPENDIX D: SOME USEFUL PHYSICAL CONSTANTS

AM1 power flux I_{AM1} = 0.107 watts/cm^2
Avogadro's number N_A = 6.02 x 10^{23} molecules/mole
Boltzmann's constant k = 1.38 x 10^{-23} joules/°K
Electric charge (magnitude) q = 1.602 x 10^{-19} Coulombs
Electronic rest mass m_{eo} = 9.11 x 10^{-31} kilograms
Permittivity of free space ϵ_o = 8.85 x 10^{-12} Farads/meter
Planck's constant h = 6.624 x 10^{-34} joule-seconds
Solar constant I_{AM0} = 0.1353 watts/cm^2
Speed of light (in a vacuum) c = 2.998 x 10^8 meters/second

APPENDIX E: SYMBOLS

A	
A*	Richardson constant
A**	Modified Richardson constant for mos junctions
A_1	Temperature coefficient for Schottky barriers
A_D	Junction area of a solar cell
Ag	Silver
Au	Gold
Al	Aluminum
$AgInS_2$	Silver indium sulfide

AlSb	Aluminum antimonide
Å	Angstroms
a	The radius of light sensitive area of a standard configuration solar cell
a-	Denotes the amorphous form of a semiconductor
a-Si	Amorphous silicon
a-SiC	Amorphous silicon carbide
a-SiGe	Amorphous silicon-germanium
a-Si:H	Amorphous silicon containing hydrogen
a-SiC:H	Amorphous silicon carbide containing hydrogen
a-SiGe:H	Amorphous silicon-germanium containing hydrogen
$\alpha(h\upsilon)$	Absorption coefficient
$\alpha(\Psi)$	Factor accounting for atmospheric losses of solar power
B	
β	Percentage of input light power lost to reflection
C	
CdSe	Cadmium selenide
CdS	Cadmium sulfide
CdS/Cu_2S	Cadmium sulfide/copper sulfide
CdTe	Cadmium telluride
CIS	See $CuInSe_2$
CO_2	Carbon dioxide
Cu	Copper
$CuGaSe_2$	Copper gallium diselenide
$CuGa_{0.25}In_{0.75}Se_2$	
	Copper gallium indium diselenide
Cu_2S	Copper sulfide
$CuInSe_2$	Copper indium selenide
c	Speed of light
c-	Denotes the crystal form of a semiconductor
D	
D	Deuterium
D	Generalized diffusion constant
D_s	Generalized diffusion constant for minority carriers in the substrate of a solar cell
D_n	Diffusion constant for electrons

D_{np}	Diffusion constant for electrons in a p-type semiconductor
D_p	Diffusion constant for holes
D_{pn}	Diffusion constant for holes in an n-type semiconductor
d	"Front Layer" thickness in a solar cell
Δ	Substrate contact thickness for a solar cell
ΔE_c	The "notch" in the conduction band lower edge in heterojunctions
ΔE_v	The discontinuity in the valence band upper edge in heterojunctions
ΔE_g	The difference in energy gap widths for the two semiconductors in a heterojunction

E	
E	Energy
E_A	Acceptor electron energy level
E_c	Conduction band lower edge
E_D	Donor electron energy level
E_F	Fermi energy level
E_g	Energy gap of a semiconductor
E_p	Energy of a phonon
E_{ph}	Energy of a photon
E_t	Energy level of a trap
E_v	Valence band upper edge
$\mathscr{E}$	Electric field
$\mathscr{E}_S$	Electric field in a substrate
ϵ	Permittivity

F	
F	Focal length of a lens
f	Fraction of light reflected from a lens, mirror or semiconductor surface
Φ	The ratio of the optically absorbing surface area to the total surface of a solar cell
ϕ_{Bn}	Schottky barrier energy adjusted for image force lowering
ϕ_{Bo}	Barrier energy for Schottky or mos junctions

G	
$Ga_{1-x}Al_xAs$	Gallium aluminum arsenide
$GaAs$	Gallium arsenide

$GaAs_xSb_{1-x}$	Gallium arsenide antimonide
GaP	Gallium phosphide
Ge	Germanium
G_L	Light driven charge carrier generation
H	
H	Hydrogen
He	Helium
Hf	Haffmium
HgTe	Mercury telluride
H_2O	Water
h	Planck's constant
$\mathcal{H}$	The hamiltonian
$\hbar$	Planck's constant divided by 2π
χ	Electron affinity for a semiconductor
I	
I	Current
I_L	Load current for a solar cell
InP	Indium phosphide
I_{ns}	Solar power density (insolation)
η_c	Carnot cycle engine efficiency
η_s	Solar cell maximum efficiency
J	
J	Current density
J_D	Solar cell normal current density
J_G	Generation-recombination current density
J_{GS}	Saturation current density for generation-recombination
J_n	Electron current density
J_p	Hole current density
J_{ph}	Photocurrent density
J_S	Saturation current density for a pn junction and the generalized saturation current density for solar cells
J_{SC}	Short circuit current density for a solar cell
J_{SH}	Saturation current desnity for a heterojunction
J_{SS}	Saturation current density for Schottky or mos solar cells
J_T	Tunneling current density

K

K$_1$	Tunneling constant
K'	Solar cell loss factor
k	Extinction coefficient
k	Boltzmann's constant

L

L	Generalized charge carrier diffusion length
Li	Lithium
L$_n$	Electron diffusion length
L$_{np}$	Electron diffusion length in p-type semiconductors
L$_p$	Hole diffusion length
L$_{pn}$	Hole diffusion length in n-type semiconductors
L$_S$	Generalized minority carrier diffusion length in the substrate of a solar cell
λ	Wavelength of a photon
λ	q/kT in expressions for diode current and resistance
λ_g	Wavelength of a photon when it has an energy equivalent to the energy band gap of a semicondcutor

M

M	Width of a metal contact ring for a standard configuration solar cell
M(T)	The ratio of the saturation current density at a temperature, T, to that at 300°K
m*	Generalized effective mass
m$_{ce}$	Effective mass for electrons in the conduction band
m$_{eo}$	Free electron mass
m$_{ve}$	Effective mass for holes in the valence band
μ	Generalized mobility
μ_{LM}	Majority carrier mobility of the "Front Layer"
μ_n	Electron mobility
μ_{np}	The mobility of electrons in a p-type semiconductor
μ_p	The mobility of holes
μ_{pn}	The mobility of holes in an n-type semiconductor
μ_S	The generalized mobility of minority carriers in the substrate regions of a solar cell
μ_{SM}	The majority carrier mobility in the substrate of a solar cell

N

N	The optical concentration level
N_A	Net acceptor impurity concentration in a p-type semi-conductor
N_c	Conduction band effective density of states
N_D	Net donor impurity concentration in a n-type semiconductor
N_L	Impurity concentration in the "Front Layer" of a solar cell
N_{SJ}	Substrate impurity concentration at the junction
N_{st}	Surface concentration of traps
N_S	Generalized impurity concentration in a solar cell substrate
N_t	Volume concentration of traps
N_v	Valence band effective density of states
n	Index of refraction
n	Ideality factor in expressions for diode current versus voltage
n	Electron concentration per unit volume
n'	Generalized carrier concentration in space charge regions
n_i	Intrinsic carrier concentration
υ	Frequency

0

Ω	The intensity of incident light

P

PCVED	Plasma enhanced chemical vapor deposition
P_d	Palladium
P_E	Electrical power delivered by a second or third stage solar cell power system
P_p	The radiated photon power density for a black body
P_T	Solar power delivered by a second or third stage solar cell power system
P_{Th}	The thermal power delivered by a second or third stage solar cell power system
Pt	Platinum
P_{aT}	Available thermal power in a second or third stage solar cell power system
P_{if}	Probability of an electron interband transition
P_{max}	Maximum delivered solar power
p	Hole concentration per unit volume

pc	Denotes that a semicondcutor is in polycrystalline form
Ψ	Angle between the sun and the normal to the earth's surface
Ψ	Schrödinger wave function
Ψ'	Average north-south angle, for a given day, between the sun and the zenith
Ψ''	Angle between the sun and the detector/collector normal
Ψ_i	The intrinsic Fermi level
Q	
Q	An amount of energy equal to 2.93×10^{14} kilowatt-hours
q	Magnitude of the charge on the electron
R	
R_L	Resistive load for a solar cell
R_l	The resistive load for a solar cell which yields maximum power transfer
R_T	The thermal resistance
r_D	Series resistance of a solar cell
ρ	Charge density
ρ_T	Thermal resistivity
S	
Si	Silicon
SiC	Silicon carbide
SiF_4	Silicon tetrafluoride
SiGe	Silicon-germanium
SiH_4	Silane
SiO_2	Silicon dioxide, also quartz or sand
s_{pn}	Surface recombination velocity for holes on an n-type semiconductor
s_{np}	Surface recombination velocity for electrons on a p-type semiconductor
σ	Conductivity
T	
T	Tritium
T	The absolute temperature
Ti	Titanium
T_c	The "cold" reservoir temperature in a heat engine

T_h	The "hot" reservoir temperature in a heat engine
T_J	The temperature of a solar cell junction
T_s	The temperature of a heat sink
ΔT_s	The temperature difference between a solar cell junction and its heat sink
T'	The light transmitted into a semiconductor
T''	The light which passes through a semiconductor
τ	Lifetime
τ^A	Lifetime due to Auger recombination
τ^R	Lifetime due to radiative recombination
τ_m	Maximum lifetime
τ_{no}	Small signal electron lifetime
τ_{np}	Lifetime of electrons in p-type semiconductors
τ_{po}	Small signal hole lifetime
τ_{pn}	Lifetime of holes in n-type semiconductors
τ_S	Generalized substrate minority carrier lifetime
Θ	East-west angle between the sun and the earth surface normal
Θ_T	Heat energy flow (in a solar cell)
U	
U	Recombination rate
V	
V	Voltage
V_L	Load voltage in a solar cell system
V_A	Externally applied voltage
V_B	Junction built-in voltage
V_D	Effective voltage across a diode junction
V_D'	The voltage across a solar cell junction when there is maximum power transfer
V_{OC}	Open circuit voltage
V_P	Photovoltage
v	Velocity of a charge carrier
V'	The voltage at which the forward generation-recombination current equals the forward diffusion current
v_{th}	Thermal velocity of a charge carrier

W
W$_c$ Width of the space charge region

X
X$_1$ The collection distance for hole-electron pairs in a semi-
                         conductor
X$_S$ Generalized space charge region width in the substrate of
                         a solar cell
X$_S'$ The maximum space charge region in the substrate of a
                         solar cell
X$_{SS}$ Substrate space charge region width for an externally
                         shorted junction
X$_T$ Empirical factor for current in a heterojunction
x$_n$ Space charge width in the n-type region of a junction
x$_p$ Space charge width in the p-type region of a junction

Y
Y   The thickness of a vertical configuration solar cell in the
                         direction of photon travel

Z
$Zn_{0.35}Cd_{0.65}O$ Zinc cadmium oxide
ZnSe   Zinc selenide
ZnTe   Zinc telluride
z   Factor accounting for surface recombination in a solar cell
ζ Energy difference between the conduction band edge
                         and the Fermi level

SUBJECT INDEX

A

Absorption 67, 120, 132-137, 139, 142, 145, 214, 219, 344, 349, 352, 354
Absorption Coefficient
see Absorption
Absorptivity
see Absorption
Acceptors 86-91, 160
Acid Rain 9, 11
Aluminum Antimonide (AlSb)
- junctions 206, 208-210, 222, 251, 252, 261, 263, 268
- properties 85, 90, 93, 103, 122, 130, 138-141, 198, 213, 214
- solar cells 139, 143, 144, 146-150, 218, 219, 227-229, 231-250, 270, 277, 284, 286, 291, 295-297, 302, 372
AMO (air mass zero) 40, 42, 43, 123, 124, 127, 129, 138-140, 143, 146-149, 177, 180, 182, 184, 186, 188, 190, 191, 223, 231-235, 239-241, 245-247, 257, 270, 273-285, 287-294, 298, 299, 303, 305, 306, 309, 311, 313
AM1 (air mass one) 40, 42, 43, 123, 124, 127, 129, 138, 139, 141, 144, 146-148, 150, 177, 181, 183, 185, 187-189, 191, 201, 223, 231, 232, 236-238, 242-244, 248-250, 257, 264, 270, 295-297, 303, 307, 310, 312, 314, 368
Amorphous 72, 92, 131, 156, 198, 329, 346, 348-358, 364, 371, 373, 388
Antireflection 67, 121-123, 145-150, 204, 341, 354

Antisite 342
Atmosphere
see Weather
Auger Recombination
see Recombination

B

Bandpass 317-320, 326
Base Load Power
see Power Plant
Battery 382-384, 386
Biological Sources 2, 8, 19, 20, 22, 26, 34--*see also* Wood, Plants, Energy -sources
Built-in Voltage 158, 163, 174, 205, 251

C

Cadmium Selenide (CdSe)
- junctions 206, 208-210, 222, 251, 261, 263, 268
- properties 85, 90, 93-95, 103, 122, 127, 130, 138-141, 198, 213, 214
- solar cells 139, 143, 144, 146-150, 218, 219, 227-250, 270, 278, 285, 286, 292, 295-297, 302, 342, 347, 372
Cadmium Sulfide (CdS)
- properties 127
- solar cells 342, 345, 347
Cadmium Sulfide/Copper Sulfide
see Copper Sulfide/Cadmium Sulfide
Cadmium Teluride (CdTe)
- junctions 206, 208-210, 222, 251, 252, 261, 263, 268
- properties 85, 90, 94, 95, 103, 122, 127, 130, 138-141, 198, 213, 214, 389

- solar cells 139, 142-144, 146-150, 218, 219, 227-229, 231-250, 270, 276, 283, 290, 293-299, 305-307, 309-314, 319-321, 325, 327, 331, 332, 342, 347, 372
Carnot 6, 24, 25, 32, 304, 310, 318
CIS *see* Copper Indium Selenide
Coal 2, 8, 9--*see also* Fossil Fuels
Collector 46-52, 55, 56, 63
Collection Distance 171, 172, 211
Computer 386, 387
Concentration 36, 57, 68, 69, 257, 258, 374, 375-380, 389--*see also* Lenses, Mirrors and Optical Concentration
Conduction Band *see* Energy Band
Conductivity 71, 72, 91, 132, 352
Conductor 71
Configuration *see* Optical Orientation
Converter 315-318, 320, 322-325
Cooling *see* Heat Sink
Cooling Systems *see* Heat Sink
Copper Indium Gallium Selenide (Cu(In,Ga)Se$_2$)
see Copper Indium Selenide
Copper Indium Selenide (CuInSe$_2$) 342, 344
Copper Indium Gallium Sulfide, Telluride, Sulfide (Cu(In,Ga)(S, Se, Te)$_2$)
see CuInSe$_2$
Copper Sulfide (Cu$_2$S) 342-344
Copper Sulfide/Cadmium Selenide (Cu$_2$S/CdS) 157, 342-344--*see also* Solar Cell
Cost 27, 42, 57, 205, 253, 257, 305-307, 340, 344, 364, 366-370, 372-381
Crystal *see* Crystalline
Crystalline 72, 73, 76, 77, 79, 81, 82, 90, 92, 93, 97, 131, 155, 198, 364, 371, 373, 388
Current *see* Current Density
Current Density 71, 91, 96, 160-162--*see also* Photocurrent Density

Czochralski 371, 388

D
Dams 2, 8--*see also* Energy-sources
"Dead Layer" 142-150, 201, 205, 214-216, 218, 219, 279
Defect Density 342
Dember Effect 156
Demographics 1, 4, 10, 12-16, 26, 29, 34, 35
Dendritic Web 388
Density of States 178
Depletion Layer Width
see Space Charge Region
Depletion Region
see Space Charge Region
Detector *see* Collector
Dichroic Mirror 333, 363, 390
Diffusion 91, 94, 166, 209, 334, 340
Diffusion Constant 160--*see also* Electron, Hole
Diffusion Length 160, 215--*see also* Electron, Hole
Donor 86-91, 160

E
Ecology 1, 25, 32, 33, 42, 389
Economics *see* Cost
Effective Mass 93, 94, 166--*see also* Electron, Hole
Efficiency *see* Energy
Electric Field 92, 108, 109, 111, 112, 114, 132, 155-157, 159, 162, 216, 217, 226, 258, 352
Electrolysis 386, 387
Electron Affinity 164, 174
Electron-Hole Pairs 97, 103, 114, 120, 125, 126, 128, 132, 139, 142, 155, 158-161, 164, 170, 177, 197, 201, 202, 207, 211, 214, 215, 217, 269, 271, 317
Electron Lifetime *see* Electron
Electron Mobility *see* Electron
Electron 82-87, 89, 93-95, 98, 101, 103, 105, 124, 133, 134, 159, 161,

211, 214

Energy
- capital *see* Energy-sources
- efficiency 5, 20, 25, 31-33, 35, 36, 130
- heat *see* Thermal
- income *see* Energy-sources
- reserves 9-14
- sources 1, 2, 7-14, 19, 26, 29, 34, 35, 253, 365
- storage 18, 20, 29-31, 364, 381, 386, 387, 390
- thermal 263, 264, 304-312, 336, 380-382
- transport 382
- uses (and consumption) 1, 3-7, 9, 10, 12-15, 26, 29, 34, 35

Energy Band 76, 82-84, 86, 87, 98, 110, 124, 125, 133-135, 155, 159, 169, 216

Energy Gap 83, 85, 87, 103, 125, 127, 130-134, 136, 145, 146, 155, 161, 162, 165, 169, 177, 179-191, 197, 198, 206, 207, 317, 349-351

Environment *see* Ecology

Epitaxy 90, 91, 94, 212, 334, 340, 388

Extinction Coefficient
see Absorption

F

Food 1, 3, 20

Fossil Fuel 2, 7-10, 14, 65, 366--
see also Energy-sources

Fresnel Lens *see* Lens

"Front Layer" 164, 165, 199, 201, 202, 206, 212-214, 216, 218, 219, 221, 226-229, 251, 252, 259

Fuel Cell 382, 384

G

Gallium Arsenide (GaAs)
- junction 166, 206, 208-210, 222, 251, 252, 261, 263, 268
- properties 83, 85, 90, 92, 94,

95, 102, 103, 122, 127, 129, 130, 137-142, 198, 213, 214, 389
- solar cells 36, 139, 144, 146-150, 218, 219, 227-229, 231-250, 270, 275, 282, 289, 293-299, 305-307, 309-314, 319-321, 324, 326, 330, 332, 342, 347, 372

Gallium Arsenide Antimonide
(GaAs$_x$Sb$_{1-x}$) 347

Gallium Arsenide Phosphide
(GaAs$_x$P$_{1-x}$) 129

Gallium Arsenide/Aluminium Gallium
Arsenide (GaAs/Al$_x$Ga$_{1-x}$As) 157, 166, 199--*see also* Solar Cells

Gallium Phosphide (GaP) 85, 90, 94, 103, 122, 129-131

Gas 2, 8, 9--*see also* Fossil Fuels

Generation-Recombination
see Recombination

Geothermal 2, 8, 11-14--*see also*
Energy-sources

Germanium (Ge)
- junction 166
- properties 83, 85, 90, 95, 103, 122, 130, 131

Glass 66-68, 341, 357

Glow Discharge 352

Grain Boundary 339-342

Greenhouse Effect 7, 9, 11, 15, 20, 21

H

Heat Energy *see* Energy, Heat, Solar
Thermal Energy

Heat Flow 264, 265, 327

Heat Sink 199, 201, 204, 265, 302, 303

Heterojunction *see* Junction

Hole-Electron Pairs see Electron-Hole
Pairs

Hole 84-87, 89, 93, 95, 99, 101, 103, 105, 133, 134, 159-161, 211, 214

Hole Lifetime *see* Lifetime, Hole

Hole Mobility *see* Hole

Hydrogen (H) 385, 386

Hydrogen Selenide (H$_2$Se) 345

Hydropower 2, 8, 14, 15, 25, 26,
  365, 366--*see also* Dams, Energy
  -sources, Tides

I
Index of Refraction 67, 68, 120,
  130, 132
Indium Phosphide (InP)
  - junction 206, 208-210, 213, 222,
  251, 252, 261, 263, 268
  - properties 85, 90, 103, 122, 130,
  138-141, 198, 214
  -solar cells 139, 143, 144, 146-150,
  218, 219, 227-229, 231-250, 270,
  274, 281, 286, 288, 295-297, 302,
  372
Indium Tin Oxide (ITO)
  see Tin Oxide
Insolation *see* Solar Insolation
Insulator 71
Intermediate Load Power
  see Power Plants
Interstitial 342
Intrinsic Carrier Concentration 84,
  87, 162, 260, 271
Inverted Configuration
  see Optical Orientation
Inverted Solar Cell
  see Optical Orientation
Ion Implantation 91, 92, 94, 209,
  212, 340

J
Junction 108-115, 170-175. 178, 197
  - heterojunction 143, 156-158, 163-
  166, 199-203, 205, 206, 210, 214,
  219-222, 227-229, 231-253, 259,
  261, 263, 270, 273-299, 305-314,
  318, 319, 323-326, 347, 348, 389
  -mos 168, 169, 201, 203, 206, 208,
  221, 260, 389
  - pn junction 109-115, 156-159,
  161-163, 178, 199-203, 205, 206,
  209, 210, 214, 219, 220, 222, 228-
  250, 253, 259-261, 263, 270, 273-

292, 295-299, 307, 389
  - Schottky 156-158, 167-169, 201,
  203, 206, 208-210, 219-222, 225,
  227-253, 259, 260, 262, 264. 269,
  270, 273-292, 295-297, 307, 308, 389
  - tunneling 357, 358

L
Lattice 73-76, 90
Lattitude 49-54, 56, 368
Lead-Acid *see* Battery
Lens 32, 33, 36, 57-60, 66, 67-69,
  257, 301, 302, 373, 379, 390
Lifestyle 1, 2, 5, 6, 7, 16
Lifetime 97, 99, 101-103, 107, 139,
  142, 155, 161, 165, 178, 188, 197,
  201, 212, 213, 222, 355
Lithium-Sulfide *see* Battery
Loss Factor 175, 176, 180-191, 197,
  229, 230, 267, 294-299, 326, 327
Luminescent 335, 336
Luminescent Concentrator
  see Luminescent

M
Mercury Telluride (HgTe) 342
Methane 22
Mirrors 32, 36, 57, 58, 60-69, 257,
  301, 302, 373, 379, 390
Mobility 92, 93, 197, 213--*see also*
Electron, Hole
Molecular Beam Epitaxy
  see Epitaxy
Momentum 133-136
MOS or MOS Solar Cell
  see Junction, Solar Cell

N
Natural Gas *see* Fossil Fuels
Nickel-Zinc *see* Battery
Nuclear
  - fission 2, 8, 11, 12, 365, 366--*see
  also* Energy-sources
  - fusion 2, 8, 16, 17--*see also*
  Energy-sources

O

Ocean 22, 24, 25
Ocean Thermal Energy Conversion 25
Oil 2, 7-9--*see also* Fossil Fuels
Oil Shale 9
Open Circuit Voltage 170, 173-175
Optical Concentration 258, 201, 267,
  269, 271, 272-285, 287-299, 302-
  307, 375-380
Optical Orientation 199, 363
  - inverted configuration 199, 204,
  205, 207, 211, 215, 218, 223, 225-
  231, 238, 241, 245-251, 253, 267,
  269, 270, 287-294, 297, 299, 305-
  314, 323-327, 389
  - standard configuration 199, 200-
  202, 205, 207, 211, 214, 216, 219,
  223-227, 229-238, 250, 253, 265,
  270, 273-279, 293-297, 298, 305-
  314, 389
  -vertical configuration 199, 200,
  202-205, 207, 214, 215, 218, 223-
  232, 235, 239-244, 250, 253, 265,
  267, 270, 280-285, 296, 298, 305-
  314, 389
Oxide *see* Quartz

P

Peak Load Power
  see Power Plants
Petroleum *see* Oil
Phonon 93, 102, 133, 135, 136
Photon 43, 44, 67, 108, 114, 115,
  119-121, 124-128, 132, 133-135,
  136, 139, 142, 145, 155, 158, 164,
  177, 186, 188, 258, 301, 319, 353,
  355, 357, 358
Photocurrent 120, 121, 128, 129,
  146-150, 155, 165, 170-172, 175-
  177, 186, 188-191, 202, 215, 217-
  220, 223, 225, 232, 235, 238, 247,
  269, 270, 321, 322, 352
Photocurrent Density
  see Photocurrent
Photosynthesis 19, 20

Photovoltage 120, 155, 159, 170,
  175, 176, 187, 189-191, 205, 206,
  229, 232, 233, 236, 239, 241, 242,
  245, 248, 251, 267, 286, 293, 295-
  299, 326, 327, 346
Photovoltaic Cell *see* Solar Cell
Planck's Constant 78
Plants 2, 19, 20, 21--*see also* Energy
  -sources
Plasma Enhanced Chemical Vapor
  Depostion 388
Plastic 66-68
PN Junction *see* Junction
Politics 1, 6-8
Polycrystalline 72, 73, 92, 131, 155,
  157, 198, 339-348, 364, 371-373, 388
Population *see* Demographics
Power Plants 365, 366, 369, 381, 386,
  387
Power Ratio 368, 369
Production 371

Q

Quartz 66, 68, 122, 168, 169
Quantum Mechanics 76, 78, 84, 127,
  133, 134

R

Recombination 97-102, 104-107, 126,
  142, 161, 162, 165, 166, 171, 172, 177,
  178, 188, 215, 218, 219, 258, 269, 271,
  301, 345
Recombination-Generation
  see Recombination
Reflection 119-123, 308
Reliability 370, 381
Resistance 165, 171, 175, 197, 212,
  216--*see also* Solar Cell Resistance
Resources *see* Ecology
Ribbon Silicon 342, 371, 388
Richardson Constant 221

S

Saturation Current
  see Saturation Current Density

Saturation Current Density 160-162, 165, 172-175, 177-188, 190, 191, 197, 201, 212, 220-223, 247, 259, 261-263, 268, 279, 319

Schrödinger 78-81

Schottky Barrier *see* Junction

Schottky Junction *see* Junction

Semiconductor 71, 72, 76, 84, 90, 101, 119--*see also* individual semiconductor listings

Series Resistance *see* Resistance or Solar Cell Resistance

Short Circuit Current *see* Short Circuit Current Density

Short Circuit Current Density 170, 172, 173

Silane (SiH$_4$) 349, 350

Silicon (Si)
- amorphous 353-358, 374-379
- junction 206, 208-210, 222, 251, 252, 261, 263, 268
- properties 77, 83, 85, 90, 94, 95, 103, 122, 130, 134, 137-141, 198, 213, 214, 349-352, 389
- solar cells 36, 139, 142-144, 146-150, 215, 219, 227-229, 231-250, 270, 273, 280, 287, 293-299, 305-307, 309-314, 319-321, 323, 326, 329, 332, 334, 343, 346, 372, 374

Silicon Carbide (SiC) 342, 349
- amorphous 353-356, 358

Silicon Dioxide (SiO$_2$) 122, 342--*see also* Quartz

Silicon Germanium (SiGe) 349, 356

Silicon Tetrafluoride (SiF$_4$) 356

Silver Indium Sulfide (AgInS$_2$) 342

Single Crystal *see* Crystalline

Solar
- solar constant 39
- solar insolation 49-51, 53, 54, 56, *see also* -solar cell energy
-solar electric 19, *see also* -solar cell energy and solar cell
-solar cell energy 2, 8, 17-19, 26, 28-35, 44, 45, 52, 140, 141, 313,

314--*see also* Energy-sources

Solar Cell 34-36, 42, 43, 57, 72, 108, 114, 120, 130, 131, 146-150, 156-158, 164, 167, 175, 188, 199-202, 207, 212-214, 227, 228, 233-250, 253, 257, 258, 264, 265, 269, 270, 293-299, 301-315, 317, 318, 321, 323-331, 335, 336, 340, 342-349, 353-358, 364, 367, 371, 384, 388, 389

Solar Cell Arrays *see* Solar Cell

Solar Cell Delivered Power Density 128, 129, 138, 139, 143, 144, 146, 148-150, 175-177, 180-187, 189-191, 197, 201, 229, 231-238, 240, 243, 246, 249, 273-292, 295-299, 326, 327, 329-332, 334--*see also* individual semiconductors

Solar Cell Efficiency 155, 169, 176, 177, 190, 191, 235, 238, 241, 244, 247, 250, 258, 272-292, 295-299, 301, 305-307, 322-326, 371, 375-380

Solar Cell Resistance 171, 175, 176, 188, 197, 201, 222, 227-230, 232, 235, 238, 247, 252, 253, 269, 271, 319, 355

Solar Spectral Irradiance *see* Sun

Solar Spectrum *see* Sun

Solar Thermal Energy 257, 258, 264, 304, 309-312, 314

Space 42, 148

Space Charge Region 159, 161, 163, 165, 205, 206, 210, 211

Spectrum Splitting 333

Sputtering 352

Stacking 202, 204, 356, 385, 389, 390

Staebler-Wronski Effect 356, 369

Stage Two Solar Cell System *see* Solar Cells

Stage Three Solar Cell System *see* Solar Cells

Standard Configuration *see* Optical Orientation

Substrate 204-207, 209, 210, 212, 213, 215-217, 221, 223, 226, 227, 229, 231-250, 253, 273-278, 280-285, 287-292, 305-307, 309-314, 318, 323-326

Sun 39-41, 43, 197, 315, 316, 363, 364, 367
Sunlight *see* Sun

T
Tandem Solar Cell 333, 334
Tar Sands 9
Temperature 12, 19, 22, 24, 26, 32, 84, 85, 92-94, 107, 135, 136, 142, 179, 198, 199, 201, 227-229, 231-250, 253, 259, 262, 263, 267, 268, 273-299, 303-307, 309-314, 322-327, 329-332, 356, 363
Thermal Efficiency 303, 328-332
Thermal Resistance 265, 266
Thermal Resistivitiy
  see Thermal Reistance
Thermoelectric 32, 33
Thermophotovoltaic 315-319, 322-327, 329-332, 363, 390
Thin-film *see* Polycrystalline
Third Stage Solar Cell System *see* Stage Three Solar Cell System
Tides 2, 8--*see also* Energy-sources
Tin Oxide (SnO) 354, 355, 357, 358
Tracking 48-56, 257, 301, 315, 374, 390
Transmission 119, 123, 385
Traps 97-99, 101, 102
Trees *see* Wood
Tunneling 166, 168, 169--*see also* Junction

U
Utility *see* Power Plants

V
Vacancies 342
Valence Band *see* Energy Band
Vertical Configuration
  see Optical Orientation

W
Waste 22--*see also* Biological
Wave Form 78-81

Wavelength 138, 139
Waves 19
Weather 5, 18, 39, 40, 44, 45, 54-56, 375-380, 382, 385
Web *see* Dendritic Web
Wind 2, 8, 19, 23, 24, 365--*see also* Energy-sources
Wood 2, 8, 14, 15, 19, 21--*see also* Energy-sources
Wronski-Staebler Effect
  see Staebler-Wronski Effect

Z
Zinc Cadmium Oxide ($Zn_xCd_{1-x}O$) 347
Zinc Oxide (ZnO) 345
Zinc Sulfide (ZnS) 127
Zinc Selenide (ZnSe) 166
Zinc Telluride (ZnTe) 342

BERRY LIBRARY
CINCINNATI STATE